CYBERSECURITY MANAGEMENT

Cybersecurity Management

An Organizational and Strategic Approach

NIR KSHETRI

UNIVERSITY OF TORONTO PRESS
Toronto Buffalo London

Toronto Buffalo London
utorontopress.com
Printed in Canada

ISBN 978-1-4875-0496-0 (cloth)
ISBN 978-1-4875-2362-6 (paper)
ISBN 978-1-4875-3125-6 (EPUB)
ISBN 978-1-4875-3124-9 (PDF)

Library and Archives Canada Cataloguing in Publication

Title: Cybersecurity management : an organizational and strategic approach / Nir Kshetri.
Names: Kshetri, Nir, author.
Description: Includes bibliographical references and index.
Identifiers: Canadiana (print) 20210229403 | Canadiana (ebook) 20210229438 | ISBN 9781487523626 (paper) | ISBN 9781487504960 (cloth) | ISBN 9781487531256 (EPUB) | ISBN 9781487531249 (PDF)
Subjects: LCSH: Computer security – Management. | LCSH: Computer crimes.
Classification: LCC QA76.9.A25 K74 2021 | DDC 005.8 – dc23

We welcome comments and suggestions regarding any aspect of our publications – please feel free to contact us at news@utorontopress.com or visit us at utorontopress.com.

Every effort has been made to contact copyright holders; in the event of an error or omission, please notify the publisher.

University of Toronto Press acknowledges the financial assistance to its publishing program of the Canada Council for the Arts and the Ontario Arts Council, an agency of the Government of Ontario.

Funded by the Government of Canada
Financé par le gouvernement du Canada

Contents

Figures, Tables, and Boxes

Figures

Tables

In Focus Boxes

Preface and Acknowledgments

Cyberthreats are among the most critical issues facing the world today. It is important to increase the awareness of and preparedness to handle such threats among national decision-makers, planners, and the public. Likewise, organizational decision-makers need a fully developed understanding of cybersecurity's substantial effects on their bottom line. A number of trends, events, and turning points have increased the importance and significance of this issue. The role of cybersecurity in the pursuit of an organizational mission, while currently underappreciated, is likely to become more evident in the future.

Due to the pervasiveness of the Internet and rapidly growing cyberthreats, cybersecurity is everyone's responsibility – not simply the sole responsibility of IT people in an organization. Readers without a technical background need a book that provides them with an understanding of the nature and extent of cyberthreats to organizations and consumers, how such threats evolve with new technological advances, and effects of cultural, organizational, and macro-environmental factors on cyberattacks and cybersecurity. This book responds to this long-standing need. The overriding goals of this book are to (i) develop an understanding of mechanisms and growth of cybercrimes and the current state of cybersecurity, (ii) analyze macro-level factors affecting cybercrime and cybersecurity, (iii) evaluate strategic and organizational issues associated with cybersecurity, and (iv) articulate and show the effects of new and evolving information and communications technologies and systems on cybersecurity and privacy issues. A related goal is to fill a strong pedagogical need in these areas.

This book is written as a textbook for upper-level undergraduate and graduate cybersecurity courses in business, public policy, and related fields. Readers are not required to have a technical background in order to understand the contents of this book. For readers who are well-versed with the technical side of cybersecurity, this book can help expand the depth of understanding of this subject by providing relevant insights about organizational and strategic issues and highlighting policy implications.

Regarding the ideas, concepts, and content presented in this book, I am grateful to several people for their comments, suggestions, support, and encouragement. I would like to express deep appreciation to Jennifer DiDomenico, Acquisitions Editor, Social Sciences, University of Toronto Press, who inspired me to undertake this project. She shepherded the project with the greatest of care and professionalism through its various phases. I am grateful to Susan Bindernagel for excellent copy-editing. I wish to thank my mentor and colleague Jeff Voas for his inspiration and support. Thanks are also due to anonymous University of Toronto Press reviewers for their useful comments and excellent suggestions. I greatly benefited from feedback and ideas from University of North Carolina at Greensboro (UNCG) students who took the Cybersecurity Management course with me over the past many years. I also received help from my talented graduate assistant Brad Gooch at UNCG. Finally, I could not have done it without the love, support, and understanding of my wife and best friend, Maya.

Abbreviations

4R	Fourth Revolution
5G	Fifth Generation
AEL	annual expected loss
AI	artificial intelligence
AICPA	American Institute of Certified Public Accountants
ANA	Association of National Advertisers
ANPD	Autoridade Nacional de Proteção de Dados
APC	Association for Progressive Communications
API	application programming interface
APT	advanced persistent threat
ARF	ASEAN Regional Forum
ARO	annual rate of occurrence
ASEAN	Association of Southeast Asian Nations
AWS	Amazon Web Services
BAS	building automation system
BD	big data
BEC	Business Email Compromise
BEUC	Bureau Européen des Unions de Consommateurs
BLOB	binary large object
BPH	bulletproof hosting
BRIC	Brazil, Russia, India, China
BRICS	Brazil, Russia, India, China, and South Africa
BRT	Bandidos Revolutions Team
BYOD	bring your own device

CAC	Cyberspace Administration of China
CAPTCHA	Completely Automated Public Turing test to tell Computers and Humans Apart
CBM	confidence-building measure
CCP	Chinese Communist Party
CCPA	California Consumer Privacy Act
CDU	Cyber Defense Unit
CEO	chief executive officer
CIA	Central Intelligence Agency
CINS	Critical Infrastructure Notification System
CIO	chief information officer
CIP	critical infrastructure protection
CISO	chief information security officer
CMO	chief marketing officer
CNIL	Commission nationale de l'informatique et des libertés
CoECoC	Council of Europe Convention on Cybercrime
CompTIA	Computing Technology Industry Association
COVID	Coronavirus Disease
CPO	chief privacy officer
CPS	cyberphysical system
CS	cybersecurity
CSA	Cloud Security Alliance
CSIS	Center for Strategic and International Studies
CSP	cloud service provider
CTB	Curve-Tor-Bitcoin
CTIA	Cellular Telecommunications & Internet Association
CTIIC	Cyber Threat Intelligence Integration Center
CTO	chief technology officer
C-UPHD	Champaign-Urbana Public Health District
DAST	dynamic application security testing
DDoS	distributed denial of service
DEC	Deep Entity Classification
DFS	Department of Financial Services
DHS	Department of Homeland Security
DL	deep learning
DNC	Democratic National Committee
DNS	Domain Name System
DoE	Department of Energy
DoJ	Department of Justice

DoS	denial of service
DPA	Data Protection Authority
DPC	Data Protection Commission
DSCI	Data Security Council of India
DVR	digital video recorder
EC	European Commission
EFCC	Economic and Financial Crimes Commission
ENISA	European Network and Information Security Agency
EO	executive order
EPIC	Electronic Privacy Information Center
ES-C2M2	Electricity Subsector Cybersecurity Capability Maturity Model
ETNO	European Telecommunications Network Operators' Association
EU	European Union
FAQ	frequently asked questions
FBI	Federal Bureau of Investigation
FCC	Federal Communications Commission
FDA	Food and Drug Administration
FIP	fair information practices
FISMA	Federal Information Security Management Act
FSU&CEE	Former Soviet Union and Central and Eastern Europe
FTC	Federal Trade Commission
GARM	Global Alliance for Responsible Media
GB	gigabyte
Gbps	gigabit per second
GCI	Global Cybersecurity Index
GDPR	General Data Protection Regulation
GGE	Group of Governmental Experts
GPS	Global Positioning System
GSR	Global Science Research
HEW	US Department of Health, Education and Welfare
HHS	Health and Human Services
HIPAA	Health Insurance Portability and Accountability Act
HRM	human resources management
IaaS	Infrastructure as a Service
IC3	Internet Crime Complaint Center
ICANN	Internet Corporation for Assigned Names and Numbers
ICO	Information Commissioner's Office
ICP	internet content provider
ICS	industrial control systems

ICT	information and communications technology
IIoT	Industrial Internet of Things
IMU	Innovative Marketing Ukraine
IoT	Internet of Things
IP	Internet protocol
ISAO	information sharing and analysis organization
ISO	International Organization for Standardization
ISP	Internet service provider
IT	information technology
ITU	International Telecommunication Union
KGB	Komitet Gosudarstvennoy Bezopasnosti
KPA	Korean People's Army
LAC	Latin America and the Caribbean
LGPD	Lei Geral de Proteção de Dados
MaaS	malware as a service
METI	Ministry of Economy, Trade and Industry
MFA	multi-factor authentication
MIAC	Ministry of Internal Affairs and Communications
ML	machine learning
MoD	Ministry of Defense
MS	millisecond
NACD	National Association of Corporate Directors
NASSCOM	National Association of Software and Services Companies
NATO	North Atlantic Treaty Organization
NCA	National Crime Agency
NCB	National Cyber Bureau
NCCIC	National Cybersecurity and Communications Integration Center
NCD	National Cyber Directorate
NCS	national cybersecurity strategy
NERC	North American Electric Reliability Corporation
NHS	National Health Service
NISC	National Center of Incident Readiness and Strategy for Cybersecurity
NIST	National Institute of Standards and Technology
NPA	National Police Agency
NSA	National Security Agency
OECD	Organisation for Economic Co-operation and Development
OEWG	Open-Ended Working Group
OPM	Office of Personnel Management
OS	operating system

PaaS	platform as a service
PC	personal computer
PFI	personal financial information
PII	personally identifiable information
PIPA	Personal Information Protection Act
PLA	People's Liberation Army
PPB	parts per billion
PPC	pay per click
RAN	radio access network
RBN	Russian Business Network
RD	Rove Digital
RFID	radio-frequency identification
ROCSI	return on cybersecurity investment
ROI	return on investment
SaaS	software as a service
SAL	single attack loss
SAS	Statement on Auditing Standards
SAST	static application security testing
SBoM	software bill of materials
SCADA	supervisory control and data acquisition
SCC	Security Consultative Committee
SCO	Shanghai Cooperation Organisation
SDN	software-defined networking
SEC	Securities and Exchange Commission
SM	social media
SME	small and medium-sized enterprises
SOX	Sarbanes-Oxley Act
STEM	science, technology, engineering, and math
Tbps	terabit per second
TIYDL	thisisyourdigitallife
UN	United Nations
UNCTAD	United Nations Conference on Trade and Development
URL	uniform resource locator
VC	venture capital
VDN	value delivery networks
WHO	World Health Organization
WISE	Women in Security Excelling
WSSC	Washington Suburban Sanitary Commission

CYBERSECURITY MANAGEMENT

PART ONE

Mechanisms and Growth of Cybercrimes and the Current State of Cybersecurity

CHAPTER ONE

Current States of Cybercrime and Cybersecurity

CHAPTER PREVIEW

In the aftermath of various catastrophic breaches in global computer networks, the severity of cyberattacks and the importance of cybersecurity need hardly be elaborated. Individuals and organizations are vulnerable to a wide variety of cyberthreats. This chapter looks at the growth in global cybercrime and the associated impacts. It gives special consideration to the nature of cyberthreats and some key challenges. Also developed is a typology of various forms of cyberattacks. First, market-based and predatory cybercrimes are discussed with examples. Then various forms of predatory cybercrimes are discussed in terms of motivation, whether a target is pre-determined, jurisdiction of the target, category of the target and the perpetrator, and whether technology or social engineering plays a dominant role in the crime. Finally, it evaluates the current state of cybersecurity (CS).

1.1 INTRODUCTION

By all accounts, the global cybercrime industry is significantly bigger than most of the major and well-known underground industries such as the illegal drug trade and human trafficking.[1] At the outset, it is important to make clear that estimating the size of any underground economy is difficult. In the 2010s, the most often cited figure for annual worldwide loss to cybercrime was US$1 trillion.[2] More recently, popular press articles have cited Juniper Research's study, which put the costs of data breaches at US$3 trillion globally in 2019 and estimated the costs will reach US$5 trillion by 2024.[3]

Costs that are often overlooked in the above estimates include damage and destruction of data; lost productivity; theft of intellectual property, personal, and financial data; post-breach disruption of companies' businesses; forensic investigation; restoration of hacked data and systems; and company devaluations associated with reputational harm. Including those costs, Cybersecurity Ventures estimated that cybercrimes cost the world US$3 trillion in 2015, which would increase to US$6 trillion annually by 2021.[4]

Data breaches carry significant economic and psychological costs to billions of consumers worldwide. According to the 2019 *NortonLifeLock Cyber Safety Insights Report*, which was based on a survey of Internet users in 10 countries, 67% of respondents reported to be "more alarmed than ever about their privacy," 66% were "very worried their identity will be stolen," and 92% expressed concerns about data privacy.[5] According to the report, 80% of Indian

Figure 1.1 Percentage of respondents ranking cyberattacks/cyberthreats as the highest or among the top five concerns to their organizations

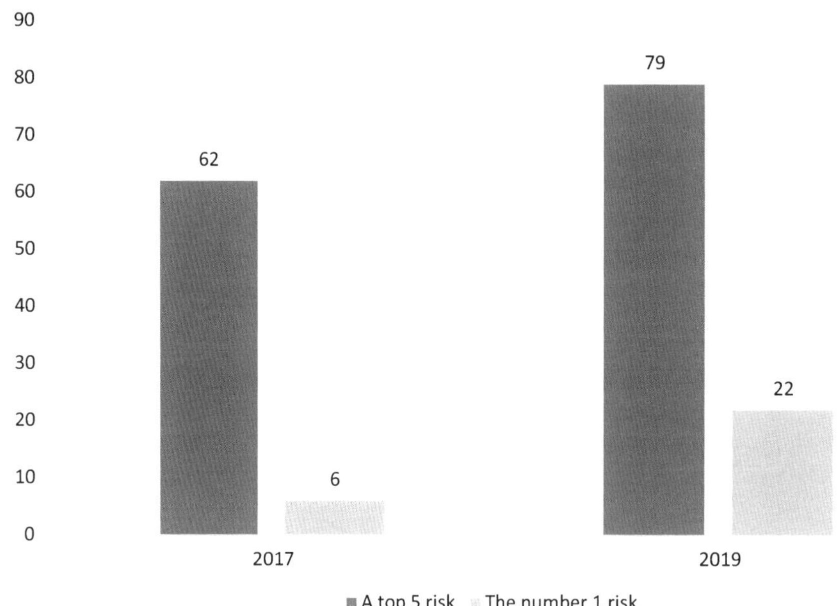

Data source: *Global cyber risk perception survey*. (2019). Marsh. Retrieved from https://www.marsh.com/us/insights/research/marsh-microsoft-cyber-survey-report-2019.html

Internet users had become cybercrime victims and 66% had experienced cyberattacks in the previous 12 months.[6]

Organizations are also increasingly concerned about cyberthreats. According to the Marsh Microsoft *2019 Global Cyber Risk Perception Survey* conducted worldwide with more than 1,500 respondents, the proportion of respondents ranking cyberattacks/cyberthreats as the highest as well as one of the top five concerns to their organizations increased drastically from 2017 to 2019.[7] (See figure 1.1.)

These worries are indeed justified. In 2018, cybercriminals breached 2.8 billion consumer data records in the US, which was estimated to cost over US$654 billion to US organizations.[8] Juniper Research's estimate suggested that 12 billion records were compromised worldwide in 2018 and will increase to over 33 billion in 2023.[9]

Data breach costs are rapidly rising. According to IBM Security's 2019 Cost of Data Breach Report, the average cost of a cyberattack on organizations was US$3.92 million.[10] For businesses with fewer than 500 employees, the average loss was US$2.5 million. While the majority of the losses were reported in the first year, the effects are felt for a long time. (See figure 1.2.) The report estimated that each lost record cost US$150. Cyberattacks have staggering economic costs that place an especially heavy burden on small and medium-sized enterprises (SMEs). According to the non-profit organization National Cyber Security Alliance, one in three small businesses become cybercrime victims every year and about 60% go out of business within six months of the attack.[11]

Figure 1.2 Distribution of data breach costs over time (%)

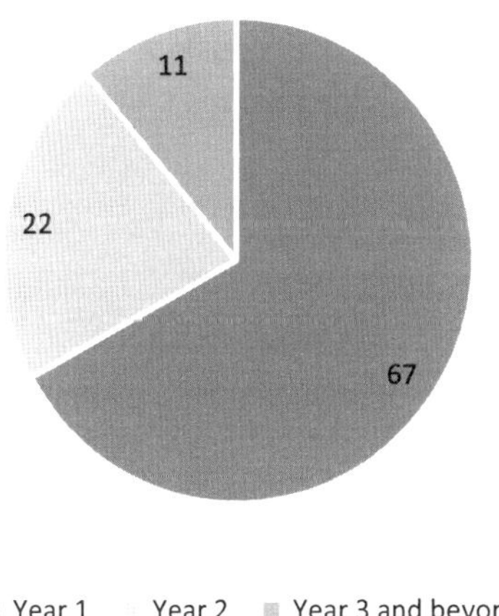

Data source: IBM Security. (2019). *Cost of a data breach report*. Retrieved from https://www.ibm.com/downloads/cas/ZBZLY7KL?_ga=2.181209767.1686129232.1586105132-321662522.1584074488

IN FOCUS 1.1: DEFINITIONS OF MAJOR TERMS

We define some of the major terms that will be used in the book.

Artificial intelligence

Artificial intelligence (AI) systems can mimic and imitate human cognition to perform tasks that seemed to be possible only with human thinking and logic before.

Big data

According to the research company Gartner, big data (BD) is defined as "high-volume, high-velocity and high-variety information assets that demand cost-effective, innovative forms of information processing for enhanced insight and decision making."[12] To Gartner's three Vs – volume, velocity, and variety – the software company SAS has suggested two additional BD dimensions: variability and complexity.[13] Recent research has indicated that the dimensions of BD identified by Gartner and SAS are tightly linked to privacy and CS issues.

Botnet

A botnet (robot network) is a network of computers infected with worms and Trojans that is controlled by a cybercriminal. A botnet is used to deliver spam and malware applications to victims and launch distributed denial of service (DDoS) attacks.

Click fraud

Click frauds involve illegitimate clicks on pay per click (PPC) ads, for which advertisers are charged on a per click basis. Human as well as technological means are used to generate invalid and artificial clicks. Some publishers click on ads on their own websites. Clicks on Internet ads also come from someone who is paid to click but has no interest in buying the product. Automated click generating programs are also used.[14]

Cloud computing

Cloud computing involves hosting applications on servers and delivering software and services via the Internet. In the cloud computing model, companies can access computing power and resources on the "cloud" and pay for services based on usage.

Cybercrime

A cybercrime is defined as a criminal activity in which computers or computer networks are the principal means of committing an offense or violating laws, rules, or regulations.[15]

Cyberphysical systems (CPSs)

CPSs are "smart systems that include co-engineered interacting networks of physical and computational components."[16] That is, in CPSs, physical entities interact with and are controlled by collaborating computational elements. CPSs process the collected information and act. CPSs facilitate the real-time collection and analysis of data related to diverse aspects such as health, environment, energy and water usage patterns, traffic, and waste disposal, and thus are likely to play key roles in smart cities.

Cybersecurity

The International Telecommunication Union (ITU) defines cybersecurity as "the collection of tools, policies, security concepts, security safeguards, guidelines, risk management approaches, actions, training, best practices, assurance and technologies that can be used to protect the cyber environment and organization and user's assets. Organization and user's assets include connected computing devices, personnel, infrastructure, applications, services, telecommunications systems, and the totality of transmitted and/or stored information in the cyber environment. CS strives to ensure the attainment and maintenance of the security properties of the organization and user's assets against relevant security risks in the cyber environment."[17] According to the ITU, general security objectives of CS comprise availability, integrity, and confidentiality.

Denial-of-service (DoS) attacks

There are two categories of DoS attacks: operating system (OS) attacks and network attacks. OS attacks entail discovering holes in the security of the OS and bringing down the system. Network attacks disconnect a network from the Internet service provider (ISP).

Distributed denial-of-service attack

In a DDoS attack, heavy incoming traffic floods a targeted system. The traffic often originates from multiple sources, often from botnets. The heavy traffic overloads the system and prevents the fulfillment of legitimate requests.

Extrinsic and intrinsic motivations of cybercriminals

Extrinsically motivated hackers look for benefits or incentives that are associated with hacking activities and are, in general, monetary compensation. The amount of financial incentives and the amount of motivation driving a hacker's behavior co-vary positively. Companies with higher digitization of values (higher potential financial incentives) are thus more likely to be attacked.

Intrinsically motivated individuals do activities for "inherent satisfactions rather than for some separable consequence" and they "act for the fun or challenge entailed rather than because of external prods, pressures or rewards."[18] For instance, hackers testing their skills and looking for fun act for purely psychological rather than monetary benefits.

Internet of Things (IoT)

The IoT is the network of physical objects or "things" (e.g., machines, devices, and appliances; animals; or people) embedded with electronics, software, and sensors that are provided with unique identifiers and possess the ability to transfer data across the Web with minimal human interventions.

Machine learning

Machine learning (ML) is a type of AI that helps increase accuracy of software applications in predicting outcomes without explicit programming. The basic idea behind ML is simple: algorithms receive input data and by using statistical analysis they predict output values within acceptable ranges. ML processes are similar to those involved in data mining and predictive modeling, which also look for patterns in data in order to adjust program actions.

Malware

A malware (malicious + software) is a software program used by cybercriminals to infiltrate or damage a computer system without the owner's informed consent. Software programs that are officially supplied by legitimate companies can also be considered malware if such programs secretly operate against the computer users' interests.

Nigerian 419 fraud

Nigerian 419 fraud (also known as advance fee fraud) is named for a section (419) of the Nigerian criminal code. Before the Internet was widespread, this type of fraud was conducted by unemployed Nigerians using snail mail, which was sent to Western businesspeople looking to make deals with oil officials. This type of fraud also originates from other countries.

Pay per click (PPC) advertising

PPC is a model used in Internet advertising in which advertisers are charged by their hosts (e.g., Google and Yahoo) only when their ad is clicked.

Phishing

Phishing involves the fraudulent acquisition of personal information by tricking an Internet user, typically by pretending to be a legitimate business or service provider. Phishers primarily use email. Information targeted by phishers includes email account credentials, bank account information, and online game account credentials.

Ransomware

Ransomware is a term to describe rogue software that blocks access to a computer or encrypts files on a computer, effectively holding the victim's computer and/or data hostage until a payment is made to the attacker.

Smart city

A smart city uses technology to gather and analyze data and take actions in order to enhance efficiency and improve quality of life.[19]

Social engineering fraud

In social engineering frauds, cybercriminals use emotional appeals such as fear, pity, or excitement to victimize Internet users by luring them to give their credentials, click malicious links, or download files containing malware. They may establish interpersonal relationships or create a feeling of trust and commitment in order to achieve these goals.

Zero-day vulnerability

Zero-day (0-day) vulnerability targets unknown flaws in software that the developer of the software has not yet had time to identify or fix. A cyberattack launched using zero-day vulnerability is called a zero-day attack.

1.2 THE NATURE OF CYBERTHREATS AND KEY CHALLENGES

1.2.1 Rapid Growth and Evolution of Cyberthreats

Organizations and individuals face multiple and diverse cyberthreats. According to *The Economic Value of Prevention*, a report from the Ponemon Institute, phishing, Domain Name System (DNS)-based attacks, viruses, bots, DDoS, and ransomware (see In Focus 1.2) are among the most common types of cyberattacks facing organizations. (See figure 1.3; also see In Focus 1.1 for definitions.)

These attacks are growing at alarming rates. For instance, by 2019, there were about 980 million malware programs, and 350,000 new malware types were detected every day.[20] In 2019, SonicWall recorded 9.9 billion malware attacks.[21] Kaspersky Lab detected more than 482 million phishing attempts in 2018, compared to 236 million such attempts in 2017.[22]

1.2.2 Difficulty of Attribution

In a cyberattack, attribution is defined as "determining the identity or location of an attacker or an attacker's intermediary."[23] The fact that the Internet was designed for ease of use without considering security gives the offense an obvious advantage over the defense. Due to the attacker's anonymity and problems associated with attribution, it is extremely difficult to use punishment as a threat to deter a potential offender from attacking.[24] Whereas retaliation and

Figure 1.3 Most common cyberattacks facing organizations (%)

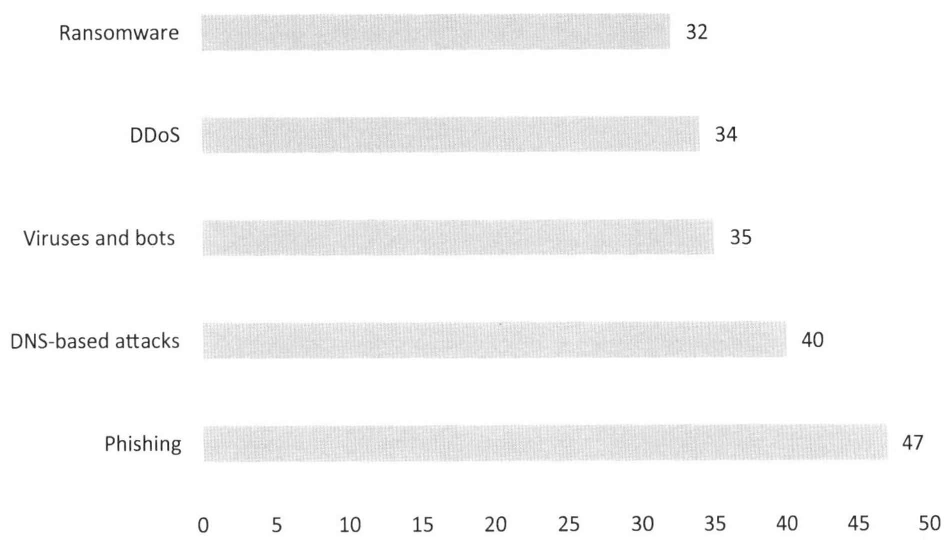

Data source: Ponemon Institute: The Economic Value of Prevention as cited in Bernard, A. (2020). *Cybersecurity prevention can save your company $682K*. TechRepublic. Retrieved from https://www.techrepublic.com/article/cybersecurity-prevention-can-save-your-company-682k/

IN FOCUS 1.2: NETWALKER RANSOMWARE TARGETS HOSPITALS DURING THE COVID-19 PANDEMIC

On March 10, 2020, cybercriminals hijacked the computer networks of the Champaign-Urbana Public Health District (C-UPHD) in Illinois, which serves 210,000 people including the University of Illinois.[25] The ransomware MailTo (Netwalker) was behind the attack.[26] A security specialist described Netwalker as "nasty" and "sophisticated."[27] Netwalker targets Windows 10 systems. It can deactivate antivirus software.[28]

C-UPHD went offline for four days. It relied on its Facebook page and local media outlets to provide updates about COVID-19 during the outage.

The district paid US$350,000 in ransom.[29] Most was paid by its insurer and the district covered its US$10,000 deductible.[30] After paying the ransom, it retrieved 99% of the files.[31]

Netwalker also targeted hospitals in Spain. The criminals sent coronavirus-themed phishing emails with PDF attachments. When the receiver opened the PDF attachment, the attackers locked down the systems.[32]

prosecution require knowing the attackers with full certainty, equivalents of DNA samples, fingerprints, and other offender information to identify a cyberperpetrator do not exist.[33] In most cases, traces lead only to botnets, which consist of businesses' and consumers' computers worldwide. Most cyberoffenders thus believe, and often rightly, that they cannot be traced and can cause widespread destruction without prompting a response from the victim.[34]

1.2.3 Increased Vulnerability of Critical and Sensitive Sectors

From a cyberoffender's perspective, different types of information have different value. Cyberoffenders are mainly interested in high-value data and information that are associated with critical, sensitive economic sectors. For this reason, government, energy, and banking industries often face high-profile cyberattacks.

According to the CS software company Claroty's *Global State of Industrial Cybersecurity* report published in 2020, 74% of CS professionals worldwide were more concerned about a cyberattack against critical infrastructure compared to an enterprise data breach. Among respondents from the US, 51% said that industrial networks are not properly protected and emphasized the importance of strengthening CS in these networks. Also, 55% of US respondents thought that US critical infrastructures are vulnerable to a cyberattack.[35]

National infrastructures that are especially critical, such as the power grid and nuclear plants, are high-value targets for financially motivated hackers, adversary governments, and terrorists. Cyberattacks on power companies and grids generate major risks, such as stopping the flow of natural gas through pipelines or causing a petrochemical facility's explosion or oil spill.[36] A major concern is that several such attacks have been reported in power companies and grids. (See End-of-Chapter Case: Cybersecurity Issues Facing Power Grids in the US.)

Banks and other financial institutions also are attractive targets because they have highly sensitive information about customers such as social security numbers and detailed records of past spending. Cyberattacks facing banks include disruption of websites, payment card fraud, and infiltration of bank networks to steal money.

Likewise, healthcare companies store financial, medical, and other types of personal information. (See In Focus 1.3.) Medical information is especially attractive to cybercriminals because the information can be used in a variety of ways and it takes victims longer to realize that their information has been stolen. For example, cybercriminals can use medical information to impersonate patients with specific diseases and obtain prescriptions for substances that are controlled by regulators such as the Food and Drug Administration (FDA).[37]

IN FOCUS 1.3: HEALTHCARE ORGANIZATIONS FACED CYBERATTACKS DURING THE COVID-19 PANDEMIC

During the COVID-19 crisis, hospitals and other healthcare organizations faced cyberattacks with various motivations. In March 2020, the US Health and Human Services (HHS) Department experienced a DDoS attack on its computer system. It is suspected that the perpetrator was a foreign actor. Analysts believe that the attack was part of a campaign of disruption and disinformation with the goal of undermining the US response to the COVID-19 pandemic. The HHS servers received millions of requests over several hours.[38] The goal of such attacks is to prevent legitimate users from accessing services.[39]

According to the US CS firm FireEye, hackers supporting the Vietnamese government launched cyberattacks against Chinese government organizations during the COVID-19 pandemic. The hacking group APT32 was reported to target email accounts of the personnel at China's Ministry of Emergency Management and the government of the city of Wuhan, where the global coronavirus pandemic started.[40]

The Czech Republic's Brno University Hospital, which was a major COVID-19 testing site in the country, needed to shut down all computers due to cyberattacks.[41] London's Hammersmith Medicines Research, which was conducting clinical trials for new coronavirus medicines, also faced cyberattacks.[42]

1.3 A TYPOLOGY OF CYBEROFFENSES

In this section, we develop a typology of cyberoffenses (see figure 1.4) to help us understand the extent, nature, causes, and consequences of cybercrimes. Such a typology suggests how cybercrimes with certain characteristics and behaviors have a certain probability of targeting a given victim and what reactions and responses they are likely to receive from various actors. A typology also has significant implications for consumers, businesses, and governments in

Figure 1.4 A typology of cyberoffenses

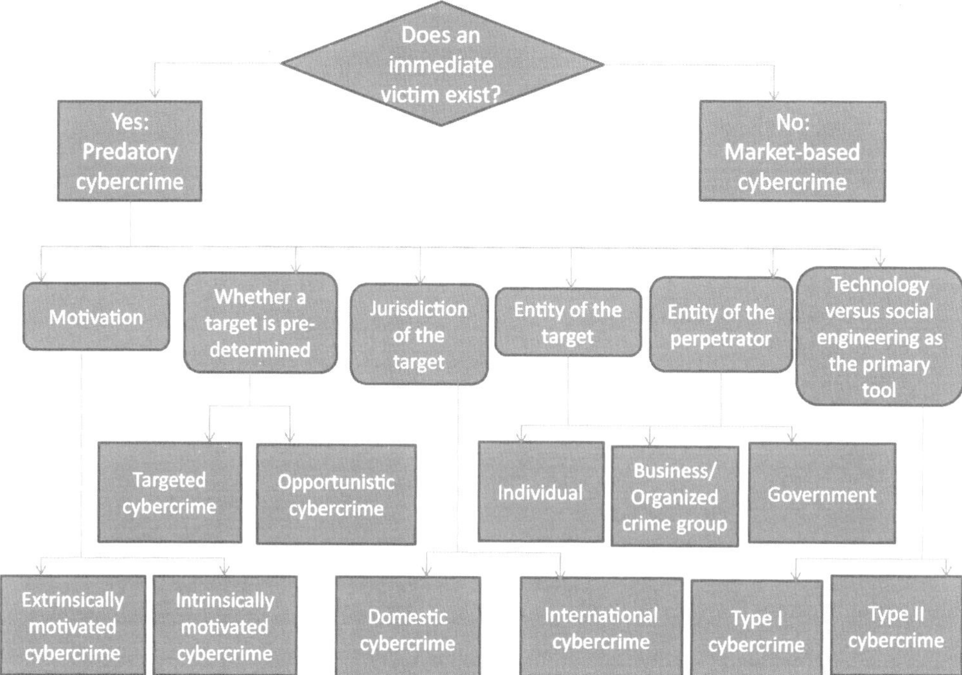

taking precautionary and defensive measures. The typology can also be used to explain shifts in the cybercrime landscape and develop a strategy to fight cybercrimes.

1.3.1 Predatory and Market-based Cybercrime

In market-based cybercrime, goods and services are illegally produced and/or distributed. As in legitimate market transactions, such actions often involve producers, distributors, retailers, and other actors on the supply side and consumers on the demand side. These actions also generate transfers, which are voluntary.[43] One example is China's export of cybercrime-enabling technologies and equipment to international groups. For instance, crime rings involved in identity theft and credit-card forgery in New York City allegedly sourced their skimming equipment, including blank credit cards, from China, Lebanon, Libya, and Russia.[44]

In this section, we focus mainly on predatory cybercrimes. We define a predatory cybercrime as a crime in which an offender inflicts harm or takes property from a victim by using computers or computer networks as the principal means.

1.3.2 Motivations Associated with Cybercrimes

The first issue is to examine the motivations associated with cybercrimes. There is a high prevalence of extrinsically motivated cybercrimes that are committed for financial benefits. For instance, running a botnet is highly profitable for botmasters and botnet controllers. One estimate suggested that a botmaster with a network of 30,000 bots can make monthly profits of US$26,000 by launching DDoS attacks, over US$18 million by engaging in bank frauds, and over US$20 million by using the bots in click frauds.[45]

An organized Russian cybercrime gang, which used Carbanak and Cobalt malware attacks to steal as much as US$1 billion from over 100 banks and financial companies in about 30 nations, serves as a high-profile example of an extrinsically motivated cybercrime.[46] The gang started its activities in 2013 using millions of credit-card details stolen from US-based retailers. After almost five years, Spanish police arrested the gang's alleged leader in March 2018. Around the same time, another member of the group was arrested in Ukraine.[47]

There are also many instances of intrinsically motivated cybercrimes, which are committed for inherent satisfaction rather than an external reward or some separable consequence. Intrinsic motivation based on obligation to the community explains the behaviors of some ideological hackers interested in political goals. Especially, intrinsic motivation based on obligation to the community, associated with China-based cybercrimes, deserves mention. The Chinese hackers, for instance, have expressed patriotic and nationalistic longings and have fought against other countries' hackers or against their businesses and government agencies.

1.3.3 Technology versus Social Engineering as the Primary Tool

Cybercrimes can also be differentiated based on whether technology or social engineering is used as the primary tool. Type I cybercrime mostly contains technological elements while Type II cybercrimes have mainly human elements and use social engineering.[48]

An example of Type I cybercrime, the WannaCry ransomware started on May 12, 2017. It infected about 300,000 computers in 150 countries,[49] costing thousands of dollars in ransom money and billions in lost productivity.[50] Hospitals in the UK's National Health Service (NHS) were forced to cancel surgeries. Ambulances were diverted and patient records were inaccessible. WannaCry used 28 languages to release ransom messages, including European and Asian dialects.[51] The extortionists demanded that individuals pay US$300–US$600 in bitcoin.

Many cybercriminals use social engineering techniques to manipulate people into divulging confidential information to gain access to systems. For instance, in a series of cyberattacks that were traced to China by Symantec, the target firms received emails that asked them to open an attachment. The emails claimed that the attachments were invitations from established business partners or security updates.[52] The 2009 China-originated attacks on Google, dubbed Operation Aurora, relied heavily on social engineering tools. The attackers had communicated with employees in US firms such as Google, Adobe, and Microsoft for a long period to gain trust. The attackers then sent messages asking them to click on websites infected with malware.[53]

Cybercrimes involving social engineering that target consumers are also growing rapidly. This can be attributed to the rapid rate of Internet and e-commerce development in China. For instance, in 2021, China is expected to add more than 31 million Internet users.[54] These novice, inexperienced users tend to lack an understanding of the dangers of phishing and are more likely to be duped by phishers' tricks.

1.3.4 Jurisdiction of the Targets

The jurisdiction in which a cybercriminal is located – domestic or foreign – affects law enforcement's ability to effectively investigate and respond to a cybercrime. It is often difficult to arrest, prosecute, and jail perpetrators who are in foreign countries. The WannaCry ransomware discussed above clearly is an example of international cybercrime. According to the US government, WannaCry hackers were based in North Korea.[55] CS researchers found evidence linking WannaCry ransomware attacks to North Korea.[56] Since North Korea maintains few diplomatic ties, it is difficult to take action against cybercriminals located there.

A substantial proportion of cybercrime activities target domestic victims. For instance, while China-based hackers have been recognized as a major source of international cyberattacks, many Chinese cybercriminals pursue domestic victims. In 2016, hackers in China were reported to compromise over 20 million accounts on Alibaba's Taobao e-commerce website.[57] The hackers used the stolen accounts to order products on Taobao.

1.3.5 Opportunistic and Targeted Attacks

The next issue concerns the predetermined and intentional selection of targets. Opportunistic attacks often entail releasing malware that spreads indiscriminately across the Internet. Opportunistic attacks pose less danger to computer networks.

Targeted attacks are often carried out by skilled hackers with expertise to do serious damage. Some are motivated by financial gain. Targeted attacks are also initiated by terrorists, rival companies, ideological hackers, or government agencies. The general trend in the global cybercrime industry is that cybercrimes are increasingly targeted. Many pursue specific high-value targets and are individualized and customized. Analysts have noted that a large proportion of the most sophisticated cyberattacks aimed at extracting high-value intellectual property, also known as advanced persistent threats (APTs), originate from China.[58] Note that APTs are characterized by a high degree of stealth. They employ sophisticated means to gain access to a network, stay hidden and undetected, and compromise data for an extended period. In order to escape observation, they act quietly, cautiously, and secretly. It is also worth noting that it is "almost impossible to secure a network against a sufficiently skilled and tenacious adversary."[59]

1.3.6 Categories of Targets and Victims (Individuals, Businesses, and Governments)

A critical issue concerns who the targets and victims of the cyberattacks are. (See table 1.1.) There is a wave of cybercrimes targeting consumers. A 2016 survey by the Internet Society of China found that four out of five Chinese Internet users had experienced some type of information leak.[60]

China-originated APTs have targeted governments as well as corporations.[61] There are instances of activists' engagement in intrinsically motivated cybercrimes for which Chinese businesses are the targets and victims. A case in point is the country's biggest dairy operator, China Mengniu Dairy. The company's website, www.mengniu.com.cn, was attacked in 2011 after the company admitted that its milk products contained a cancer-causing substance.[62] Several Chinese news sites, such as Sina.com and People's Daily, displayed Mengniu's defaced

Table 1.1 Some examples of cybercrimes associated with various categories of perpetrators and targets

	Perpetrator		
Target	Government	Business	Individual
Government	• Cyberwarfare • International spying activities	• Organized cybercrime groups targeting government networks	• Intrinsically motivated cybercrimes targeting government networks
Business	• Spying on businesses • Government attacks on business websites (e.g., China)	• Industrial espionage (e.g., stealing trade secrets) • Online extortion targeting businesses	• Intrinsically motivated cybercrimes (e.g., committed by disgruntled employees) • Fighting for ideology (e.g., targeting of multinational companies' websites by ideologist hackers fighting against capitalism)
Individual/ Consumer	• Spying on citizens • Cyberattacks targeting citizens (e.g., the Ethiopian government targeting dissidents)	• Illegitimate companies targeting individuals • Online extortion targeting individuals • Sending email spam	• Intrinsically motivated cybercrimes (e.g., cyberbullying) • Extrinsically motivated cybercrimes (e.g., eBay auction frauds)

website. In a message left on Mengniu's website, a hacker called the company "a disgrace to our nation."[63]

It was noted above that click fraud is the most profitable use of botnets. Some illicit media sites buy fake traffic to defraud advertisers. It was reported that advertising frauds cost the *Financial Times* as much as US$1 million a month. The company managed to stop the frauds by persuading 24 advertising exchanges to drop the ads associated with the frauds.[64]

Finally, government agencies have been victimized. For instance, in October 2001, a hacker replaced a Chinese government agency's website with pornographic content.[65]

1.3.7 Categories of the Perpetrator (Individuals, Organizations/Organized Crime Groups, and Government Agencies)

Another way to classify cyberattacks is by perpetrator. Many cyberattacks may well be the work of individual, non-organized cybercriminals. As an example of an individual cyber-offender, a lone cyberbully may send vulgar messages or insulting remarks to victims.

Organized crime groups employ hackers. In its annual crime-threat report published in March 2017, Europol, which coordinates law-enforcement agencies across the EU, noted that it spends most of its time and resources combating the digitization of organized crime. As a first sign of this new phenomenon in Europe, in 2013, Belgian police arrested some members of a Dutch-Turkish drug gang that had employed two hackers. These hackers used a phishing attack to penetrate a port's computer system. Gaining illegal access to the port's computers, the gang tracked all shipping containers coming into and going out of the port. They were also able to issue permits to take containers that were carrying cocaine from South America.[66]

Likewise, according to the National Police Agency of Japan, about 90% of bank accounts in Japan that received fraudulently transferred money online were opened under Chinese names. The agency suspected that Chinese organized crime groups were behind these frauds.[67]

Some companies have also engaged in cyberattacks on competitors. In 2009, a Chinese online gaming company's attack on rival companies' servers led to an Internet outage in many Chinese cities.[68]

Finally, some authoritarian regimes have engaged in cyberattacks targeting domestic entities. For instance, Chinese government agencies allegedly sent a virus to attack banned websites.[69] Likewise, according to the Toronto-based research institute The Citizen Lab, the Ethiopian government used spyware to target dissidents living in about 20 countries. The victims included Ethiopian media professionals, a lawyer, and a student. Dissidents received phishing emails that contained spyware pretending to be Adobe Flash updates and PDF plugins.[70]

1.3.8 Combining Different Dimensions in the Typology

The different dimensions can be combined to identify and categorize cybercrimes so that a crime in each cell (or quadrant) exhibits the characteristics of both dimensions. As an example, various categories of cybercrimes are plotted onto a 2 × 2 matrix (see table 1.2) that illustrates the location of the target or victim (domestic versus international) on the horizontal axis against the motivation (extrinsic versus intrinsic) on the vertical axis.

As illustrated in table 1.2 and discussed above, digitization of the Chinese economy has increased the opportunities for extrinsically motivated cybercrimes and some domestic industries such as online gaming are attractive (cell I). As illustrated in cell II, intrinsically motivated cybercrimes against domestic targets have various combinations of perpetrators and victims involving individuals, businesses, and governments.

Extrinsically motivated cybercrimes that originate in China and pursue international targets are found to focus primarily on industrial and economic espionage activities involving intellectual property and trade-secret thefts. Nonetheless, cybercrimes characterized by quick monetization that involve data- and credential-stealing malware aimed at committing financial frauds are also traced to China (cell III). Finally, intrinsically motivated cyberattacks that originate in China and pursue international targets are allegedly associated with political espionage activities as well as Chinese nationals' engagement in international cyberwars (cell IV).

Table 1.2 A 2 × 2 matrix of cyberattacks representing jurisdiction of the target/victim and motivation: An illustration of China

	Victims/target	
Motivation	Domestic	International
Extrinsic	[I] ▪ Some industries (e.g., online gaming) are attractive targets ▪ Chinese cybercriminals with a lack of organizational capability to internationalize may focus on the domestic market ▪ Chinese systems with weak defense mechanisms are targeted	[III] ▪ Activities include industrial and economic espionage ▪ Crimes involve data- and credential-stealing malware aimed at committing financial fraud or selling the data to others
Intrinsic	[II] ▪ Politically motivated attacks on Chinese companies ▪ Cyberattacks on Chinese government agencies' websites ▪ Chinese government's attacks on non-complying websites	[IV] ▪ Activities involving political espionage ▪ Chinese nationals' engagement in international cyberwars

1.4 CURRENT STATE OF CYBERSECURITY

Many organizations and individuals have not taken appropriate steps in preparation for rapidly growing cyberthreats. A Harris-IBM survey of 690 city and county employees in the US conducted in early 2020 found that half of them had not done anything in 2019 to increase their preparedness for ransomware attacks. More than a quarter of the respondents had not received any CS training.[71] This is of significant concern due to the increasing prevalence of ransomware attacks. According to Emisoft, state and local governments in the US faced more than 113 ransomware attacks, and K–12 school districts and universities faced 89 such attacks in 2019.[72]

Likewise, the lack of effort to fight cyberattacks targeting power grids is a major concern. While a security breach could have catastrophic consequences, some power providers tend to view security upgrades as "sunk costs." There is little incentive to increase the level of CS. The executive director of a Brazilian company that specializes in securing industrial control systems put the issue this way: "The [power] industry has 20-year-old devices – we have to think of other kinds of tools."[73] Some companies rely on outdated control protocols that fail to take advantage of well-known security practices such as encryption and authentication.[74]

A report by the European Commission (EC) noted that a key challenge is the lack of skilled personnel. The Commission emphasized improving coordination among stakeholders to build necessary expertise to design, build, and maintain smart grid systems that are secure. The Commission's report concluded that education and awareness among electrical power executives needs to be heightened.[75]

A further concern is the lack of comprehensive, integrated, holistic efforts to fight cyberattacks. For instance, it was reported that the German company Siemens and the Swiss company ABB, which dominate the global market for power grid and industrial equipment, are taking measures to make their designs secure. But infrastructure companies have been slow to upgrade their equipment.

Having noted the above CS challenges, it is also worth emphasizing that many organizations and individuals are intensifying efforts and are devoting more and more resources to improving their defenses against rapidly growing cyberthreats. According to Juniper Research's report *The Future of Cybercrime & Security: Enterprise Threats & Mitigation 2017–2022*, global CS spending was US$93 billion in 2017 and will reach US$135 billion in 2022.[76] Cybersecurity Ventures estimates even higher global CS spending: US$1 trillion for the five-year period from 2017 to 2021 or an average of US$200 billion annually.[77] (See figure 1.5.)

1.4.1 Drivers of CS Spending

Organizations' awareness of emerging threats has led to increased CS spending. Various other factors can be considered as additional causes of this rapid increase in CS spending.[78] First, major countries have issued new regulations that emphasize strong CS measures among

Figure 1.5 Global cybersecurity spending (US$, billions)

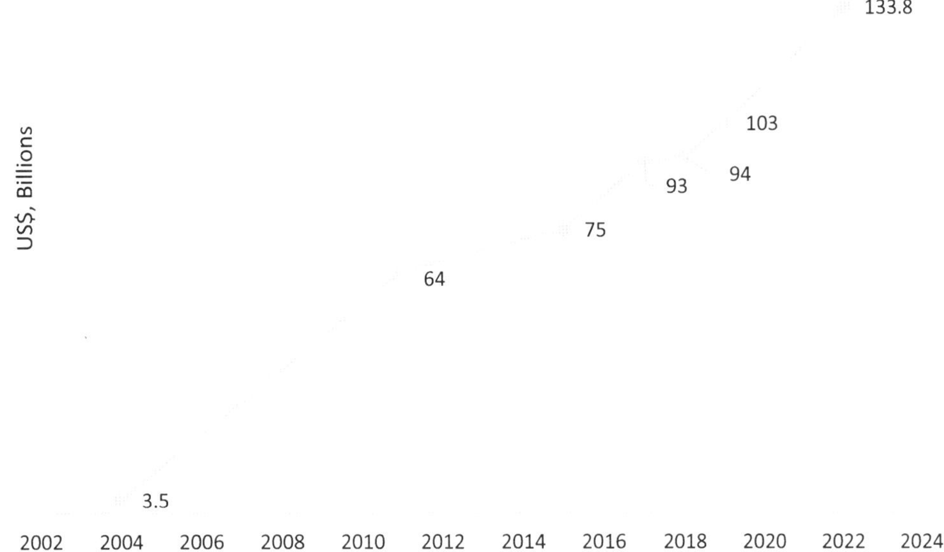

Data sources: Data from 2004 and 2011 from Ross, A. (2016, April 25). Want job security? Try online security. *Wired*. Retrieved from https://www.wired.co.uk/article/job-security-cybersecurity-alec-ross; data for 2017 from Moar, J. (2017). *The future of cybercrime & security: Enterprise threats & mitigation 2017–2022*. Retrieved from https://www.juniperresearch.com/researchstore/innovation-disruption/cybercrime-security/enterprise-threats-mitigation; data for 2018, 2019, and 2022 are estimates of International Data Corporation as reported in Dignan, L. (2019). *Global security spending to top $103 billion in 2019, says IDC*. ZD Net. Retrieved from https://www.zdnet.com/article/global-security-spending-to-top-103-billion-in-2019-says-idc/

organizations. For instance, China's Cybersecurity Law enacted in 2017 requires financial services firms to have IT infrastructures that meet various specifications. They also need to pass standard CS tests and have certifications. The failure to comply can result in fines of over US$150,000 and even criminal charges.[79]

A second factor is shifting customer mindset. According to Mine, a company based in Tel Aviv, Israel, the average Internet user is in the databases of 350 companies, and the top 5% of users had 2,834 companies accessing their data.[80] Today's consumers expect companies to give a high priority to CS and protect their data. According to the *RSA Data Privacy & Security Report*, based on a survey of consumers in France, Germany, Italy, the UK, and the US, 62% of the respondents said that they would blame the company, not the hacker, if their data were breached.[81] Likewise, a survey of 1,000 UK consumers commissioned by FireEye indicated that 72% of consumers would stop purchasing from a company if a CS breach were found to be linked to the boardroom's lack of priority on CS.[82] More than half of consumers warned that they would take legal action against companies if their personal data were stolen or used for criminal purposes due to a data breach and 62% of consumers reported that they would share fewer personal details with companies. For 10% of the respondents, security of their data was the main consideration when purchasing products.[83] In a 2018 global survey conducted by Salesforce Research, 87% of consumers believed that CS had already transformed, was actively transforming, or would transform their expectations of companies by 2023. In terms of changing customer expectations, the survey found that CS was viewed to have a higher transformative capacity compared to the so-called Fourth Revolution (4R) technologies such as AI, IoT, cloud computing, virtual/augmented reality, and blockchain.[84] Customers are thus far more likely to trust a company that has proper CS measures in place to protect their information.

Finally, the evolution to a digital business strategy has stimulated CS spending.[85] Companies are gathering more and more data on consumers. Effective measures to secure sensitive information and maintain privacy are important to build and retain trust in brands.[86]

1.4.2 Cyberinsurance as a Tool to Protect against Future Losses Related to Cyberattacks

Cyberinsurance is emerging as an important tool among organizations to protect themselves against future cyberattack-related losses. (See In Focus 1.4.) During 2018–26, the global cyberinsurance market is expected to increase by a factor of six. (See figure 1.6.)

Figure 1.6 Global cyberinsurance market size (US$, billions)

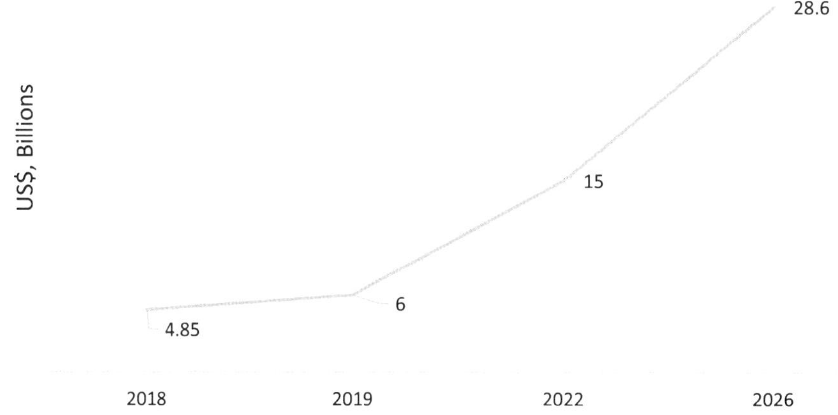

Data sources: Data for 2018 and 2026 from Global Newswire. (2020). *Cyber insurance market is expected to grow $28.60 billion by 2026: Says AMR*. Allied Market Research. Retrieved from https://www.globenewswire.com/news-release/2020/03/31/2009314/0/en/Cyber-Insurance-Market-Is-Expected-to-Grow-28-60-Billion-by-2026-Says-AMR.html; data for 2019 and 2022 from TRC Capital Markets' estimate as reported in Ralph, O. (2019, November 19). Data hacks and big fines drive cyber insurance growth. *Financial Times*. Retrieved from https://www.ft.com/content/751946b2-fb0a-11e9-a354-36acbbb0d9b6

Cyberinsurance expresses a cyberrisk in terms of a dollar value. The cyberinsurance underwriting process can help identify CS gaps and provide opportunities for improvement. Understanding how cyberinsurance functions and the costs of cyberinsurance premiums will help organizational decision-makers increase the effectiveness of the CS budgeting process.[87]

In the US alone, about 200 insurers offered cyberinsurance in 2019.[88] A number of insurance technology (insurtech) companies utilize AI to underwrite cyberinsurance. The startup insurtech Cowbell Cyber offers AI-powered cyberinsurance for SMEs.[89] Likewise, Israel's Sayata Labs uses artificial intelligence and data science to underwrite cyberinsurance for SMEs. Sayata's solutions help insurers assess cyberrisks and offer recommendations to policyholders. Sayata works with insurers such as Axa.[90]

IN FOCUS 1.4: WHAT IS CYBERINSURANCE?

Cyberinsurance provides coverage for the theft or loss of first-party and third-party data. For the loss or theft of first-party data, an insurer may cover expenses related to notifying clients regarding the data breach, purchasing credit-monitoring services for affected customers, and launching a public relations campaign to restore the company's reputation.

Third-party cyberinsurance is important to businesses that are responsible for a client's CS. Cyberinsurance helps them to pay for lawsuits if their actions or lack of action leads to a data breach on a client's system.[91] Some examples of companies that need third-party cyberinsurance include web hosting businesses, IT consultants, software developers, app developers, security consultants, and website designers.[92]

1.5 CHAPTER SUMMARY AND CONCLUSION

Organizations and individuals face complex and diverse cyberthreats. The escalation of cyberthreats has made CS an increasingly critical function. From the typology developed in this chapter, it is clear that cybercriminals with different profiles and personal characteristics follow diverse modus operandi and different structures. Perpetrators also differ in terms of the goals of cybercrime activities. Variations can be found in the characteristics of potential targets for criminal activities, the nature and extent of the damage inflicted on the victims, and the implications to and responses elicited from various actors. Governments and other actors interested in fighting cybercrimes need to adapt their responses to the nature of the cybercrimes and criminals involved.

Some cybercrime activities can be minimized by changing the metrics used in purchasing technologies and services. For instance, many marketers provide incentives for agencies to buy cheap traffic and offer a lower PPC rate. While a given budget can buy a large number of clicks, the quality of those clicks may not be sufficiently high to ensure conversion. The advertiser may be affected by fake clicks. The problem can be managed if advertisers focus on conversions as a metric.[93] For instance, by conducting a PPC audit and analyzing the conversion rates, advertisers can find if the clicks are fake.

Regulators can play a key role in strengthening the CS of organizations and individuals. In addition to new regulations requiring stronger CS measures, governments may need to provide economic incentives, especially to strategically important players such as power providers and equipment suppliers.

1.6 DISCUSSION QUESTIONS

1. What types of costs might a firm encounter if it faces a cyberattack?
2. How is the cyberthreat landscape changing?
3. What are predatory and market-based cybercrimes? Explain with examples.
4. Describe the current state of CS.

1.7 END-OF-CHAPTER CASE: CYBERSECURITY ISSUES FACING POWER GRIDS IN THE US

The US Department of Homeland Security (DHS) has labeled 16 critical infrastructure sectors as vital.[94] A study by the US Cyber Consequences Unit indicated that the costs of a single wave of cyberattacks on US critical infrastructures could exceed US$700 billion, approximately the same as that associated with 50 major hurricanes. US military leaders employ scenario planning to better understand the risks of such cyberattacks – for example, in a confrontation over Taiwan, China might try to cut off the electricity to Fort Bragg, California, to ground US airborne forces.[95]

As table 1.3 shows, there have been several cyberattacks against power companies and grids over the past decade. In a 2019 survey of 1,726 utility professionals from the Asia-Pacific region, Europe, Latin America, the Middle East, and North America, 54% expected an attack on critical infrastructure in the next 12 months.[96]

Of the approximately 30 hacking groups targeting critical infrastructure sectors and industrial control systems that are tracked by the industrial CS firm Dragos, 11 were reported to be focusing on North American power grids.[97] When the US launched a drone strike to kill the Iranian general Qasem Soleimani, tensions between the two countries escalated. Analysts suggested that physical infrastructures such as the electrical grid were a major potential target for Iranian cyberwarriors.[98]

Table 1.3 Examples of cyberattacks on power companies and grids

Date	Cyberattack on Power Companies or Grid Networks	Remarks
2008	▪ A Brazilian energy company's website was defaced.	
February 2011	▪ The Conficker worm infected a Brazilian power plant's system.	
2013–14	▪ A hacking group known as Dragonfly or Energetic Bear, which was believed to operate from Russia, compromised networks of firms in the energy sector and other critical infrastructures. Over 1,000 firms in 84 countries including France, Germany, Italy, Poland, Spain, Turkey, and the US were affected.[99]	▪ The group was first identified in January 2014 by the CS firm CrowdStrike. It used a Stuxnet-like cyberweapon to access power plant control systems. The Finnish security firm F-Secure and others believe that Dragonfly is a Russian hacking group[100] linked to the Russian government.[101]
December 2015	▪ Two Ukrainian power distribution companies – Prykarpattyaoblenergo and Kyivoblenergo – reported that hackers hijacked their systems. ▪ About 225,000 people lost power for up to six hours.[102]	▪ The malware was named Industroyer. ▪ Ukrainian officials blamed Russia for the attack.[103]
December 2016	▪ A transmission facility at the Pivnichna substation outside Ukraine's capital city Kiev experienced a cyberattack. Customers in part of Kiev and a surrounding area lost power for an hour.[104]	▪ Ukrainian security researchers believed this attack and the 2015 attack were launched by the same perpetrators. ▪ FireEye, Dragos, and other security firms attributed the attacks to a hacker group known as Sandworm, believed to operate from Russia.[105]
May 2017	▪ At least four offices of West Bengal State Electricity Distribution Company in India were attacked by WannaCry ransomware.[106]	▪ The offices served about 800,000 households.
January–June 2017	▪ The CS company Symantec reported that a hacking group, which it called Dragonfly 2.0, targeted dozens of energy companies in the first half of 2017. The hackers gained access to more than 20 companies' networks. They were believed to obtain operational access to some power firms in Turkey and the US.[107]	▪ Symantec tracked Dragonfly 2.0 attacks as early as December 2015, but its activities became more aggressive in 2017, especially in Switzerland, Turkey, and the US.[108]
March 2019	▪ Hackers exploited firewall vulnerabilities to cause "blind spots" for a power grid company in the western US for about 10 hours.	▪ The hackers disabled control systems for over 500 megawatts of wind and solar power.

On March 5, 2019, hackers were able to exploit firewall vulnerabilities to cause "blind spots" periodically for a power grid company operating in the western US (California, Utah, and Wyoming) for about 10 hours. This meant that the grid operators were unable to monitor the electricity flow and other activities in the affected area. It was arguably the first known cyberattack that had caused this type of disruption.[109] The hackers disabled control systems for over 500 megawatts of wind and solar power, which is sufficient capacity to serve several hundred thousand homes.[110] The attack is a serious concern because when wind turbines do not generate enough power during low wind periods, utilities must be able to communicate rapidly with other grid operators so that fill-in electricity flows.[111]

1.7.1 Vulnerability of Power Grids

Hackers can use various types of vulnerabilities to launch attacks against power grids. Power-grid attacks are also increasing in severity. ESET security researchers regarded the Industroyer malware that caused a blackout in Kiev in December 2016 to be the biggest threat to industrial control systems since Stuxnet, which did substantial damage to Iran's nuclear program.[112]

Industroyer used a variety of communication protocols.[113] For instance, the malware ran on machines using Windows but also sent commands to specialized hardware products that interacted with the substations.[114]

Malware like Industroyer is particularly dangerous because it enables hackers to do more than carry out industrial sabotage: it gives them operational access to power companies' networks, which means they can directly control the interfaces used to send commands to equipment such as circuit breakers, switches, and disconnectors and halt electricity flow at will. Such malware could be secretly planted and exploited at an opportune time such as during a conflict.[115]

One factor contributing to the vulnerability of power grids is that industrial communication protocols are often standardized across different infrastructures, which limits security. Malware used against one type of industrial control system can simply be "tweaked" to attack a power grid.[116]

Hackers do not need to directly hack the grids to cause substantial damage. For instance, hackers can launch attacks on multiple high-wattage electric vehicle charging stations in a city, which can potentially cause a blackout.[117] One estimate suggested that electric vehicle sales will account for 28% of global light-duty vehicle sales around 2025.[118]

1.7.2 Countermeasures

Regulators, energy companies, and power-grid equipment suppliers and other actors have taken steps to address the problem of power-grid attacks. In the US, the Department of

Energy (DoE) and DHS developed the Electricity Subsector Cybersecurity Capability Maturity Model (ES-C2M2) in 2012 to guide the implementation and management of CS measures and sharing of best practices by utility companies.[119] In May 2017, former president Trump issued an executive order outlining actions for federal agencies to strengthen the CS of power grids and other critical infrastructures. It instructs the DOE and DHS to work with state and local government agencies to identify risks to the US power grid and assess the potential consequences of cyberattacks.[120]

In recent years, US lawmakers have also proposed legislation to study cyberthreats to the power grid, the impact (especially on the military) of a major blackout, and potential solutions, including lower-cost "analog" approaches that involve taking the grid offline.[121] Two bills introduced in 2017 include S. 79 (Securing Energy Infrastructure Act)[122] and H.R. 3855 (Securing the Electric Grid to Protect Military Readiness Act of 2017).[123]

Some argue that launching destructive attacks against the power grid sector will be a difficult task for nation-state hackers because this sector is well-prepared to identify and defeat attackers.[124] The critical infrastructure protection (CIP) standards have also strengthened the CS of power grids.[125] The North American Electric Reliability Corporation critical infrastructure protection plan aims to improve the North American power system's security through a range of efforts such as standards development, compliance enforcement, and assessments of risk and preparedness. Additional efforts undertaken include the dissemination of critical information and creation of awareness regarding key CS issues.[126] Penalties for non-compliance include fines, sanctions, or other actions.

The system for California is broken up into many components. The operators can isolate components readily in order to control the system, which makes it more difficult to take the grid down.[127]

1.7.3 Conclusion

Past cyberattacks were often launched by criminals seeking economic gain or by political activists. In recent years, state intelligence agencies have emerged as the dominant and the most dangerous adversaries. They have the resources, personnel, and tools to infiltrate critical infrastructure networks such as power grids and launch devastating attacks. In the wake of recent alarming incidents, government and industry stakeholders have become aware of the problem but progress remains slow. Although it is not possible to defend against every cyberthreat, the probability of a successful attack can be reduced. In addition to mandating stronger CS, governments should provide additional economic incentives to energy providers and power-equipment suppliers to implement strong CS measures.

CHAPTER TWO

The Relationship and Differences between Privacy and Security

CHAPTER PREVIEW

Internet users are giving increasing importance to data privacy. In some jurisdictions, such as the EU, privacy is viewed as a basic human right. This chapter explains how privacy is related to and different from confidentiality and security. It gives special consideration to fair information practices to address privacy concerns. It takes a look at some organizations' engagement in deceptive privacy practices. It discusses the development of laws related to data privacy worldwide and examines the General Data Protection Regulation in detail. It also gives an overview of various levels of emphasis on privacy and security in different countries.

2.1 INTRODUCTION

In the context of personal information, privacy is tightly linked to and often spoken of interchangeably with confidentiality and security. However, privacy, confidentiality, and security are different concepts. Put simply, privacy refers to the use and governance of personal data. It is concerned with individuals' rights to control their personal information and how such information is used. Personal information includes personally identifiable information (PII) (e.g., name, social security number, address, biometric information), financial information, and information about career, education, health, family, or criminal history.[1]

Some view privacy as a basic human right with a high intrinsic value. This means that privacy has value "in itself" or "in its own right." Many philosophers consider intrinsic value tightly linked to moral judgments.[2] For instance, it can be argued that it is morally wrong to violate individual privacy based on the idea that privacy is objectively valuable in itself and is an essential component of human well-being.[3]

Security, on the other hand, is related to how information is protected.[4] As stated in chapter 1, it includes technical safeguards used to ensure confidentiality, integrity, and availability of data.[5]

In order to understand the difference between privacy and security, let us take an example. Patients share personal information with healthcare providers in order to get services. If the hospital uses a patient's information only to provide the services and protects that data, it maintains both privacy and security. The hospital might sell the patient's information to a pharmaceutical company without the patient's permission. In this case, the patient's privacy is compromised. Security measures are irrelevant in such invasions of privacy by parties that have authority to access the data.[6]

If the hospital faces a cyberattack and cybercriminals steal the patient's information, security is compromised. In this case, if the patient's stolen information is sold on the dark web, privacy is also violated.[7]

2.1.1 Confidentiality

In an exchange relationship, confidentiality involves ensuring that personal information is not disclosed to third parties.[8] In a physician–patient relationship, for instance, a confidentiality clause prevents physicians from disclosing a patient's information to others who are not authorized to see it. That is, disclosing personal information inadvertently or in an unauthorized manner is a breach of confidentiality.[9] However, exceptions to physician–patient confidentiality may occur when confidentiality may result in danger to public health and safety. Since World War II, the trend in many Western countries has been toward increased disclosure if health and safety may be jeopardized when such data is not disclosed.[10] For instance, it will be in the public interest to breach confidentiality if an attorney knows that their client plans to commit murder, suicide, or child abuse.[11] Likewise, in order to ensure public safety from intoxicated or impaired pilots and bus drivers, a physician has a duty to report such information.[12]

2.2 FAIR INFORMATION PRACTICES AND PRIVACY CONCERNS

The fair information practices (FIPs) are an established set of principles for addressing privacy concerns on which modern privacy laws are based. In 1973, an advisory committee of the

US Department of Health, Education and Welfare (HEW) started the process of formally codifying FIPs. In 1980, the Organisation for Economic Co-operation and Development (OECD) used the HEW's work to create a set of eight FIPs.[13] Key principles of FIPs are presented in table 2.1.

Many examples exist of companies that have violated FIPs. For instance, several of Google's services have violated personal privacy. One example is Google Street View.[14] As of December 2019, Google had covered 98% of the world's places where people live and produced 10 million miles of Street View imagery.[15] Google's Street View imagery provides information about homeowners. For instance, researchers have found that different features that are visible on a house's picture (e.g., detached/terraced house, block of apartments, age, condition, etc.) are associated with different probabilities of accident. This information is interesting to insurance companies who can maximize profitability by more precisely predicting car accidents by profiling homeowners.[16]

Table 2.1 Key principles of FIPs

FIP Principle	Explanation
Collection limitation	There should be limits to personal data collection. Such data should be obtained by lawful and fair means and with the data subject's knowledge and consent.
Data quality	Data should be relevant to the purposes, accurate, complete, and up to date.
Openness/ transparency	A general policy is required with respect to the openness of personal data (e.g., existence/nature of personal data, the main purposes of use, and who controls it).
Security safeguards	Reasonable security safeguards should be provided to protect personal data against unauthorized access, destruction, modification, disclosure, or other risks.
Purpose specification	The purpose of data collection should be specified before or at the time of data collection.
Use limitation	Data should not be disclosed or used for purposes other than those specified at the time or before the data is collected except with the subject's consent or to fulfill legal requirements.
Individual participation	Individuals should be given the right to know whether or not organizations have data relating to them. The data subject should be able to challenge such data and have the data erased or amended if the challenge is successful.
Accountability	Organizations should be accountable for complying with measures that ensure the above provisions.

Source: The OECD Guidelines on the Protection of Privacy as reported in Dixon, P. (2006). *A brief introduction to fair information practices*. World Privacy Forum. Retrieved from https://www.worldprivacyforum.org/2008/01/report-a-brief-introduction-to-fair-information-practices/

In 2013, Google acknowledged that its Street View mapping project violated privacy when its cars involved in the project collected Wi-Fi data such as people's passwords and email while gathering information to improve location-based searches. In order to settle a case filed by 38 US states about the project, Google agreed to have its employees follow privacy rules as well as explicitly inform the public how they can defend themselves against such privacy violations.[17]

Google has also been accused of secretly collecting sensitive healthcare data without patients' knowledge and giving its employees access to such information. The accusation is concerned with the initiative "Project Nightingale" carried out with the St. Louis-based Catholic organization Ascension, which is the second largest health system in the US. Ascension has 2,600 hospitals, doctors' offices, and other facilities across 21 US states. Google allegedly participated in the data-sharing scheme without the knowledge of Ascension's doctors or patients. Patients' sensitive information collected and shared under the initiative included lab results, doctor diagnoses, hospitalization records, patient names, and dates of birth. The stated goal of the initiative was to improve medical outcomes with cloud-based AI services. Tens of millions of patients' data were accessed by at least 150 Google employees.[18] Some Ascension employees expressed concerns that Google employees' access to patients' private health information violates privacy.

After media reports of the potential misuse of patients' sensitive information, Google published an FAQ[19] that argued that the initiative followed all relevant regulations. However, such practices of collecting large amounts of personal information about patients without their knowledge or consent and for purposes that they don't expect or understand obviously lack an ethical foundation.[20] Such practices also violate the transparency principle.

Google also reportedly found a flaw with its social-networking site Google Plus, which exposed personal data such as birth dates and contact information of hundreds of thousands of subscribers. The company, however, decided not to disclose the flaw to users due to fears of regulatory scrutiny. In this case also, Google argued that it followed relevant legal requirements in its decision not to inform users.[21]

Likewise, there have been accusations that Facebook violated fair uses of personal data. It was reported to have data-sharing agreements with at least 60 device makers, such as Apple, Samsung, and many other companies. Sensitive data about users' friends, such as relationship status, religion, and political leaning were allegedly shared.[22] In April 2014, Facebook announced that it would not allow third-party developers to collect data about the app users' friends. However, it failed to fulfill this commitment. Developers were to stop collecting data as of April 2015 but it was reported that user data was shared via third-party apps until June 2018.

Facebook's content-labeling projects undertaken to develop AI also allegedly violate users' privacy. As of mid-2019, Facebook was reported to have 200 such projects that employed thousands of people globally. In one such project, about 260 contract workers in India of the outsourcing firm Wipro were given access to millions of Facebook users' photos, status updates, and other content that had been posted since 2014. The items posted were categorized in terms of various dimensions such as the subject of the post, the occasion on which it was posted, and the author's intention.[23]

2.3 THE GDPR: THE MOST COMPREHENSIVE DATA PRIVACY LEGISLATION

As noted above, FIPs provide basic principles for addressing privacy concerns. However, these principles are vague and ambiguous and thus hard to implement. Consequently, many companies exploit regulatory loopholes. The European Consumer Organization's (Bureau Européen des Unions de Consommateurs [BEUC)]) analysis of 14 of the largest Internet companies' policies found that they were too vague and did not comply with the General Data Protection Regulation (GDPR). The BEUC argued that technology companies such as Facebook, Google, and Amazon use language that is unclear. While these companies required users to agree with their policies, they provided insufficient information for users to judge what they were agreeing to.[24] In order to address such concerns and to improve legal clarity and legal certainty around data privacy, many jurisdictions are revamping their regulatory systems.

The EU's General Data Protection Regulation was adopted by the European Parliament in April 2016 and went into effect on May 25, 2018. It replaced the outdated Data Protection Directive enacted in 1995. The GDPR arguably was enacted to apply to any technology. (See In Focus 2.1.)

IN FOCUS 2.1: THE GDPR AND BLOCKCHAIN

While the GDPR was enacted to apply to any technology, there has been quite a lot of confusion as to how solutions based on blockchain can be designed to comply with the GDPR. Blockchain solutions developers currently operate in an uncertain regulatory environment.

The European Parliament's resolution on distributed ledger technologies and blockchain (the "Blockchain Resolution"), passed in October 2018, affirmed that blockchain applications pseudonymize users but cannot anonymize them. Note that anonymization makes it impossible to connect the data to a natural person. Pseudonymization, on the other hand, makes it possible to re-identify the data and, with additional information, connect the data to a natural person. The fact that data on blockchain is only pseudonymized, but not anonymized, makes data stored in blockchain's ledgers subject to the GDPR.

The GDPR assumes that there is a data controller. Data subjects enforce their data protection rights against the controller. Blockchain's decentralization feature means that there is no single center of control. There has been a lack of regulatory guidance as to how the data controller is determined in a blockchain network.[25] When personal data is abused, it is difficult to determine who should be legally accountable.[26]

Blockchain's immutability feature is also of concern. When a block is added, it is extremely difficult or even impossible to delete or modify data in the block. The difficulties of deleting blockchain data violates data minimization and purpose limitation provisions of the GDPR. The idea here is that personal data should not be held longer than needed to achieve the purpose for which the data is collected. This is clearly stated before the data is collected. Furthermore, data must be collected for "specified, explicit and legitimate purposes and not further processed in a manner that is incompatible with those purposes."[27]

Blockchain's functioning is also incompatible with the "right to be forgotten" article. This provision allows a data subject to request the deletion of personal data or limits the duration for which it can be stored.

The recommendation of the French data protection watchdog Commission nationale de l'informatique et des libertés (CNIL) has been against the use of permissionless blockchain. Avoiding such blockchain networks, parties can come to a careful agreement regarding various members' data responsibilities. The CNIL has also suggested developers look for "other solutions" if possible until the agency provides legal clarification.

The GDPR is viewed as the most significant change to EU data privacy legislation. It has dramatically changed the processes that organizations need to follow in order to track consumers' online behaviors and process that data. Under the GDPR, companies are required to obtain specific legal bases in order to use customers' data or track their behaviors. Many companies choose consent as the legal basis.[28] The GDPR requires companies to have an explicit opt-in consent from customers in order to keep personal data.

By January 27, 2020, 160,921 personal data breaches were reported in the EU, which resulted in over US$126 million in fines.[29] Figure 2.1 presents data for select countries.

Figure 2.1 GDPR violations and fines imposed in select EU countries (May 25, 2018 to January 27, 2020)

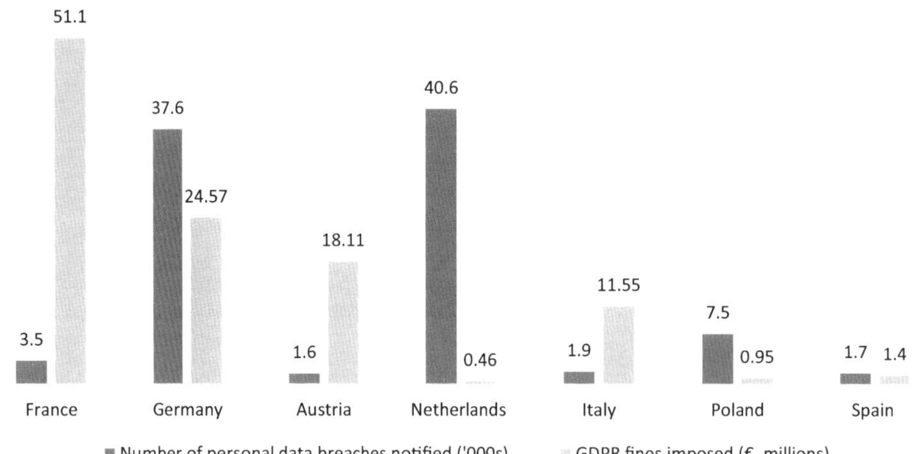

Data source: *DLA Piper GDPR data breach survey*. (2020, January). Retrieved from https://www.dlapiper.com/en/us/insights/publications/2020/01/gdpr-data-breach-survey-2020/

There is a wide variation in GDPR enforcement across EU countries. Austria, France, and Germany are the top three EU countries in terms of GDPR fines. As of January 2020, several countries, such as Estonia, Finland, Iceland, Liechtenstein, Luxembourg, and Slovenia, had not issued financial penalties.[30] Especially, Ireland and Luxembourg, where many technology companies' international headquarters are located, are reported to lack the resources to handle excessive caseloads of GDPR violations.[31]

2.3.1 Key Provisions of GDPR

According to the European Commission, a wide range of information is considered personal data. This includes information related to an individual's private and professional life. Some examples of personal data are names; government ID numbers; physical and email addresses; credit-card and banking information; physical health, mental health, and genetic data; biometric data; racial, cultural, or ethnic information; and data related to online and offline activities such as location information and IP addresses that allow marketers to track users.[32]

GDPR gives website visitors various rights. They include the right to receive specific and up-to-date information on the type of data collected, the purpose of the data collected, and where the data is sent. Website visitors also have the right to prevent companies from collecting, storing, and processing data about them.[33]

The GDPR has also made it mandatory to notify the relevant supervisory authority of any data breach within 72 hours of becoming aware of such breach. The "right to be forgotten" article provides data subjects with the right to request that a data controller erase their personal data. Data controllers are also required to stop further processing of such data and potentially stop third parties from doing so. The "privacy by design" article requires data controllers to hold and process only data that is absolutely needed to perform their duties.

GDPR aims to harmonize the governance of personal data (such as credit-card, banking, and health information) across the EU member states. In order to comply with the GDPR, organizations are required to have greater oversight and control over where and how personal data is stored and transferred. Access to personal data also needs to be rigorously policed and audited.

The GDPR has also substantially increased penalties for data breaches. For breaches that are serious, the fines could be as much as 4% of the concerned companies' global annual revenue, or €20 million, whichever is greater.[34]

Consent may not be needed for cookies that collect non-sensitive personal data such as those used to track items in a shopping cart. However, cookies used to collect personal data that are tied to users require user consent.

The GDPR has certain provisions and principles. Implicit consent was considered sufficient in the pre-GDPR era to process personal data. Such consent is implicitly granted by the data subject's actions and the circumstances of a situation (e.g., person's silence or inaction). Under the GDPR, data subjects need to be informed in explicit terms as to what data will be collected and why. Approaches that do not give users a free choice to say "yes" or "no" may be considered "forced consent."[35]

The principle of transparency emphasizes that information addressed to a user needs to be in clear and plain language, concise, accessible, and easy to understand. In addition, if it is appropriate, information visualization must be used to help users to understand easily. Even after a website has obtained valid consent, visitors should be provided with an easy way to withdraw consent.[36]

Regulators, such as the UK Information Commissioner's Office (ICO) and the French Data protection authority Commission nationale de l'informatique et des libertés (CNIL), have been clear that a cookie notice needs to avoid lengthy, technical details.[37] Another key provision is that entry to a website should not require a cookie consent. A cookie wall requiring a consent to enter is not "freely given" consent.[38]

The attention so far has mostly been confined to cookies and other online tracking mechanisms through traditional computers and browsers, as well as users' devices such as wearables and smart televisions. For instance, the ICO and CNIL have made it clear that the E-Privacy Directive applies to any technique used to read from or write to "terminal equipment." The directive also covers device fingerprinting, the practice of combining different pieces of information such as operating system, browser, fonts, and clock information to identify a unique device. However, there is a lack of practical recommendations as to how firms should obtain informed consent on those devices.[39]

The GDPR may have significant effects on companies that use wearable devices and other methods of monitoring employees. For example, a business that employs fleet tracking will see changes in the company's right to record data on employees' movements and performance. Employees need to be informed in explicit terms as to what data will be collected and why. The GDPR requires legitimate reasons to process employees' personal data. The data can be used only for the purpose that was specified before the data collection. Companies must ensure that employees understand their right to ask for a copy of data in which they can be clearly identified. Such information must be supplied within 30 days.[40] Likewise, employers need to conduct detailed privacy impact assessments when they use other mechanisms such as chip implants to track their employees.[41]

Article 45 gives the European Commission the power to determine whether a country outside the EU provides an adequate level of data protection. If a third country meets the adequacy standard, data flow from the EU (and Iceland, Liechtenstein, and Norway) to the third country is treated in the same manner as intra-EU transmissions of data.[42] That is, data can flow to the country without additional safeguards. As of December 2020, the EC recognized that the following jurisdictions provided adequate data protection: Andorra, Argentina, Canada (commercial organizations), Faroe Islands, Guernsey, Israel, Isle of Man, Japan, Jersey, New Zealand, Switzerland, and Uruguay.[43]

2.3.2 Examples of Breaches and Fines Imposed under the GDPR

As mentioned, a key element in the GDPR is the role of consent or the user's approval to process personal data, which must be freely given rather than obtained from direct or indirect coercion. The process should also be unambiguous, transparent, specific, informed, and simple to understand.[44] Many companies' consumer tracking practices fail to address several of these major issues and thus they lack the legal basis to process personal data. As a result, official complaints against many companies have been filed by privacy activists, consumer protection agencies, and government regulators since the first day the GDPR went into effect.[45]

In table 2.2, we list and describe some examples of complaints and formal regulatory actions against some companies that have been initiated by regulators in several EU countries as well as GDPR-related breaches and fines. Critics have questioned these companies' legal basis for processing personal data because they lacked mechanisms to obtain consents that must be "specific, informed and freely given." This table also illustrates FIPs violated as discussed in the previous section.

Concerns have been expressed about Microsoft telemetry data, which is collected from remote or inaccessible points and automatically transmitted to receiving equipment for the

Table 2.2 Examples of some organizations' engagement in deceptive privacy practices and GDPR-related complaints filed against them

Company	Unfair and Harmful Practices Involving Private Data	Complaints and Regulatory Actions
Facebook	• Facebook allowed third-party developers to collect data about app users' "friends" violating an agreement to not collect "friend" data (collection limitation principle). • Private information of about 87 million users was shared illegally with Cambridge Analytica (use limitation principle). • To facilitate content labeling projects to develop AI programs, contract workers of an outsourcing firm were given access to millions of users' personal information (use limitation principle). • In December 2019, more than 267 million Facebook users' personal information was stolen (security safeguards principle).	• Facebook was criticized for its "take it or leave it" position in consent: it allegedly blocked accounts of users who did not give consent (no free choice). • As of August 2019, there were 11 cases against the company in Ireland. • In July 2019, the European Court of Justice ruled that Facebook plugins such as "Like it" in third-party websites violate the GDPR.
Twitter	• In May 2019, a "bug" in Twitter's iOS app enabled sharing of user location data with an advertising partner[46] (use limitation principle). • Twitter refused to give users information about how they are tracked (openness/transparency principle).	• In September 2018, a GDPR complaint was filed with Ireland's DPC arguing that Twitter exposed users' information to advertisers when users visit a website. The process, known as a "bid request," fails to protect personal data against unauthorized access. • Irish privacy authorities investigated Twitter's refusal to give users information about how they are tracked when links in tweets are clicked.
Microsoft	• Telemetry data collection allegedly collected sensitive data inappropriately. It did not provide users an option to turn off the feature (openness/transparency principle).	• In July 2019, the German state Hesse banned Office 365 in schools.[47] • In July 2019, the Dutch government asked government institutions to avoid Office Online and the Office mobile apps.[48]

(Continued)

Table 2.2 Continued

Company	Unfair and Harmful Practices Involving Private Data	Complaints and Regulatory Actions
Deutsche Wohnen	▪ Deutsche Wohnen was retaining tenants' personal data for an unlimited period without considering the legitimacy and necessity of the retention (use limitation principle).	▪ Berlin DPA issued a US$16.1 million fine.
Google (see the end-of-chapter case: France Issued the Biggest Fine under the GDPR against Google)	▪ The Street View mapping project violated privacy (collection limitation principle). ▪ Project Nightingale was a secret data-sharing scheme with Ascension in which sensitive patient information was accessed by at least 150 Google employees (collection limitation principle). ▪ A flaw in Google Plus exposed personal data of hundreds of thousands of subscribers (security safeguards principle).	▪ In January 2019, the CNIL announced a fine of €50 million.

purpose of monitoring. Microsoft allegedly collects data about the use of its Office apps (e.g., Word, Excel, and PowerPoint) and records and stores them without informing people. The collected data include sensitive personal information such as email addresses, email subject lines,[49] and requests for translation service through the Office software.[50] Such data are produced by the system-generated event logs.[51] According to a report prepared by the firm Privacy Company that was commissioned by the Dutch government, Microsoft did not provide users an option to turn off the feature. This violates a key provision of the GDPR.

Similarly, Irish privacy authorities investigated Twitter regarding its refusal to give users information about how they are tracked when links in tweets are clicked. When users put links into tweets, Twitter applies its link-shortening service, t.co, to them. Twitter's stated goal of the link-shortening service is to measure the number of times the link has been clicked. The company argues that such information helps to fight malware. Privacy advocates have argued that Twitter actually gets more detailed information when its users click the shortened links. It is suspected that Twitter might leave cookies in users' browsers and use the information to track people when they visit other websites.[52]

Likewise, in late 2019, the Berlin Commissioner for Data Protection and Freedom of Information (Berliner Beauftragte für Datenschutz und Informationsfreiheit [Berlin DPA]) issued a €14.5 million (US$16.1 million) fine on the largest private real estate owner in Berlin, Deutsche Wohnen.[53] Berlin DPA had conducted onsite inspections of Deutsche Wohnen's data retention policy in June 2017 and March 2019. Deutsche Wohnen was managing 111,000 apartments[54] and retained tenants' personal data for an unlimited period without considering the legitimacy and necessity of the retention. Tenants' financial and personal data such as bank statements, training contracts, tax, social insurance, and health insurance information were affected. The data archiving system used by the company lacked mechanisms to remove data when such data were no longer required for the purpose for which they were collected.[55]

2.4 DATA PRIVACY REGULATIONS IN THE REST OF THE WORLD

There are some old as well as new privacy laws that have been put in place in countries worldwide. Many new privacy laws have been inspired, at least in part, by the GDPR.

2.4.1 The US

Compared to the EU, the US provides more autonomy to businesses regarding the way they disclose and store personal information. The US government does relatively little to restrict companies from tracking people and selling information to third parties. US companies also have more autonomy to decide how much information they want to reveal about their data practices.[56]

Organizations' data privacy and security frameworks are subject to a number of federal laws and regulations enacted to protect the privacy, security, and confidentiality of specific categories of data and information. A further observation is that there is heterogeneity in the implementation and interpretation of CS across the US states. In addition to federal laws, there are about 47 different state laws in the US regarding how people should be notified in case of a data breach involving personal information.[57]

The Federal Trade Commission (FTC) is the key federal agency that regulates how consumers' personal information is handled under Section 5 of the FTC Act. The Act gives the FTC authority to investigate "unfair and deceptive acts and practices in or affecting commerce."[58] Other important laws include the Health Insurance Portability and Accountability Act (HIPAA) to protect sensitive health information; the Sarbanes–Oxley Act (SOX) to protect shareholders and the public from accounting errors and frauds in enterprises and to improve the accuracy of corporate disclosures; the Gramm-Leach-Bliley Act to regulate financial information; the Fair Credit Reporting Act and the Fair and Accurate Credit Transactions Act to cover information used in credit, insurance, and employment decisions; and the Children's Online Privacy Protection Act to regulate personal information obtained online from children under the age of 13.[59] For instance, HIPAA requires healthcare providers to have measures in place to protect patient data privacy, integrity, and availability. Those not complying with the HIPAA standards might face up to US$1.5 million in fines and 10 years in prison.

2.4.1.1 The California Consumer Privacy Act (CCPA)

Among the most notable recent developments is the CCPA, which became effective on January 1, 2020. The CCPA is largely modeled after the GDPR but is less stringent. For this reason, the CCPA is often referred to as GDPR Lite.

Both the GDPR and CCPA strongly emphasize transparency. For example, in order to comply with the CCPA, a business is required to include a section in its privacy policy that describes the rights of consumers in California. It must also provide clear instructions regarding how consumers can opt out of the sale of their personal information.

The consent process is one of the major differences between the two laws. The GDPR requires businesses to have a legal basis to collect and process data. Most businesses rely on user consent as a legal basis. This means that they must first obtain the user's consent before the data are collected. Under the CCPA, businesses can collect data from consumers without acquiring consent while consumers can opt out of the sale of their information after it is collected.[60]

2.4.2 Asia

The overall landscape of privacy legislation in Asia is heterogeneous and fragmented. As of February 2020, Japan was the only Asian country that was granted an Adequacy Decision by the EC.[61] In other Asian countries, privacy laws do not offer the same level of safeguards as in the EU and Japan.[62]

South Korea enacted the Personal Information Protection Act (PIPA) in 2011. In addition, the country has sector-specific legislation to protect data privacy.[63] The PIPA requires private- and public-sector entities which collect information that identifies a specific person to meet strict compliance requirements. Individuals also have data ownership rights as well as the right to be forgotten. Government agencies, however, can collect and use personal data without obtaining consent in order to use them for public interest purposes.[64]

In China, a key driver of data privacy regulations is the public's increased awareness of their right to privacy. The Employment Services and Management Regulations require employers to keep certain employee data confidential. Similarly, Certain Regulations on Standardizing the Order of the Internet Information Service Market (2012 Regulations) provided a legal definition of personal information. Likewise, the 2012 Online Data Protection Regulation bans the sale and distribution of personal information without the owner's consent. It requires ISPs to ensure the security of personal data and prevent misuse as well as provides consumers the right to seek deletion of personal data posted without consent and to sue for violations.

On the other hand, the Chinese government's use of state power to get unlimited access to citizens' personal information for surveillance purpose has been a concern.[65]

As of April 2020, India lacked a comprehensive data privacy law to protect personal data. In December 2019, India's lower house of the bicameral Parliament, Lok Sabha, introduced the Personal Data Protection Bill, 2019. It specifies how data of Internet users are stored, processed, and transferred. The bill criminalizes illegitimate re-identification of user data but gives the government the power to exempt its agencies from the law.[66] This means that if Indian government agencies ask social media companies for the personal information of their customers, the social media companies must comply.[67] As of December 2020, a Joint Parliamentary Committee of both the Houses was reviewing the bill and 228 amendments had been submitted to the Committee.[68]

2.4.3 Latin America

Several Latin American countries had data protection policies in place for a long period. However, after the GDPR's implementation in the EU, significant amendments and improvements in privacy laws were made by some of these countries. For instance, Argentina passed its Personal Data Protection Law (Law No. 25,326) in 2000, which regulates personal data's processing and protection.[69] Subsequently, Argentina took initiatives to change its data protection laws to align with the GDPR. In 2018, a bill to replace Law No. 25,326 was proposed. The bill contains key provisions of the GDPR, such as a requirement for governmental agencies processing sensitive and big data to appoint a data protection officer and standards for the lawfulness of data processing.[70] Likewise, the Brazilian General Data Protection Law (Lei Geral de Proteção de Dados [LGPD]) was ratified in the Congress in mid-2018. The LGPD became effective on September 18, 2020.

Some countries in the region already meet the EU standard for adequate data privacy protection. As of February 2020, the European Commission had recognized Argentina and Uruguay as jurisdictions that provide adequate protection.

2.4.4 Africa

In most countries in Africa, underdeveloped regulatory regimes fail to provide adequate data privacy protection. As of mid-2018, only 23 out of 55 nations in Africa had passed or drafted personal data privacy regulation. Only nine of them had data protection authorities.[71]

In 2014, the African Union's convention on CS and personal data was adopted. As of February 2020, 14 nations had signed it and 5 had ratified and/or accessed the convention.[72]

Kenya is ahead of other African countries in the development of data privacy regulations. In November 2019, Kenya passed its new data protection law, which has many elements of the GDPR. For instance, it requires organizations to inform users about why their data is being collected, for what purpose that data is used, and for how long the organization will store the data. The bill also has a provision that gives consumers the right to request organizations delete their data. Any violation of the law will be investigated by an independent office, and violators may face a maximum fine of 3 million shillings (US$29,283) or two years in jail.[73]

The new law is a significant step toward comprehensive data protection. Like other East African countries, digital lending apps in Kenya that engage in predatory lending practices are accused of violating consumers' privacy. These apps determine potential borrowers' creditworthiness by accessing data from their phones such as SMS, call logs, bank balance messages, and bill payment receipts.[74] The new law makes such practices illegal.

In addition, the law requires organizations to have a certain level of security standards for storing data. Its data localization provision makes it illegal to send Kenyans' personal data outside the country. Critics have argued that if this provision is implemented, the Kenyan economy will not be able to benefit from the economic gains that arise from cross-border data flows.[75]

2.5 LEVELS OF EMPHASIS ON PRIVACY AND SECURITY

Countries around the world have placed various levels of emphasis on privacy and security. Political, social, cultural, and historical contexts shape the formation of privacy-related institutions. An observation is that while the use of radio-frequency identification (RFID) to automate the tracking and monitoring of people is becoming a serious concern in the West, privacy concerns are less prominent in Asia. In general, there is lower expectation of privacy in Asian countries compared to the West.

In the EU, individuals' privacy and freedom of expression have almost the same level of protection. The European Convention on Human Rights states the EU's fundamental principle of privacy. Especially due to decades of state surveillance by the Nazis and the East German communists, Germans are more anxious and fearful about the misuse of personal information. A researcher at Deutsche Digital Institut described this as "a deep collective experience" rather than "paranoia."[76] The EU has set a baseline common level of privacy in order to protect privacy rights of its citizens irrespective of the data location. On the other hand, the US, where freedom of expression is given more weight than privacy at the federal level, follows a self-regulatory approach and has sector-specific regulations for sensitive data.

In the EU, increased consumer activism has been a key force strengthening privacy protection. This can be contrasted with China where nongovernment entities, special interest groups, and the civil society are organized loosely. There is little room for these groups to influence national policy-making.

Chinese consumer fear of and anger against corporate misuse of personal data is arguably used as a convenient distraction by the Chinese government while it intensifies its own data

Figure 2.2 China's CS spending (US$, billions)

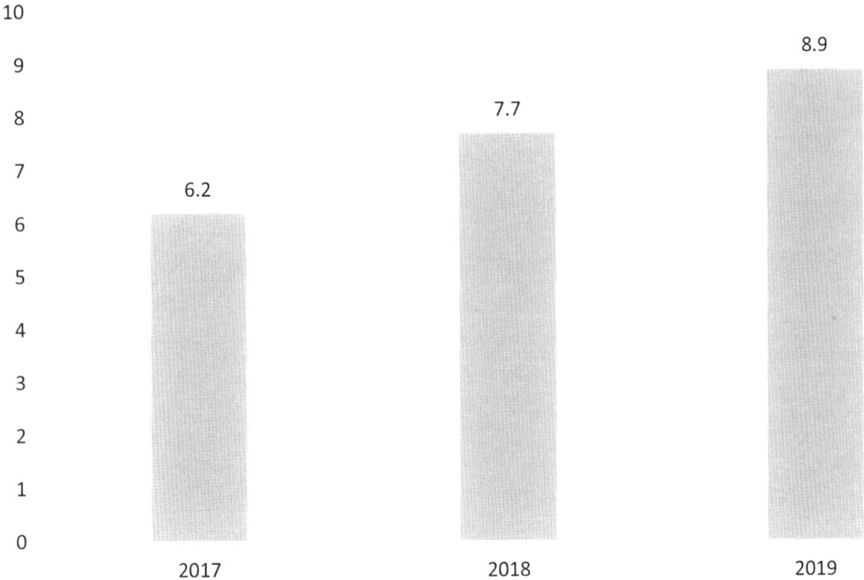

Data source: China Academy of Information and Communications Technology (CAICT) as reported in Ferreira, L. (2019, September 19). *Cybersecurity in China is a business worth $8.9 billion*. CGTN. Retrieved from https://news.cgtn.com/news/2019-09-19/Cybersecurity-in-China-is-a-business-worth-8-9-billion-K6RRVvwmgU/index.html

collection initiatives for surveillance.[77] Especially with the help of its technology companies, the Chinese government has elevated the notion of AI-based surveillance to a new height.

Yet, despite the lower level of protection of consumer rights in China, the country's CS performance has been impressive. As of 2019, China had about 3,000 CS companies.[78] According to FireEye's Cyber Trendscape 2020, 39% of Chinese organizations had mature cyberattack and breach event response and communications plans. China had the second highest proportion of organizations with such plans, just behind the US.[79] The country's CS spending is growing rapidly. (See figure 2.2.)

2.6 CHAPTER SUMMARY AND CONCLUSION

While there is a considerable overlap between privacy and security, these are two distinct issues. From an individual's perspective, personal privacy is an issue that is associated with intrinsic value. Regulators face an increased demand from citizens to enact and enforce privacy laws. Different countries have enacted a number of regulations and established various institutions to tackle data privacy issues. CS, on the other hand, is important because of possible economic and national security consequences. Most jurisdictions are thus enacting CS regulations and strengthening enforcement mechanisms.

There is a wide variation in the emphasis placed by various regulatory agencies on these two issues. Regulations such as the GDPR place major emphasis on privacy. Others, such as New York's regulations for the financial sector, mainly focus on CS.

2.7 DISCUSSION QUESTIONS

1. How do privacy, confidentiality, and security differ?
2. What key principles and provisions of fair information practices address privacy concerns?
3. Which provisions of FIPs are most likely to be violated by companies?
4. What are the key features of the GDPR?
5. How do social, cultural, and historical factors affect the development of laws, regulations, and policies related to privacy?
6. What are some examples of organizations that engaged in deceptive privacy practices? How did those practices violate the principles of FIPs?
7. What is the current status of privacy regulations in African and Latin American countries?

2.8 END-OF-CHAPTER CASE: FRANCE ISSUED THE BIGGEST FINE UNDER THE GDPR AGAINST GOOGLE

In January 2019, the French Data Protection Authority CNIL announced a fine of €50 million (US$57 million) to Google "in accordance with the General Data Protection Regulation … for lack of transparency, inadequate information and lack of valid consent regarding the ads personalization."[80] The CNIL had been investigating Google's privacy practices for some time. In September 2018, for instance, the CNIL studied the information that Google makes available to users to create a Google account on an Android phone. Users were presented with most of the information required by the GDPR, such as the purposes of data processing, duration of data storage, and categories of personal data. But the consent process it followed was not up to the mark based on some general requirements such as transparency and simplicity in order to obtain valid consent.[81]

To access complementary information, users are required to click on many buttons and links. The CNIL found that in some cases, as many as five or six actions were needed to access the relevant information. The CNIL also noted that some information lacked clarity and comprehensiveness.[82] When new users signed up for an account, they needed to click through a special section to learn about the way their data is processed by Google. The information was disseminated across many documents.

2.8.1 Legal Challenges in Other Jurisdictions

France is not the only jurisdiction where Google has faced a legal challenge. Norway's Consumer Council and other groups have argued that Google lacks a legal basis to track users through Location History and Web & App Activity, and then to process personal data. These settings are integrated into Google accounts. Complaints have been filed on the grounds that Google's use of the Location History and Web & App Activity menus, especially for users of Android smartphones, are misleading.[83] Even if users turn off the Location History option, Google continues to collect location data and track consumers through services such as Google Maps, weather updates, and browser searches. In order to stop being tracked, users need to turn off Web & App Activity through settings.[84] Users find it difficult to avoid being surveilled.[85] The plaintiffs argued that Google uses "forced consent" and thus presses/coerces consumers into consenting to processing data without understanding the details. Critics such as European consumer organizations have raised important objections to this approach, saying that the setting fails to meet GDPR standards. The relevant information about Location History should not be hidden, should not be in submenus, and should not require extra clicks. According to Norway's Data Protection Authority, an additional issue was lack of clarity regarding how the collected data are used.[86]

2.8.2 Privacy Activism

The EU's consumer activists have also been pressuring regulators to clamp down on the violators of the GDPR. In November 2018, the BEUC had filed complaints about "deceptive" location tracking.[87] In November 2019, the BEUC sent another letter to the Irish Data Protection Commission (DPC) expressing concerns about the regulator's lack of priority in enforcing the GDPR: "One year after these complaints were filed, it has yet not been decided whether Google infringed the GDPR." [88]

In 2018, a complaint against Google was filed before the Irish (DPC) by the chief privacy officer (CPO) of the privacy-centric web browser Brave, the Open Rights Group, and University College London. Google's Ad Exchange system allegedly leaked personal data to more than 1,000 companies without users' consent or their ability to take actions to stop the practice.[89] The complaint labeled it as a "massive data breach."[90] In February 2020, the Irish DPC announced a probe into the legality of Google's processing of location data and the transparency issues related to the processing.[91]

2.8.3 Google's Response

Google has made some significant steps toward improving its privacy policy. One such change has been in its policy about third-party cookies, which are created by websites other than the one a consumer is visiting directly. Such cookies are used for activities such as cross-site tracking and ad serving. Currently Google relies on third-party cookies for delivering ads on other sites. Google announced that it would phase out third-party cookies from its popular Chrome browser by 2022.[92] It plans to launch what it refers to as Privacy Sandbox as the alternative. Privacy Sandbox centralizes data by using an application programming interface (API) in Chrome. The data will be accessible to marketers only when the browser determines that user activity will be anonymous.[93]

CHAPTER THREE

The Economics of Cybercrimes

CHAPTER PREVIEW

Technology and skill intensiveness, a high degree of globalization, and newness make cybercrimes structurally different than conventional crimes. This chapter analyzes the workings of cybercriminals' minds in order to understand the real and perceived costs and benefits that underlie their misbehavior. Special consideration is given to international institutions; rules that affect the political economy of cybercrime are also analyzed. The chapter also tries to describe how cybercriminals in different economies face different cost-benefit considerations in their decisions to engage in criminal activities. The chapter suggests that characteristics of cybercriminals, cybercrime victims, and law enforcement and judiciary systems reinforce each other and form a vicious circle. The chapter then proposes some mechanisms to break the vicious circle of cybercrimes and to alter a hacker's cost-benefit structure.

3.1 INTRODUCTION

In order to identify the best ways to protect their information assets, organizations need to understand what their cyberadversaries are after. Especially, there is the need for a clearer articulation of the structures of costs, benefits, and attractiveness of cybercrimes and basic mechanisms that underlie cybercriminals' behavior.

Three factors contribute to the structural uniqueness of cybercrimes: technology and skill intensiveness, a higher degree of globalization than conventional crimes, and newness. First, unlike conventional crimes against persons or property such as arson, burglary, and murder, most cybercrimes are very skill intensive. Even persons who lack the expertise to write their own scripts or codes and thus use existing tools to hack into computers (known as script kiddies), often to commit victimless and/or marginal cybercrimes, possess more skills than their conventional world counterparts. It is thus important to know the distribution of skills required to engage in cybercrime activities and economic opportunities that individuals with such skills face.

Second, given the global nature of the Internet, there are important procedural and jurisdictional issues associated with cybercrimes. Although some progress has been made in the international harmonization and standardization of CS norms, there is a long way to go. For instance, the Council of Europe's Convention on Cybercrime (CoECoC), more commonly known as the Budapest Convention, is the first international treaty on cybercrimes.[1] As of March 2020, 65 countries had signed as well as ratified the convention in accordance with their national constitutional or legal requirements, making it enforceable. Three additional countries had signed the CoECoC but had not ratified.

Third, mostly due to the newness of cybercrimes, law enforcement authorities across the world are relatively inexperienced in dealing with these crimes. Another implication of newness is that some countries have not yet enacted laws related to cybercrimes. Cybercriminals find it attractive to operate from countries with weak laws and enforcement. Still another dimension of newness is a lack of established mechanisms, codes, policies, and procedures. Due to these factors, offenders committing cybercrimes are likely to feel less guilt compared to conventional criminals.

An implication of the above is that differential labor market rewards and countries' varied approaches to international relations and international cooperation imply that would-be hackers face different cost-benefit calculations related to their engagement in cybercrime activities. While the West faces significant jurisdictional challenges in investigating and prosecuting international cybercriminals, such challenges are even more pronounced in dealing with economies with low degrees of modernization of institutional frameworks and/or a low degree of cooperation and integration with the West. Cybercriminals tend to jurisdictionally shield themselves in such countries. This is known as *jurisdictional arbitrage*. For instance, in Estonia and Romania, which are among the countries most integrated with the West, cybercriminals are jurisdictionally less shielded than those in Russia.

Many Russian hackers are believed to have ties to the country's government agencies or oligarchs. If Russian cybercriminals are arrested in other countries at the request of the US government, the Russian government tries to stop their extradition to the US using a variety of techniques.[2] (See In Focus 3.1.) In several cases, Russia has pushed for extradition of the cybercriminals to Russia rather than the US. For instance, Russian hacker Aleksei Burkov was charged by the US for operating a credit-card fraud forum called CardPlanet. Burkov was arrested in Israel in 2015. The Russian government fought for four years to stop Burkov's extradition by Israel to the US. It even proposed a prisoner swap with an Israeli arrested in a Moscow airport on cannabis charges for Burkov's extradition to Russia.

IN FOCUS 3.1: VLADIMIR ZDOROVENIN WAS AN EXTREMELY UNLUCKY RUSSIAN CYBERCRIMINAL

The extradition of alleged Russian cybercriminal Vladimir Zdorovenin in 2012 by Swiss authorities was a big victory for the FBI. Zdorovenin was arrested in Zurich in March 2011 and then extradited to the US. According to US prosecutors, Zdorovenin and his son Kirill were engaged in various crimes involving cyberfraud, identity theft, and illegal operations with securities since 2004. Commenting on Zdorovenin's extradition, an FBI assistant director warned that cybercriminals cannot "hide behind the safety and anonymity of a Russian IP address" and that they would be brought to justice.[3]

In January 2013, Zdorovenin pleaded guilty in a Manhattan federal court to a count of conspiracy and a count of wire fraud. Zdorovenin apologized to the judge and asked for leniency, saying, "The lessons I have learned will be with me for the rest of my life." He was sentenced to three years in prison, which was much shorter than the sentence sought by prosecutors. US District Judge Paul Gardephe justified the lesser term on the fact that the defendant did not have the technical expertise to commit the crimes. His son, Kirill Zdorovenin, whose hacking skill was used to commit the fraud schemes, remains on the loose.

Vladimir Zdorovenin was among a few extremely unlucky Russian cybercriminals to be arrested and extradited to face criminal charges in the US. Most Russian cybercriminals, like Kirill Zdorovenin, are never arrested, prosecuted, or convicted.

In light of the above observations, this chapter examines the environment and structure of cybercrimes and assesses the cost-benefit structure of cybercriminals. We offer a simple economic analysis to provide insights into factors that encourage and energize cybercriminals' behaviors.

3.2 THE ENVIRONMENT AND STRUCTURE OF CYBERCRIMES: THE VICIOUS CIRCLE

3.2.1 The Environment

Research on crimes in the conventional world has indicated that sociocultural practices and political and economic systems are tightly linked to crimes. In this section, we provide a brief overview of the effects of these factors on cybercrimes.

3.2.1.1 Political and Legal Environment

The quality of the political and legal environment strongly influences cybercriminals' cost-benefit calculus. For instance, cybercriminals find it appealing to engage in criminal activities from countries that lack or have weak cybercrime laws and enforcement mechanisms. As of 2018, 17 of the 194 ITU member countries lacked cybercrime legislation. Moreover, about a third to half of nations worldwide lack legal frameworks that criminalize extraterritorial cybercrimes.[4]

Some countries, such as China and Russia, allegedly ignore cybercrimes unless such crimes are against their national interests. A member of Honker Union ("hong ke" means "red hacker"), who ran the group's website when members attacked many foreign websites for political causes, reported that a provincial police official "discouraged" him from attacking the US websites but his group attacked anyway and was not punished.[5]

There is an accusation that Russian cybercriminals are controlled or co-opted by the country's intelligence agencies. The Russian government allegedly ignores Russia-originated cybercrimes as long as the hackers help to achieve the government's political agendas when they are called. For instance, while Russia has strict laws controlling the media, *Hacker Magazine* published an article in 2010 on how to crack the NATO website, with screenshots and step-by-step instructions.

The lack or ineffectiveness of law enforcement allows hackers to engage in criminal activities without any trouble and get away with their crimes. Law enforcement agencies are also accused of giving various pretexts not to prosecute cybercriminals. Prosecutions related to cybercrime in Russia are extremely low at five to six people per year.[6]

3.2.1.2 The Availability and Effectiveness of Resources

The system capacity argument maintains that the legal response to a suspected crime is a function of organizational resources and caseload pressures.[7] Resource limitations are of particular concern for white-collar crimes; due to their complexity and hidden nature, substantial investigative and prosecutorial effort is required.[8] State capacity to respond to white-collar crime is constrained by the limited capacity of criminal justice agencies and resulting system overload.[9]

Similar constraints apply to cybercrimes. Most countries experience a severe shortage of law enforcement resources for cybercrime investigation.

3.2.1.3 International Institutions, Rules, and Relations

A significant proportion of cybercrimes are international. Estimates suggest that globally about 70% of cybercrime cases cross national borders.[10] For example, about half of all cybercrimes victimizing people in the UK are committed from abroad.[11]

Many cyberfraud schemes involve sophisticated international rings. For instance, some Japanese gangs reportedly hire Russian hackers to attack Japanese law enforcement agencies' databases. Likewise, some Australian swindlers have established links with Russian and Malaysian organized crime networks to transfer stolen money from overseas banks they have cracked into.[12]

International institutions and rules and international relations issues involving bilateral cooperation and multilateral initiatives are thus of high relevance. However, progress on this front has been limited. For instance, as noted above, the CoECoC, the only multilateral treaty on CS, is not accepted by many countries.

These ineffective treaties and international organizations are exacerbated by poor bilateral relations between major countries. For instance, allegations and counterallegations have been widespread in the US–China discourse on the governance of cyberspace. There has been too little cooperation and too much conflict in this area.

Russia and the US also engage in what some consider "a digital version of the cold war." Although Russia has signed agreements with the US to help with investigating a number of crimes, cybercrimes are not among them. In the past, the US Department of Justice (DoJ) complained that the agency received no response when it requested the help of Russian authorities.

3.2.1.4 Sociocultural Factors

It is also important to look at how societies view cybercrime victimization. In some societies, law enforcement agencies and others tend to blame cybercrime victims.[13] In addition to this victim-blaming attitude, societal negative stereotypes of such victims are pervasive. Cybercrime victims are sometimes perceived as greedy.[14] In order to provide further insights and a deeper understanding of this issue, it is important to consider primary and secondary victimization. Primary victimization describes a victim of the crime directly. Impacts involved in primary victimization include physical and psychological suffering and financial losses.

Secondary victimization, on the other hand, takes place because of the victim's social environment. Key mechanisms involved in secondary victimization include stigmatization, social isolation, or intrusive and humiliating questioning.[15] Secondary victimization also occurs due to journalists' faulty and insensitive practices in gathering or reporting news or due to inappropriate actions of the criminal justice system.[16] As noted earlier, law enforcement agencies' unsupportive attitudes and unwillingness to help victims have contributed to a low reporting rate of cybercrime incidents.[17]

In some societies, hackers have been able to project a positive social image of hacking activities. Russian hackers have cultivated a Robin Hood image by such deeds as hacking software programs of Western technology firms and distributing them for free. They view such activities as a donation to their society. Hacking and cyberattack activities in Russia thus carry less negative connotations than in the West.

3.2.1.5 The Economic Environment Facing Would-be Cybercriminals

The cybercrime industry provides a high economic incentive for would-be criminals. (See In Focus 3.2.) Most of the former Soviet Union and Central and Eastern Europe (FSU&CEE) economies are too small to employ their existing talent, which has forced the educated workforce to the electronic underground. They lack an equivalent of the US Silicon Valley.

A lack of English language proficiency limits access of most FSU&CEE technical experts to the Western labor market.[18] In Russia and other FSU&CEE economies, top university graduates are reportedly paid by organized crime groups up to 10 times more than legitimate jobs.[19, 20] A self-described hacker from Moscow confessed, "Hacking is one of the few good jobs left here."[21] Regarding computer attacks originating from Romania, the US-based Internet Fraud Complaint Center noted, "Frustrated with the employment possibilities offered in Romania, some of the world's most talented computer students are exploiting their talents online."[22]

IN FOCUS 3.2: MEXICO'S BANDIDOS REVOLUTIONS TEAM

Mexico's cybercrime group Bandidos Revolutions Team (BRT) was believed to be in operation since 2014. Héctor Ortiz Solares was the leader of the BRT, which had eight other members. The group was estimated to make between US$5.2 million and US$15.7 million per month.[23] The group made US$10 million on a single hack.[24] Each member of the gang thus made many times more than the average salary of an IT specialist in Mexico, which is less than US$350/week.[25]

The group also recruited experts to create malware and viruses that were used to launch cyberattacks against financial institutions. The gang then transferred money to false accounts that they controlled. Their associates went to automated teller machines (ATMs) to withdraw the cash.

In March 2018, authorities started tracking the group after ATMs in Léon and Tijuana started spitting out cash. The group drew police attention in April 2019 when they transferred about US$26 million to 849 false accounts.[26] Mexico's law enforcement agencies arrested members of the BRT in June 2019.

3.2.2 The Structure of Cybercrime

The environmental conditions presented above lead to a number of structural features of cybercrime. Characteristics of cybercriminals, cybercrime victims, and law enforcement and judiciary systems reinforce each other and form the vicious circle of cybercrime. Key elements of the vicious circle are presented in figure 3.1.

3.2.2.1 Law Enforcement and Judiciary Systems

As noted above, cybercrime investigations are highly complex as well as resource and expertise intensive. In most countries, law enforcement agencies such as police forces are inexperienced with cybercrime investigation and prosecution. Technological illiteracy and low levels of cybercrime awareness in the law enforcement community hinder efforts to arrest, prosecute, and incarcerate smugglers.

Moreover, cybercrimes are increasingly sophisticated; forms and methods change rapidly. Law enforcement agencies, lacking resources, fail to catch up with technologies enabling such crimes. Many countries thus do not investigate all reported cybercrimes. Law enforcement agencies' lack of ability to solve cybercrimes reinforces the confidence of cybercriminals as well as the unwillingness of victims to report such crimes. (See figure 3.1.)

A high proportion of cybercrime investigations have significant jurisdictional issues. In many cases, cybercrimes crossing borders create paperwork and slow down responses to such crimes. In addition, police forces in most countries are highly localized and are not well equipped to deal with the global nature of cybercrimes. National boundaries thus create serious obstacles for law enforcement agencies. Collaboration and cooperation between law enforcement agencies in different jurisdictions are far from sufficient to solve cybercrimes.

Figure 3.1 The vicious circle of cybercrimes

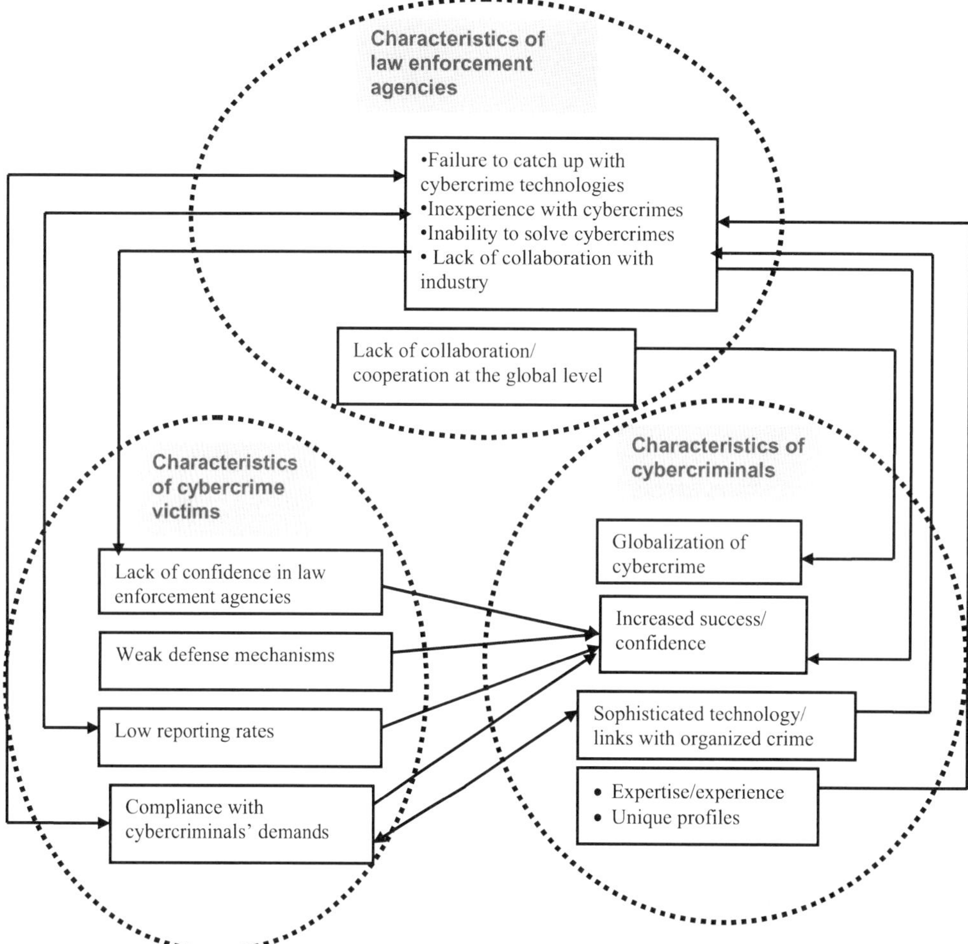

A lack of industry–government collaboration has also hampered law enforcement agencies' ability to solve cybercrimes. For instance, most of the critical infrastructures in the US are owned by the private sector. Law enforcement agencies have expressed concern over service providers' unwillingness to cooperate in cybercrime investigations.

3.2.2.2 Cybercriminals

In the conventional world, most crimes are committed close to home. Criminals travel far only if there are sufficient incentives to leave familiar territory.[27] Some crimes, such as kidnapping or robbing a bank, are attractive enough to do so. These crimes require much more planning. Crimes in the digital world differ significantly on this dimension. Information and communications technology (ICT) has drastically increased the porosity of national borders. The increased porosity and anonymity of the Internet superimpose in a complex interaction that enables criminal and violent groups, transnational terrorist organizations, and companies engaged in espionage to expand their operations globally without leaving home.

Evidence indicates that criminals' skill, intelligence, and experience co-vary positively with the odds of getting away with crimes. Some serious cybercriminals are highly skilled and thus face very low odds of getting caught. For instance, Russian Mafia hack rings are reportedly operated by former KGB agents.[28] There is evidence that some less skillful criminals get help from experienced hackers and transnational organized crime groups, which minimizes the probability of getting caught.

Increased success is making cybercriminals brasher. It was reported that some international hackers do not even conceal their real identities and the origin of their mailings. What is more, many organized criminals have invested illegally earned incomes in new technologies and in globalizing their operations, making it even more difficult to solve cybercrimes.

3.2.2.3 Cybercrime Victims

Cybercrime victims' unwillingness to report such crimes to law enforcement agencies further encourages cybercriminals' behavior. (See figure 3.1.) Indeed, cybercrimes are among the most underreported forms of criminality. One estimate by the New Zealand Police suggested that only 17% of companies that were victims of cybercrimes reported cybercrime-related losses to law enforcement. In the US, only 15% of cybercrime victims report the crimes to law enforcement agencies.[29] Likewise, about 1 in 10 Indian adolescents become cyberbullying victims every year but half of them do not report.[30]

Negative social image and self-image of cybercrime victims also lead to low reporting rates. Many victims are unwilling to report cybercrimes because they think going to law enforcement does not prevent an attack. Some victims feel that they have themselves, rather than cybercriminals, to blame for their victimization.[31] Other factors contributing to low reporting rates could be embarrassment; the fear of losing customer trust; damage to corporate credibility; and potential stock price decreases. Banks, other financial institutions, and other businesses that deal with sensitive data are especially reluctant to turn over the investigation to authorities. According to the Seventh Annual Computer Crime and Security Survey, 70% of those not reporting cybercrimes cited negative publicity as a reason. Difficulties related to documentation and proof further discourage businesses from reporting cybercrimes.

Effective risk management and incident prevention demand implementation of a well-proven security standard. Put differently, weakness of defense mechanisms co-varies positively with the likelihood of attack. A large proportion of Internet users fall prey to phishing scams because they are unable to distinguish between real and fraudulent emails.

Moreover, some companies choose to negotiate with cybercriminals by paying ransom. According to the FBI New York office, between April and July of 2015, over 100 companies in the financial sector experienced cyberattacks tied to ransom requests. Many banks reportedly paid the ransom.[32] One estimate suggested that targeted institutions pay ransoms of US$5 for every US$100 worth of damage they could suffer if the extracted data were published.[33] To take an example, DDoS attacks were launched by extortion group Armada Collective against banks and other organizations. According to CloudFlare, more than US$100,000 was sent to Armada-linked bitcoin addresses during March and April of 2016.[34] Firms' compliance with ransom demands helps cybercriminals develop and sustain their operations.

A large proportion of cybercrimes victimize businesses. Such crimes may be perceived as "victimless." In this regard, a strong parallel can be drawn between cybercrimes and shoplifting. One estimate suggested that 90% of shoplifting cases are never reported.[35] A report of the British Retail Consortium asserted that shoplifting tends to be perceived as a "victimless" crime. As a result, there is a lack of seriousness and urgency on the part of law enforcement agencies to respond to such crimes. The low reporting rate can arguably be attributed to the lack of confidence retailers have in the law enforcement response to shoplifting.[36]

3.3 A CYBERCRIMINAL'S COST-BENEFIT CALCULUS

In this section, we integrate key elements of the simplified framework representing the vicious circle discussed above and some additional characteristics of cybercrimes to examine a perpetrator's perceived cost-benefit structure associated with committing such crimes. Following the economic approach, a cybercriminal weighs benefits and costs to make a decision about engaging in a crime.[37] A cybercrime is committed if

$$M_b + P_b > I_c + O_{1c} + P_c + O_{2c}\,\pi_{arr}\,\pi_{con} \tag{1}$$

where

M_b = the monetary benefits of committing the crime;

P_b = the psychological benefits of committing the crime;

I_c = direct investment costs;

O_{1c} = opportunity costs to engage in cybercrime;

P_c = psychological costs of committing a cybercrime;

O_{2c} = monetary opportunity costs of conviction;

π_{arr} = the probability of arrest;

π_{con} = the probability of conviction.

The product term on the right side in the equation, $O_{2c}\,\pi_{arr}\,\pi_{con}$, is also referred to as the expected penalty effect.

3.3.1 The Benefits

Cybercrimes provide substantial monetary benefits to cybercriminals. Furthermore, monetary benefits are not the only incentives that motivate various actors engaged in hacking and other cybercrimes. Indeed, a number of high-profile and very costly cyberattacks were not motivated by economics. From this discussion, two categories of benefits for cyberattackers emerge: monetary benefits (M_b) and psychological benefits (P_b).

3.3.1.1 Monetary Benefits (M_B)

The cybercrime landscape is rapidly changing in terms of hackers' financial motives. An article published in IDG News Service quotes a Russian hacker employed as a security expert: "There is more of a financial incentive now for hackers and crackers as well as for virus writers to write for money and not just for glory or some political motive."[38] An estimate suggested that Russian cybercriminals annually made in the range of US$2.5–US$3.7 billion, or 35% of global cybercrime revenue in the early 2010s.[39]

When selecting victims, cybercriminals are more likely to target individuals and organizations that will give them a higher reward for the level of risk they are taking. According to Check Point Research's Brand Phishing Report for the first quarter of 2020, due to the high value of Apple IDs, Apple was the brand most targeted by phishers. About 10% of all phishing attempts globally had targeted Apple devices.[40]

3.3.1.2 Psychological Benefits (P_B)

Psychological benefits can be better explained in terms of intrinsic motivation. Some hackers may be socialized into acting appropriately and in a manner consistent with the norms of a group. The respect of one's peer hackers acts as a source of psychological benefit for some hackers. Some psychologists' conceptualization of intrinsic motivation is broader.[11] According to them, acting on the basis of principle is also a form of intrinsic motivation.

Many ideological hackings fall into this category. In some cases, hackers with a common moral framework launch cyberattacks to achieve their goals. In April 2020, about 25,000 email addresses and passwords allegedly belonging to employees at the Gates Foundation, World Health Organization (WHO), and National Institutes of Health were breached and posted online by "anonymous activists." These organizations became targets of some extremist groups during the COVID-19 pandemic. The lists first emerged on 4chan, the online message board where users can post anonymously.[42] Some consider Anonymous to be vigilantes and argue that in some situations a reasonable moral framework justifies their actions in the absence of effective legal alternatives to enforce norms that the group shares.[43]

3.3.2 The Costs

In economic terms, there are four components of cybercrime's cost structure: (i) direct investment costs needed to commit cybercrime (I_c); (ii) opportunity costs to engage in cybercrime (O_{1c}); (iii) psychological costs of committing a cybercrime (P_c); and (iv) expected monetary opportunity costs of conviction (O_{2c}, π_{arr}, and π_{con}).

3.3.2.1 Direct Investment Costs to Engage in Cybercrime (I_c)

Cybercrimes often require more investment than most other crimes, including significant time commitment and skill development on the part of the criminal. Direct investment costs are a function of the sophistication of cybercrime operations. Consider the ransomware industry, which was estimated to make US$1 billion in 2016.[44] According to Trustwave, the costs of a ransomware payload[45] (e.g., Curve-Tor-Bitcoin [CTB] Locker[46]), infection vector (e.g., RIG exploit kit), camouflaging services (encryption), and traffic (20,000 visitors) totaled US$5,900 per month for criminals.[47]

Some cybercrime operations, on the other hand, have extremely low investment costs. Consider a botnet kit called "Aldi Bot," which was first available for sale in 2011 in underground electronic forums for about US$7. A criminal could buy the kit and use a botnet to perform large-scale cyberattacks.[48]

Malware as a service (MaaS) is becoming popular among cybercriminals. Instead of developing their own crime tools, they can pay a subscription fee and outsource the development and maintenance of the malware.[49] MaaS offerings have become more powerful and less expensive.[50]

Highly organized groups and states are at the high end of the cyberattack capability spectrum and make heavy investments in time and resources to develop cyberattack capabilities. One example is the Stuxnet worm, which was designed to damage Iran's uranium processing capacity. Based on the worm's unusual sophistication and complexity, analysts suggested that it might have been created by well-funded computer experts with possible support from a national government. Microsoft estimated that creation of the worm took 10,000 person-days of work by top-ranked software engineers.

3.3.2.2 Opportunity Costs of Engaging in Cybercrime (O_{1c})

Opportunity costs vary for cybercriminals from different countries depending on the availability of jobs related to their skills and how much the jobs pay. As shown in figure 3.2, the average legitimate monthly wages in countries considered to be cybercrime hotspots are much lower than wages in the West.

Figure 3.2 A comparison of average monthly wages in the US and countries considered to be cybercrime hotspots (US$)

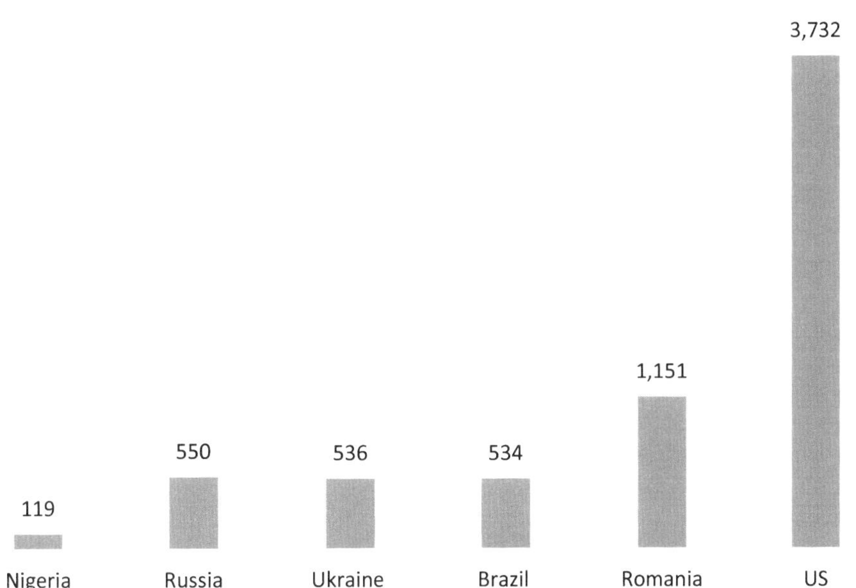

Data sources: Nigeria's wage is for 2018 from *Nigeria living wage individual.* (n.d.). Trading Economics. Retrieved from https://tradingeconomics.com/nigeria/living-wage-individual; Russia's wage is for 2019 from *Half of working Russians earn less than $550 a month.* (2019, July 19). The Moscow Times. Retrieved from https://www.themoscowtimes.com/2019/07/19/half-working-russians-earn-less-than-550-usd-month-a66487; For Ukraine, the figure is for Kiev in February 2019 from Unian Information Agency. (2019, March 29). *Average salary in Ukraine grows to US$348 in Feb.* Retrieved from https://www.unian.info/economics/10497684-average-salary-in-ukraine-grows-to-us-348-in-feb.html; Brazil's figure is for November of 2019 from *Brazil real average monthly income.* (n.d.). Trading Economics. Retrieved from https://tradingeconomics.com/brazil/wages; Romania's is for October 2019 from *Average wage in Romania up 15% yoy in October 2019.* (2019, December 10). Romania Insider. Retrieved from https://www.romania-insider.com/average-wage-ro-october-2019; US figure is for 2018Q4 from Fiorillo, S. (2021). *What is the average income in the US?* The Street. Retrieved from https://www.thestreet.com/personal-finance/average-income-in-us-14852178

As noted above, a large segment of the educated workforce in most of the FSU&CEE economies has been forced to the electronic underground due to small sizes of the economies to employ the existing talent. According to some estimates, 10,000–20,000 people in Russia work in diverse forms of cybercrime activities such as bank frauds, selling scareware, and sending fake pharmacy spam.[51]

In some economies, the lack of employment opportunities for would-be hackers results in a low perceived monetary opportunity cost of conviction. The combination of over-educated and under-employed network specialists has made Russia and other Eastern European countries fertile ground for hackers.

3.3.2.3 Psychological Costs of Committing a Cybercrime (P_c)

Psychological costs are associated with the psychological and mental energy needed in committing cybercrimes. They result from fear of punishment, guilt, etc. Compared to most physical crimes, cybercriminals are less likely to feel guilty due to the relative newness of such crimes and the lack of easily identifiable victims in some cases.

Responding to the question of whether he ever felt guilty, ex-cybercriminal Brett Johnson, who was on America's most-wanted list, said: "Generally, no. Although once I got to the point where I was stealing massive amounts of money, I would sometimes send some victims their money back if I believed their story was sad enough. That wasn't me feeling guilty [though], it was me justifying my crimes."[52]

The feeling of guilt is not equally pervasive across hackers of different sociocultural backgrounds. The psychological cost of a cybercrime is thus a function of the hacker's sociocultural background. It was reported that many Indonesian hackers feel that cyberfraud is "wrong" but acceptable, especially if a victim is rich and not an Indonesian.

3.3.2.4 Monetary Opportunity Costs of Conviction (O_{2c})

The opportunity cost of conviction is the monetary income forgone when serving a criminal sentence. For instance, a hacker who is sentenced to a three-year prison term and who could legally earn US$20,000 per year faces an opportunity cost of US$60,000. In recent years, many countries have enacted stricter laws against cybercrimes, which have increased the opportunity costs of conviction. Nonetheless, many countries have no laws at all against cybercrimes – zero opportunity costs of conviction.

3.3.2.5 The Probability of Arrest (π_{ARR}) and the Probability of Conviction (π_{CON})

As discussed in the last section, only a small proportion of cybercrimes are reported. Among reported cybercrimes, arrest rates are very low. Arrest entails identifying the pool of potential suspects and narrowing the pool by eliminating innocents. The structure of cybercrimes discussed in the previous section makes it difficult to identify the pool of potential suspects.

International cybercrimes in general have alarmingly low prosecution and conviction rates. Some FSU&CEE economies lack law enforcement agencies with high-tech crime-fighting abilities, which has led to cybercriminals' high likelihood of getting away with cybercrimes. Despite their motivation and willingness, some countries that are signatories to the CoECoC are often unable to control cyberattacks originating in these countries.

The conviction phase of a cybercrime is equally complex. Difficulties related to documentation and proof compound the problems at this phase. The newness of cybercrimes has also presented a challenge to the court system. For small cases, it is difficult to find an attorney who takes cyberfraud cases. Experts also say that explaining Internet-related crimes to judges is difficult.

It was reported in 2015 that, in the UK, only one person every month was convicted under the Computer Misuse Act during the previous 23 years.[53] In 2017, the UK cybercrime conviction rate was estimated at 0.36%.[54]

3.4 COMBATING CYBERCRIMES BY ALTERING THEIR COST-BENEFIT STRUCTURE

In the absence of appropriate measures, the elements of the vicious circle reinforce each other and lead to more favorable benefit-cost ratios for cyberperpetrators. Such a condition, in turn, is associated with increased confidence of cybercriminals and higher public distrust in government and law enforcement agencies, which may result in more and more serious cybercrimes. Where should we start to break the vicious circle of cybercrime and alter its cost-benefit calculus?

Obviously, there is no pure technological solution for security-related problems involving technologies. Micro- and macro-level measures combining technological and non-technological fixes are needed.

3.4.1 Macro-level Measures

At the macro level, appropriate policy response is to change cybercriminals' cost-benefit calculus so that cybercrimes become less attractive. Development of national technological and personnel capabilities, enactment of new laws, a higher level of industry–government collaboration, and international coordination are critical for combating these frauds.

Fighting cybercrime requires global collaboration and cooperation. Although some signs of success have materialized, nations have far to go. There has been limited success in establishing formal international CS treaties. More efforts and resources should be directed to create stronger bilateral and multilateral partnerships and cooperation in order to fight cybercrimes. Cooperation of governments in countries that have high concentrations of cybercrime is especially critical to fighting cybercrimes that originate in these countries.

Investments in the skills of law enforcement authorities enhance national capabilities to fight cybercrimes and increase the probability of arrest and conviction. In light of the lack of law enforcement capacity, especially in many developing countries, more attention needs to be focused on strengthening system capacity. For instance, while some of the signatories of the CoECoC have demonstrated willingness and responsibility to cooperate with other members in fighting cybercrimes, they lack resources to do so. Western countries may provide CS-related resources, training, and expertise to enhance the system capacity of developing countries. For instance, eBay was reported to be engaged in initiatives to educate Romanian prosecutors about cybercrimes, including explaining cybercrimes to a judge using layperson's language.[55]

Public–private partnerships can also tackle some of the above issues. For instance, in 2008, Microsoft was reported to have been working with Nigeria's Economic and Financial Crimes Commission (EFCC) and Paradigm Initiative Nigeria to fight cybercrimes as well as to educate Nigerians on cybercrimes and to create economic opportunities.[56] Such measures alter the cost-benefit calculus for cybercriminals because would-be criminals have opportunities to work in the legitimate economy. The EFCC announced in 2009 that it shut down 800 websites associated with cybercrimes and arrested 18 fraudster groups. Efficient law enforcement systems provide a negative sociopolitical feedback mechanism to cybercrimes. Cybercrime awareness may lead to organizations' and individuals' adoption of strong CS measures, which would make it more difficult for attackers to succeed. Increased legitimate economic opportunities would lead to a lower predisposition of young people to engage in cybercrime activities.

Timely information sharing would facilitate a better understanding of past cyberattacks in order to prevent future ones, as perpetrators often use the same malware to infiltrate multiple targets. Realizing this, some measures and efforts are underway to facilitate information sharing. In the US, the Executive Order (EO) 13691, Promoting Private Sector Cybersecurity Information Sharing signed in 2015 by President Obama, lays out a framework to share cyberthreat information among private sector organizations and with government agencies. It gave the private sector "targeted" liability protection to share information, that is, the private sector will not be legally liable for sharing some categories of personal data of consumers. The government will also disclose more classified threat information to the private sector.

A key provision of EO 13691 is the establishment of information sharing and analysis organizations (ISAOs), which will comply with voluntary standards envisioned by the EO. The DHS and the National Cybersecurity and Communications Integration Center (NCCIC) are given the authority to share data with ISAOs, so organizations will be able to access classified CS data.[57] The Cyber Threat Intelligence Integration Center (CTIIC) was also created in 2015 to carry out "coordinated cyberthreat assessments" based on information received from various sources. The CTIIC aims to provide "all-source analysis" of cyberthreats to policy-makers and assist relevant agencies in dealing with cybercrimes that impact numerous industries. The proposals would make it easier for companies to share intelligence with the NCCIC, including various cyberthreat indicators such as attempts to access restricted files, how a website runs, and how a company utilizes user data.

Greater information sharing has also been a central theme in emerging countermeasures in the banking and financial sector. For instance, one problem for banks that outsource DDoS monitoring and mitigation is that such services "do not necessarily communicate with other companies that have not bought the same technology – or with other technology the same company has bought."[58] To address this problem, the not-for-profit Financial Services Information Sharing and Analysis Center[59] has launched a new software platform, the Critical Infrastructure Notification System (CINS), that distills cyberthreat information into actionable intelligence to member companies worldwide nearly simultaneously. CINS should make it more difficult for hackers to deploy the same malware against multiple financial institutions. More initiatives of this type are needed.

3.4.2 Micro-level Measures

At the micro level, ensuring that both technological and behavioral factors are considered in the design and implementation of a network is crucial. Technological measures range from simply disconnecting databases from the Internet if they contain sensitive information to deploying sophisticated antifraud technologies.

By using stronger defense mechanisms, organizations and individuals can make it more difficult for cybercriminals to commit crimes (thereby increasing the costs to engage in cybercrime activities). Although CS is often cast as a largely technical problem, employees are often the weakest links. Simple behavioral measures, such as a training program aimed at improving the ability of consumers, employees, and the public to distinguish a fraudulent email from a real one, may reduce a significant proportion of phishing-related cybercrimes.[60]

An appropriate response to organized criminal groups' investments in cybercrime technologies is to increase the development and deployment of sophisticated CS systems and software. Organizations that give CS a low priority increase their chance of facing cyberattacks. For instance, consider the December 2013 high-profile cyberattacks on the US retailer Target. The attacks occurred during the peak holiday shopping season from November 27 to December 15. The malware tried to steal card data during peak customer visit times (10 am to 5 pm local times) to Target stores.[61] Target had disabled some security features during the period to deal with high traffic flow.[62] It also ignored crucial warnings from CS firm FireEye, which had provided network monitoring tools to the company. Target had been told that there was an intrusion to its system and that criminals had exfiltrated the data.[63]

Finally, in the conventional world, research indicates that the time it takes a person to report a crime is one of the most important factors that increases the probability of arrest. This is especially important for crimes for which the proper preservation of evidence is critical for a successful prosecution. Reporting a cybercrime to law enforcement authorities in a timely manner would help to preserve physical as well as digital evidence required in the successful prosecution of offenders.

3.5 CHAPTER SUMMARY AND CONCLUSION

The structure of cybercrimes provides an attractive cost-benefit assessment for cybercriminals. Some societies' sociocultural practices provide some degree of legitimacy to cybercrimes and thus provide fertile ground for these crimes. Cybercriminals find it attractive to operate from economies with a lack of law enforcement capacity and/or a low degree of integration with the West. While some governments may be willing to fight international cybercrimes that originate in their countries (e.g., CoECoC signatories such as Ukraine), they face institutional and law enforcement constraints.

By taking appropriate measures and initiatives at various levels, the existing cybercrime structure can be changed so that the perceived benefits are lower and the perceived risks and costs are higher for cyberperpetrators. There have been some positive signs on this front, the most important being the enactment of cybercrime legislation in most countries. Technology and e-commerce companies such as eBay and Microsoft have engaged in activities, such as providing training to law enforcement and judiciary officials, creating jobs for IT graduates, and increasing cybercrime awareness, that might change cost-benefit tradeoffs faced by cybercriminals. This approach to altering the cost-benefit equation for the would-be cybercriminal offers promise.

3.6 DISCUSSION QUESTIONS

1. Why do some individuals find it attractive to engage in cybercrime?
2. How are the costs and benefits to engage in cybercrime changing for a cybercriminal?
3. How do different environmental factors affect cybercrime?
4. What measures can be taken to make cybercrime economically less attractive for hackers and cybercriminals?

3.7 END-OF-CHAPTER CASE: INNOVATIVE MARKETING UKRAINE AND THE SCARY SCAREWARE INDUSTRY

Innovative Marketing Ukraine (IMU) was a pioneer of fake antivirus software, also known as scareware, founded by a Swedish, a Canadian, and a US developer. IMU exhibited many features of a legitimate company. The organization and its employees had LinkedIn profiles.[64] It was incorporated in Belize and its main offices were located in Ukraine's capital, Kiev. McAfee estimated that IMU employed more than 600 employees in Kiev and subsidiaries in Argentina, Canada, India, Poland, the US, and other countries in posts such as receptionists, financial managers, webmasters, and R&D engineers.[65] McAfee also analyzed professional records of 180 of the 396 employees' names found online. 100 of them worked for at least a year. Most were college students.[66]

Business models of scareware programs center on infusing fear and anxiety to sell fake software. Typically, scareware pretends to scan a computer for malware and tells the user that the machine is infected. In IMU's case, the goal was to persuade the victim to voluntarily provide credit-card information to pay US$50–80 for the scareware.

3.7.1 The Modus Operandi

IMU's products were superficially similar to genuine antivirus products. For instance, its Win Antivirus looked like Microsoft security software. Another product, DriveCleaner, identified 179 visits to adult websites no matter which computer it was installed on. The fake software was designed to tell users that their personal computers (PCs) were working properly once they had paid. IMU relied on fear and intimidation, rather than product sophistication, to assure compliance.

IMU invested heavily in call center facilities in India, Ukraine, and the US, which responded to 2 million calls in 2008.[67] When people made calls to complain, agents would "guide" them through the steps needed to make the scareware messages, such as "Warning: Your computer is infected Detected spyware infection! Click this message to install the last update of security software," disappear. Often that required disabling legitimate antivirus software. A McAfee researcher listened to digitized audio recordings of customer service calls that IMU kept on its servers. The researcher found that most customers seemed to be happy and satisfied at the end of the call.[68]

IMU hired many young employees who did not seem to care about ethical behaviors, practices, and standards. They knowingly refused to acknowledge the scareware's harm to consumers. A former IMU employee, who later joined a Kiev bank, noted, "When you are just 20, you don't think a lot about ethics. I had a good salary and I know that most employees also had pretty good salaries."[69]

According to a McAfee researcher, IMU received approximately 4.5 million orders in 2008, which amounted to revenues of US$180 million at the price of US$40 each. IMU sold programs in at least two dozen countries.

IMU also created dummy ad agencies to place bogus, innocent-looking ads on popular websites for reputable businesses, including Career Builder, eharmony, the *Economist*, Major League Baseball, the National Association of Realtors, the National Hockey League, Priceline, and others.[70] A click on such ads triggered automatic bogus virus scans, which showed that a user's PC was virus infected. The bogus software made sales pitches involving false promises for a clean-up and directed users to purchase IMU's scareware.[71]

3.7.2 IMU's Affiliates and Business Partners

In an attempt to "recruit" business partners (e.g., credit-card processors), IMU created subsidiaries that were designed to hide its identity. A high proportion of IMU's victims complained to their credit-card companies to obtain refunds on purchases, which deteriorated the relationships between IMU subsidiaries and merchant banks that processed the transactions. IMU was forced to switch from a bank in Canada to one in Bahrain.

In 2005, the Bahrain-based Bank of Bahrain & Kuwait terminated ties with IMU. Then IMU had no credit-card processor for five months. Following that, it established a relationship with Singapore's DBS Bank, which showed willingness to handle IMU accounts. DBS Bank processed IMU's tens of millions of dollars in backlogged payments.[72]

For infecting each machine, IMU paid affiliates 10 cents and generated average returns in the range of US$2–US$5 through software, sales, and product promotion. Affiliates loaded software by hijacking legitimate websites, setting up corrupt sites for spreading viruses, attacking social networking sites, and other methods. One affiliate recruiting site, www.earning4u.com, reportedly paid US$6–US$180 for every 1,000 infected PCs. IMU and its affiliates also rewarded the top performers. Panda Security reported that it found pictures of a party organized in March 2008 in Montenegro by KlikVIP, an IMU affiliate, to reward scareware installers. One picture showed a briefcase full of euros given to the top performer.[73]

3.7.3 The Fallout

More than 1,000 people filed complaints against IMU with the US FTC.[74] The FTC's investigation lasted more than a year, which led to a federal lawsuit to shut down IMU. Reportedly, IMU's servers were not password-protected. Therefore McAfee researcher Dirk Kollberg was able to collect more than 67 gigabytes (GB) of data from IMU servers.[75] Kollberg forwarded the information to the FTC and the FBI, which helped build the case against IMU.

According to a May 2010 announcement by the US Attorney's Office, the three people charged in the IMU case allegedly cheated customers in 60 countries out of approximately US$100 million.[76] The US government retrieved US$117,000 by settling charges against one of the defendants, who ran a customer support center in Cincinnati.[77] Profiles posted on LinkedIn indicated that after IMU was shut down, some of the former employees started working at leading banks, consulting companies, and other Kiev-based antivirus companies.[78]

CHAPTER FOUR

Increasing Returns, Externality, and Rise in Cybercrimes

CHAPTER PREVIEW

The Internet was designed without considering security, which gives the offense an advantage over the defense. This feature leads to the asymmetric nature of cyberattacks. Entities with limited financial and technical resources can compromise high-value targets. Nations lacking military strength may resort to cyberattacks, which require less resources. In order to provide insight into factors that might encourage cybercriminals' activities, this chapter focuses on various positive or self-reinforcing feedback systems to examine increasing returns in activities related to cybercrime. It considers how inefficiency and congestion in the law enforcement system lead to cybercriminals' positive assessment of risks related to cybercrimes. The chapter explains how diffusion of cybercrime expertise and technology have increased cybercriminals' ability to commit cybercrimes. It provides details about the uses of new tools and techniques such as formjacking, blockchain-based cryptocurrencies, and AI. This chapter also gives special consideration to factors that have led to increased predisposition toward cybercrime.

4.1 INTRODUCTION

The Internet was designed without considering security, which puts the offense at an advantageous position over the defense. This feature leads to the asymmetric nature of cyberattacks, which means that entities with limited financial and technical resources or militarily weaker actors can compromise high-value targets.[1] The Aldi Bot example in chapter 3 supports this theory.

Positive feedback mechanisms encourage hackers to engage in cybercrime activities. A study found that by investing US$183,000 in fake traffic, cybercriminals can make as much as US$4.6 million in profit.[2] Positive feedback to cybercriminals could also arise from the weak law enforcement environment for cybercrimes in most countries as described in chapter 3. In some societies, some cybercrime activities are accepted, which again would provide positive feedback to cybercriminals.

This chapter focuses on various positive or self-reinforcing feedback systems to examine increasing returns in cybercrime activities. In this way, this chapter looks at the economics of cybercrimes from another angle in order to provide a better and more in-depth understanding of cybercriminals' activities.

4.2 INCREASING RETURNS AND FEEDBACK LOOPS IN CYBERCRIMES

In many ways, cybercriminals' businesses can be viewed as highly successful. Increasing returns explain how innovation, industries, and the environment influence each other and drive the success of firms. The law of increasing returns argues that economies of scale, decreasing costs, and feedback mechanisms lead to further success of already successful entities. This chapter explores evidence of increasing returns in cybercrime activities.

There are three types of self-reinforcing feedback systems: economic, sociopolitical, and cognitive.[3] Cybercrimes' significant financial benefits provide positive economic feedback to cybercriminals. A low probability of being caught and prosecuted and a low severity of punishment give them high positive economic feedback.[4]

Sociopolitical feedback is related to formal and informal institutions.[5] Social feedback is linked to informal measures, such as sanctions applied by a social group to exclude a cybercriminal from one's circle of friends. Political feedback, on the other hand, is applied by regulatory institutions.

Cognitive feedback systems are associated with mental maps of hackers. These systems influence the lens through which criminals view cybercrimes.[6] As discussed in chapter 3, enjoyment from and low guilt about cybercrimes serve as forms of cognitive feedback.

4.3 MECHANISMS ASSOCIATED WITH EXTERNALITY IN CYBERCRIMES

In the cybercrime environment, increasing returns could manifest in many ways. For instance, cybercriminals may "invent" sophisticated new tools. Cybercriminals could also operate from countries with weak cybercrime laws.

We examine three mechanisms that may give positive feedback to cybercriminals: inefficiency and congestion in the law enforcement system, acceleration of the diffusion of cybercrime expertise and technology, and increase in potential criminals' predisposition toward cybercrimes.[7] From the victims' perspective, there is arguably a vicious circle of cybercrimes linking characteristics of cybercriminals, cybercrime victims, and law enforcement agencies and a corresponding virtuous circle for cybercriminals. These externality mechanisms strengthen the elements of the vicious circle for victims.

Table 4.1 presents how the externality mechanisms and the feedback systems described above are intertwined.

4.3.1 Inefficiency and Congestion in the Law Enforcement System

Congestion and inefficiency in law enforcement systems arise from factors such as newness of cybercrimes, low government priority, lack of cross-border and industry–government cooperation, and victims' unwillingness to report. Cybercriminals and victims tend to be scattered across the country and the world, posing logistical challenges. At the same time, while large law enforcement agencies such as the FBI have developed capacity to deal with cybercrimes, local police forces are not equipped to deal with cybercrimes due to staff shortages.

4.3.1.1 Law Enforcement Agencies' Lack of Resources

Sufficient resources have not been committed to fight cybercrimes. For example, in 2005 the FBI spent US$150 million, only 3% of its US$5 billion budget, on fighting cybercrimes. Ten years later in 2015, the FBI's cyberbudget was US$470 million, amounting to 5.7% of the agency's US$8.3 billion budget request that year.[8] While this is a significant increase, this level of resources is still insufficient to fight all types of cybercrimes. The FBI is devoting increased priority to prevent attacks linked to terrorism. A key focus has been in developing an informant network that can provide information about plots to FBI agents and connect Internet-radicalized plotters with people working for the FBI.[9]

Table 4.1 Externality mechanisms and feedback systems producing increasing returns in cybercrime-related activities

	Externality Mechanisms		
Feedback System	Inefficiency and Congestion in the Law Enforcement System (Assessment of Risks Related to Cybercrimes)	Diffusion of Cybercrime Expertise and Technology (Ability to Engage in Cybercrimes)	Increased Predisposition toward Cybercrime (Willingness to Engage in Cybercrimes)
Economic	▪ Law enforcement agencies' lack of resources ▪ Sophistication in cybercrimes ▪ Lack of cybercriminal database ▪ Difficulties in explaining cybercrime in courts	▪ Easily available and efficient hacking tools and payment mechanisms such as cryptocurrencies (*relative advantage*)* ▪ Schools teach skills related to hacking in some countries (*lower complexity*) ▪ High digitization (mechanisms by which cybercrimes are committed are compatible) ▪ High return on investments in cybercrime tools (*observability*)	▪ Lack of employment opportunities ▪ Victims' tendency to pay ransoms ▪ Increasing financial incentive for hackers ▪ Diffusion of cryptocurrencies ▪ Internet users' ignorance
Sociopolitical	▪ Weak cybercrime laws in some countries ▪ *Jurisdictional arbitrage* ▪ Lack of industry–government collaboration ▪ Lack of international cooperation ▪ Barriers to developing CS personnel	▪ Less skillful criminals get help from experienced hackers/crime groups (*lower complexity*) ▪ Information sharing among hackers (*lower complexity*) ▪ Cybercriminals are less likely to be caught and prosecuted (*relative advantage*)	▪ Intrinsic motivations of ideological hackers based on obligation to the community ▪ Cybercrimes acceptable in some societies
Cognitive	▪ Victims' lack of confidence in law enforcement: unwilling to report ▪ Cybercriminals' confidence	▪ Easy-to-use hacking tools (*lower complexity*) ▪ Money-back guarantees for the purchases of "zero-day" malware (*trialability*) lowers risk	▪ Enjoyment-based intrinsic motivations: a sense of pride and accomplishment ▪ Compliance with cybercriminals' demands: high confidence ▪ Low guilt

*Terms in italics are discussed later in this chapter.

Figure 4.1 Number of complaints reported to IC3, 2014–19

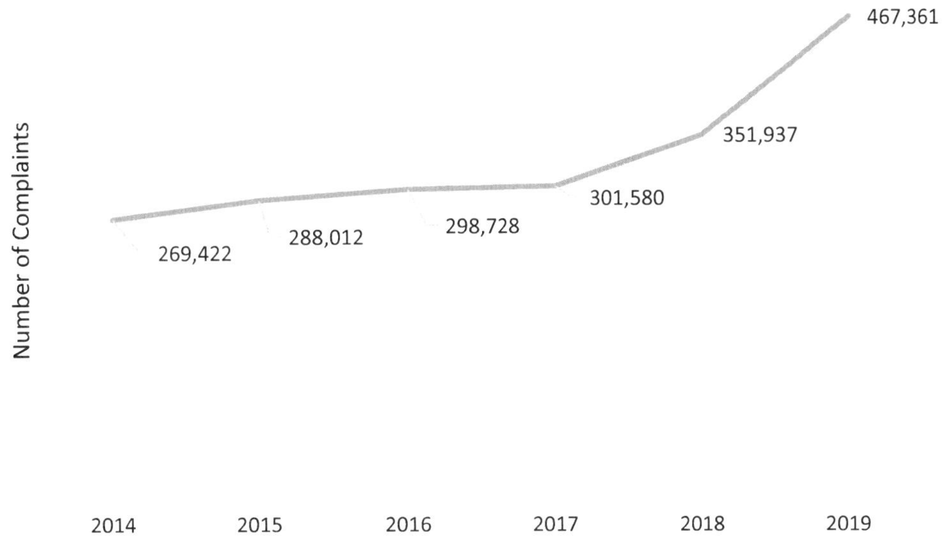

Data source: IC3. (2020). *2019 Internet crime report.* Retrieved from https://www.ic3.gov/Media/PDF/AnnualReport/2019_IC3Report.pdf

Governments' cybercapabilities are far less than what are required to address growing cyberthreats. For instance, the Internet Crime Complaint Center (IC3) was established by the FBI in 2000 to handle cybercrime. By 2019, about 4.9 million complaints were reported to the agency. (See figure 4.1.) In 2019 alone, IC3 received 467,361 complaints. Agencies such as the IC3 have resources only to focus on the biggest and most serious threats.[10]

Insufficient staff and financial resources hinder law enforcement agencies' efforts to fight growing cyberthreats. In 2020, there were over 314,000 open CS positions in the US.[11] Likewise, in 2018, over 13,000 open positions existed within US state and federal agencies.[12]

The situation is worse in developing countries. The cybercrime awareness level is low among India's law enforcement community. For instance, it was reported that when a police officer was asked to seize a hacker's computer in an investigation of a cybercrime in India, he brought the hacker's monitor. In another cybercrime case, the police seized the CD-ROM drive from a hacker's computer instead of the hard disk.[13]

Conventional crimes divert law enforcement agencies' attention, especially in countries where crimes related to drugs, gangs, and violence are serious problems, such as some countries in Latin American and the Caribbean (LAC). Some observers argue that it is more urgent for local police forces to control street gangs committing violent crimes than to investigate possible bank thefts from a foreign country.[14] For instance, the Mexican government is forced to spend an overwhelming amount of resources on gang and drug problems, and the drug Mafia arguably has better technologies.[15]

Likewise, only 5%–8% of crimes are solved in Brazil compared with 80% in France.[16] In 2017, the number of murders in Brazil's Rio de Janeiro state was 40 per 100,000, which was 14 times the rate in New York state.[17] This remarkable situation is a direct consequence of the scarcity of law enforcement resources. A prevailing culture of violence in cities such as São Paulo, Rio de Janeiro, and Brasília consume most of the available law enforcement resources, leaving little to enforce cybercrime laws. Beyond all that, police corruption has also been a problem.[18] In many countries, the law enforcement system and resources experience inefficiency and congestion, which generate positive externalities for criminals.[19]

Cybercrimes are increasingly sophisticated and new forms and methods develop rapidly. Law enforcement agencies have failed to catch up with the constant progressive nature of such crimes.

Cybercriminals' unique profiles are significantly different from those of conventional criminals (see chapter 3). In Russia, for instance, most hackers are young, highly educated, and work independently and thus do not fit the conventional police profiles of criminals.[20]

Cybercriminals are also more difficult to catch and prosecute than conventional ones. In most countries, the lack of a cybercriminal database has hampered the ability to solve cybercrimes. Collection and retention of evidence has been another critical challenge facing law enforcement agencies. Consequently, cybercrime conviction rates have been low. (See figure 4.2.)

4.3.1.2 Inefficient Judicial Systems

Cybercrimes' newness has also presented challenges to the court system. For small cybercrime cases, it is difficult to find an attorney. In many cases, explaining cybercrimes to judges and juries is also difficult.

India's only Cyber Appellate Tribunal, which started functioning in 2006, had not adjudicated a single case during 2011–2014 due to the unavailability of the chairperson and judicial members.[21] The position of chairperson was not filled until December 2018.[22]

4.3.1.3 International Nature of Cybercrimes

As noted in chapter 3, most cybercrimes are international. Cybercriminals thus benefit from jurisdictional arbitrage – that is, taking advantage of the differences between judicial systems in different countries. Due to newness, jurisdictional arbitrage is higher for cybercrimes than conventional crimes. Additionally, national boundaries create law enforcement barriers. Collaboration and cooperation among law enforcement agencies in different jurisdictions are insufficient. (See In Focus 4.1.) Some countries ignore cybercrimes unless such crimes jeopardize their national interests.

Figure 4.2 Cybercrime conviction rates in select jurisdictions (%)

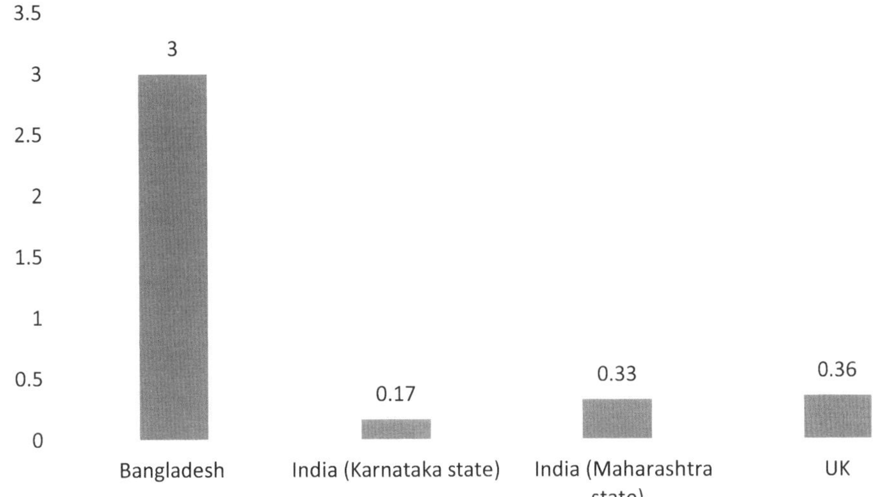

Data sources: Data for Maharashtra (2012–17, July) as reported in Naidu, J. (2017, August 21). 10,000 cybercrime cases, only 34 convictions in Maharashtra between 2012 and 2017. *Hindustan Times*. Retrieved from https://www.hindustantimes.com/mumbai-news/10-000-cybercrime-cases-only-34-convictions-in-maharashtra-between-2012-and-2017/story-IfzZbXmzVlqrG8dclkwzeO.html; data for Karnataka (2014–19) as reported in Ram, T. (2019, October 1). *Why conviction rate for cyber crime cases in Karnataka is abysmally low.* The News Minute. Retrieved from https://www.thenewsminute.com/article/why-conviction-rate-cyber-crime-cases-karnataka-abysmally-low-109803; data for the UK (2017) as reported in Gopalakrishnan, C. (2019). *Cyber-crime thrives on legal inefficiency & business leaders turning a blind eye.* SC Media. Retrieved from https://www.scmagazineuk.com/cyber-crime-thrives-legal-inefficiency-business-leaders-turning-blind-eye/article/1586202; data for Bangladesh (February 2013–February 2019) as reported in Salman, S. (2019, March 31). *Conviction rate 3pc.* Daily Sun. Retrieved from https://www.daily-sun.com/printversion/details/381494

IN FOCUS 4.1: EVIL CORP'S FOUNDER DEFRAUDED VICTIMS OF HUNDREDS OF MILLIONS OF DOLLARS BUT IS LIVING SAFELY IN RUSSIA

In December 2019, law enforcement authorities in the UK and the US reported that they had identified the two Russian kingpins behind the "most dangerous cybercrime group in the world": Maksim Yakubets and Igor Turashev. Yakubets ran the criminal hacking outfit Evil Corp.[23] Turashev was Yakubets' administrator and controlled the Dridex malware. Dridex first appeared in 2011 and then was further developed as a targeted ransomware. In March 2020, Check Point Research ranked Dridex as the third most prevalent malware.[24]

Victims in the UK lost "hundreds of millions of pounds."[25] Evil Corp had been victimizing Internet users since 2009. The US and UK authorities also targeted 21 entities that were associated with Evil Corp, including a network of money launderers.

The National Crime Agency (NCA), the FBI, and the National Cyber Security Centre collaborated in the investigation. The NCA and the FBI were able to disable the Dridex botnet for a short time in 2015. Evil Corp, however, adapted the malware and infrastructure and continued to engage in cybercrime activities.

Yakubets was reported to drive a customized Lamborghini car. His personalized number plate translates to "Thief." He spent more than US$300,000 on his wedding.

The US State Department has offered a reward of US$5 million for information that leads to the arrest of Yakubets. This is the largest reward ever offered for a cybercriminal. It is next to impossible to arrest and prosecute them. The NCA noted: "If Yakubets ever leaves the safety of Russia he will be arrested and extradited to the US."[26]

4.3.1.4 Barriers to Developing CS Human Resources

Economic and institutional factors in developing countries present idiosyncratic challenges that act as a barrier to developing CS human resources. For instance, Cameroon, which is among the countries worst affected by cybercrime in Africa, faced a dilemma. In 2016, policy-makers in the country were in the process of launching CS skill development programs. They feared, however, that after completing the training program, the trainees could use the skills gained to commit cybercrimes.[27]

4.3.1.5 Victims' Unwillingness to Report Cybercrimes

As noted earlier, proportionally fewer cybercrimes than conventional crimes are reported. For instance, in the Netherlands only about a quarter of cybercrime cases were reported to law enforcement agencies in 2016.[28] Some cyberextortion victims face additional fears in reporting cybercrimes because they have criminal records. For instance, Internet gambling sites such as sports books and online casinos in LAC countries such as Antigua, Aruba, and Costa Rica are lucrative targets for online extortionists. Reports suggest that gambling sites in the LAC region alone pay out hundreds of millions of dollars in extortion money each year.[29] Police in these countries are poorly equipped to fight sophisticated international cybercrimes.[30] Moreover, some casino operators face indictments in the US on charges of illegal gambling. Law enforcement agencies such as the FBI do little to battle attacks against these online gambling sites.[31]

4.3.1.6 Cybercriminals' High Degree of Confidence in Carrying Out Criminal Activities

In jurisdictions with weak regulative institutions, cybercriminals show a very high degree of confidence in carrying out criminal activities. A computer forensics expert in Brazil noted that Internet crime gangs in the country do not use techniques to hide themselves.[32]

4.3.2 Diffusion of Cybercrime Expertise and Technology

How do cybercrime expertise and technology diffuse? What factors lead to broader and deeper cybercrime adoption among criminals? Diffusion of cybercrimes can be explained in terms of *relative advantage, compatibility, complexity, observability,* and *trialability.*[33]

4.3.2.1 Relative Advantage

Relative advantages are perceived benefits of a technology over previous technologies.

Cybercrimes' principal relative advantage stems from the fact that such crimes are less likely to be caught and prosecuted. Cybercrimes can be committed without leaving home. This is contrary to most conventional crimes, for which criminals leave a known territory only for sufficient incentives.

Cybercriminals continuously explore new techniques, methods, and tools as well as new ways to circumvent CS protection to victimize consumers and businesses. (See In Focus 4.2.) For instance, PhishLabs identified more than 29,000 phishing kits in 2017, of which over a third had used techniques to evade the existing detection systems.[34]

IN FOCUS 4.2: FORMJACKING

Formjacking can be viewed as an online form of credit-card skimming. While a skimmer attached to a gas pump or ATM to steal credit/debit card information was popular in the past, in formjacking the information is skimmed through online forms using JavaScript. For instance, when an e-commerce customer enters information such as their name, address, and payment card details and clicks "submit" on a malware-infected website, the submitted information goes to the cybercriminal's server.[35]

According to Symantec, about 5,000 websites became victims of formjacking per month in 2018. The company identified over 30 popular websites affected by formjacking, such as fashion stores, educational websites, and websites selling sports products.[36] Symantec's study of underground markets indicated that data from a single credit card is sold for up to US$45. With this rate, if the hackers were able to steal only 10 credit cards from each compromised website, their monthly income from the 5,000 websites they hacked could be about US$2.25 million (US$45 × 10 × 5,000).[37]

To further illustrate the rapid growth and evolution of new victimization tools, we consider ransomware (see In Focus 4.3; see also figure 4.3), which is among the newest types of malware. Finnish CS firm F-Secure's *State of Cyber Security* 2017 report stated that there was only one ransomware "variant" in 2012. This increased to 35 in 2015 and 193 in 2016.[38] In the third quarter of 2017 alone, new ransomware increased by 36%.[39] In the second quarter of 2019, more than 232,292 unique users were targeted by such attacks.[40]

IN FOCUS 4.3: TRAVELEX FACES A RANSOMWARE CYBERATTACK

The London-based foreign currency exchange Travelex faced a ransomware cyberattack on the eve of New Year 2020. Travelex does business in 26 countries. It operates more than 1,000 stores and 1,000 ATMs. It allowed money transfers using cash or prepaid card. Following the attack on December 31, 2019, affected Travelex locations processed transactions manually. Online services were disconnected to prevent the spread of ransomware.[41]

The attacks affected the company's customer services for many weeks.[42] Travelex's bank and wire service partners, such as Virgin Money and Sainsbury Bank, reportedly also experienced outages due to the attack.[43]

A ransomware gang called Sodinokibi (also known as REvil) was reported to tell *BBC News* that it hacked Travelex's computer networks and wanted the victim to pay US$6 million. The gang claimed that it had gained access to the company's networks six months before and downloaded 5 GB of sensitive customer information.[44] Travelex allegedly paid US$2.3 million (about 285 bitcoins) to the criminal gang in order to regain access to its data. The company needed to negotiate with the criminals for many weeks.

Figure 4.3 Increasing pervasiveness and costs of ransomware

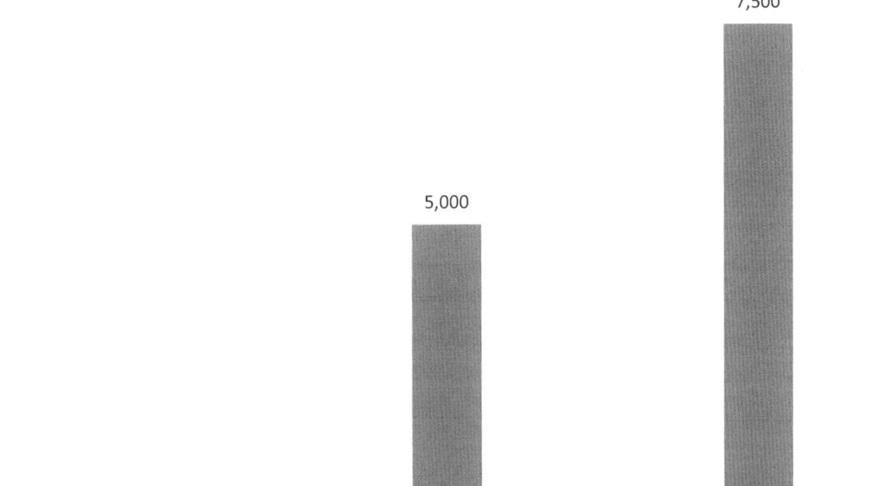

Data sources: Data for costs from Cybersecurity Ventures (2015 and 2017) as reported in Morgan, S. (2019). *Global ransomware damage costs predicted to reach $20 billion (USD) by 2021*. Cybercrime Magazine. Retrieved from https://cybersecurityventures.com/global-ransomware-damage-costs-predicted-to-reach-20-billion-usd-by-2021/ and Emsisoft (2019) as reported in Hope, A. (2020). Ransomware costs in 2019. *CPO Magazine*. Retrieved from https://www.cpomagazine.com/cyber-security/ransomware-costs-in-2019/; data for variants from *F-Secure State of Cyber Security 2017* (2015) and Recorded Future (2017) as reported in Liska, A. (2018, March 7). *2018 ransomware trends*. IS Buzz News. Retrieved from https://www.informationsecuritybuzz.com/articles/2018-ransomware-trends/; data for 2019 as reported in Fuks, L. (2019, April 28). *10 ransomware attacks you should know about in 2019*. Allot. Retrieved from https://www.allot.com/blog/10-ransomware-attacks-2019/

In recent years, the use of cryptocurrencies has taken the relative advantage of some forms of cybercrimes to a higher level. Before cryptocurrencies, extortionists usually asked victims to send money via money transfer agencies or by direct deposit to bank accounts. However, law enforcement could trace such transfers. Cryptocurrencies are an online extortionist's dream come true. For instance, it is almost impossible to pinpoint perpetrators based on bitcoin addresses. Bitcoins can be converted into cash secretly through third parties. Pinpointing the extortionist is of little value if that person is in a jurisdiction that does not cooperate in detection. For instance, as noted in chapter 1, WannaCry ransomware attacks originated from North Korea. Given that North Korea maintains few diplomatic ties, it is difficult to take actions against hackers that operate from such an isolated country.

Extortionists can further reduce the probability of detection using next-generation cryptocurrencies such as Monero, Dash, and Z-Cash. (See In Focus 4.4.) These have built-in anonymity features. For instance, Monero is arguably the most anonymous cryptocurrency. In 2016, AlphaBay, which might be the most widely used darknet market, started accepting Monero.[45] Monero is less widespread than bitcoin. For instance, as of April 10, 2021, Monero's total value was around US$5.2 billion compared to bitcoin's US$1128.3 billion.[46] Moreover, converting Monero into real-world cash is more difficult. However, this problem might not be as significant in the future if there is a higher degree of diffusion and more competition in cryptocurrencies.

IN FOCUS 4.4: SODINOKIBI SWITCHES TO MONERO

Sodinokibi is a ransomware as a service provider and is described by CS firm CyberReason as "The Crown Prince of Ransomware." It is believed to be linked to the gang that created the successful GandCrab ransomware.[47] Sodinokibi offers customized versions of the ransomware to its affiliates. If the affiliate successfully infects a system and a ransom is paid, the affiliate keeps 60%–70% of the profits.[48] Sodinokibi (REvil) ransomware targeted many healthcare organizations during the COVID-19 crisis.[49]

Sodinokibi was reported to switch from bitcoin to Monero in order to better hide the identities of the criminal gang members. In February 2020, a strategy analyst for the European Union Agency for Law Enforcement Cooperation (Europol) explained that the agency could not trace the funds and IP addresses due to the use of TOR and Monero by the gang. The agency had made some progress when Sodinokibi used bitcoin because bitcoin blockchain was more visible.

In a hacker and malware forum, the hackers running the Sodinokibi ransomware announced their switch to Monero: "we inform you that after a while the BTC will be removed as a payment method. Victims need to begin to understand the new cryptocurrency, as well as other interested parties who work with us."[50] In a two-week period in March 2020, Monero's price increased by about 11%.[51]

Ransomware involving cryptocurrencies might increase because cryptocurrencies are becoming more widely adopted. According to the World Economic Forum, 10% of global GDP will be stored on blockchain by 2027[52] compared to 0.025% in 2016.[53]

4.3.2.2 Compatibility

Compatibility refers to the degree to which a technology and the tasks it performs are perceived as being consistent with the existing values, beliefs, past experiences, and needs of potential adopters of the technology. The Internet's rapid diffusion and digitization of economic activities have led to increased cybercrime activities.

This is because the mechanisms by which cybercrimes are committed are compatible with digitization. Consider smart toys. The global smart toy market size was US$5.7 billion in 2019 and is expected to reach US$13.8 billion by 2024.[54] Smart toys employ sensors, cameras, microphones, data storage, voice recognition, GPS, and more. These technologies make toys more fun and engaging but also provide more vectors for cyberattacks.

Traditional bank robbery skills are incompatible as banks increasingly digitize their activities. In general, the traditional ways that organized crime groups committed crimes have become incompatible with the digitization of value and organizations' increased dependence on digital technology. As a result, as noted in earlier chapters, organized crime groups are developing expertise in cybercrime.

4.3.2.3 Complexity

The natures of the technology and of hacking communities and organized crime groups have greatly reduced the complexity of cybercrime expertise and technology. Most hacking tools are widely available online and require little or no expertise. Less skillful criminals also get help from experienced hackers.

Information sharing in the cybercriminal community also reduces complexity. Members in the community help fellow hackers in accessing a router and getting through a firewall. There are also reports that US-based low-end criminals get cybercrime-related help from Russian and Eastern European professional criminals.

Moreover, in some countries, specialized schools reportedly teach hacking skills. While learning network hacking skills requires programming skills, there are also some training institutes that teach would-be cybercriminals other skills that do not require technical sophistication. For instance, in China, there are reportedly many "sales brushing" schools. Note that "sales brushing" (*shuadan* in Chinese) involves placing fake orders with legitimate vendors who sell on e-commerce websites such as Alibaba-owned Taobao. Online vendors engage in "sales brushing" in order to give potential customers a false impression of high sales volume or good reputation. The "sales brushing" schools offer training programs that teach how to mimic a normal buyer's behavior so that fake orders are not detected and caught. The processes participants learn in the course include browsing websites before buying a product, asking the seller for information, and comparing prices of similar items. Such steps are likely to help escape Alibaba's detection system. Operators of the brushing groups reportedly charge a training fee in the US$6–US$12 range. Online training is provided through video. The would-be brushers are also required to take written tests to prove their proficiency in effective brushing techniques. Once the tests are passed, users graduate from ordinary member status to higher levels, which also gives them greater powers over managing the group.[55]

4.3.2.4 Observability

Observability is the degree to which the results of cybercrime activities can be seen by individuals who are likely to engage in such activities. Cybercrimes also induce a high degree of observability for criminals because cybercrimes are easy to commit and rewards are high. Some criminals in the conventional world are cashing in on the trend of increased sophistication in cybercrime technologies. For instance, Russian hack rings are reportedly operated by Mafia and former KGB agents. As discussed in chapter 1, investments in cybercrime tools such as botnets provide a high return on investment for cybercriminals.

Several North Macedonia–based fake news websites that publish fabricated or misleading articles were reported to attract large number of readers, millions of Facebook likes, and thousands of dollars of income every day.[56] This phenomenon increases the observability of their actions. In an interview with CNN,[57] an owner of a fake news site based in North Macedonia said that they were able to make more than US$2,000/day from news items that had about a million Facebook likes. Some reported that they earned US$100,000/month.[58]

4.3.2.5 Trialability

Online availability of hacking tools offers risk-free trials for would-be hackers. Recently, the quantity, availability, and quality of hacking tools have increased. Some sources of externalities thus exist in the technology. Evidence also indicates that many college students pirate software and gain illegal access to a computer system to browse and/or exchange information. Some malware development organizations were reported to offer money-back guarantees for the purchases of "zero-day" malware, meaning that if the malware were caught by antivirus programs, the purchaser would get a full refund.[59] Such experiences provide "trialability" and help purchasers get their foot in the door of the cybercrime world.

4.3.3 Increased Predisposition toward Cybercrime

This section looks at various factors that increase predisposition toward cybercrime.

4.3.3.1 Lack of Employment Opportunities

First, crime rates are linked to economic opportunities. Gary Becker comments on crimes committed by teenagers: "[L]ow earnings are a factor behind crime, and teenagers have lower earnings and fewer opportunities."[60] In some economies, lack of employment opportunities has led to an increase in cybercrimes. In Russia and Eastern Europe, students who are good in mathematics, physics, and computer science cannot easily find jobs. Likewise, creation of fake news has been an attractive economic activity for younger people in some economies burdened with high unemployment rates. For instance, in the first quarter of 2017, North Macedonia had an unemployment rate of 22.9%.[61] For those who work, the average monthly salary in North Macedonia is US$37.[62] Between August and November of 2016, a Macedonian teenager was reported to make around US$16,000 from two fake news websites.[63]

4.3.3.2 Social Acceptance of Cybercrimes

Cybercrimes are justifiable in some societies. An IDG News Service article describes how a Russian hacker-turned-teacher and his friends hacked programs that they distributed for free: "It was like our donation to society, it was a form of honor; [we were] like Robin Hood bringing programs to people."

4.3.3.3 Intrinsic Motivations Based on Obligation to the Community

Behaviors of ideological hackers interested in political goals can be explained by intrinsic motivations based on obligation to the community. Chinese hackers, for instance, fought cyberwars with Taiwanese, Indonesian, Japanese, and US hackers. Chinese hackers involved in cyberwars argued that they were patriotic and didn't do anything wrong. Patriotism and nationalism thus provided cognitive legitimacy for these hackers' activities. Ideological hackers are also motivated to fight against global capitalism and people of other religions.

4.3.3.4 Internet Users' Ignorance

Cybercriminals take advantage of Internet users' ignorance. For instance, many Internet users do not understand the importance of secure passwords. According to the UK government's agency for protecting the country from cyberthreats, National Cyber Security Centre, about 23.2 million people worldwide used "123456" as their password; another 3.8 million used "qwerty," which are the first six letters on the top left of a standard keyboard; and "password" was used as the password by 3.6 million accounts worldwide.[64]

A large proportion of Internet users lack the ability to distinguish fake from real news headlines. Estimates suggested that in the final three months of the 2016 US presidential campaign, the top-performing fake "election" news stories on Facebook attracted more views than top stories from major news outlets such as *Huffington Post, NBC News,* the *New York Times*, and the *Washington Post*. During that time, 20 top-performing false election stories from fake news sites generated over 8.7 million shares, reactions, and comments on Facebook compared to just over 7.3 million shares from 19 major news websites.[65] According to individuals experienced in running fake news sites, ad networks, when deciding on what sites to publish ads, do not care about the quality of sites but only that the sites meet minimum thresholds (e.g., no pornography). The main requirement for most advertising networks is that the traffic comes from people, not bots.[66]

At the same time, children's online activities are not sufficiently monitored. Many parents don't know about the parental control options available with the latest versions of operating systems. For instance, Windows 10's "family" feature can be used to compile reports about a young user's session activities. This feature creates "activity reports" and emails them to a parent's account. The reports include the lists of the games played, apps used, websites browsed, and keywords used on search engines.[67] Parents can also limit the content available to the child without getting permission.[68] There are also other parental control programs such as Net Nanny and Qustodio. These offer broader options to protect and monitor young users, though they are not free.

Surveys have found extremely low adoption rates of such tools. For instance, according to a 2019 study conducted by the online game platform Roblox, the proportions of parents using digital monitoring tools such as parental controls for tracking kids' online activities were 55% among millennial parents, 42% among Generation X, and 47% among baby boomers. The study also found that 60% of teens had rarely discussed inappropriate online behaviors with their parents.[69] This is a serious concern, because cyberoffenses targeting teens and young adults are growing rapidly. For instance, in a survey conducted by the Pew Research Center, 59% of US teens reported being cyberbullied.[70]

4.3.3.5 Victims' Tendency to Pay Ransoms

Some victim companies pay ransom to cybercriminals. While paying ransom might be wrong from a societal and national perspective, it might be the right strategy for the individual entity that is the subject of the ransom. When an organization's critical data are at risk, it may require an immediate response. (See In Focus 4.3.)

4.3.3.6 Diffusion of Cryptocurrencies

One study reported that 33% of UK-based corporations had bought bitcoins so that they would be prepared to pay ransoms. More than 35% of large firms (defined as those with more than 2,000 employees) were willing to pay as much as US$65,000 to unlock files containing critical data such as intellectual property.[71] A university in the US reportedly created an account to be prepared for ransomware attacks.[72] Increased success sends positive cognitive messages and makes cybercriminals disrespectful of law enforcement.

4.3.3.7 Less Guilt Compared to Conventional Crimes

As noted in chapter 3, various factors lead to less guilt in cybercrimes compared to conventional ones. It is also argued that norms guiding actions are based on the notion of face-to-face relations. Compared to conventional crimes, people involved in cybercrimes are less likely to see their actions' negative impacts.

4.4 CHAPTER SUMMARY AND CONCLUSION

The rapid rise in cybercrime has been an issue of pressing concern to our society. In order to better manage and control cybercrimes, it is important to understand the factors that fuel such crimes. The discussion in this chapter points to several mechanisms that have contributed to cybercrime escalation.

There are synergies between increasing return activities in cybercrimes. Economic, sociopolitical, and cognitive legitimacy for cybercriminals influences the degree of increasing returns to these criminals. Low costs of hacking tools and low barriers to entry make cybercrime more attractive for cyberoffenders. Over-educated and under-employed computer specialists in some countries find opportunities in online dark web activities. Weak legal and enforcement systems encourage cybercriminals' behaviors. Social and cultural acceptance of such crimes in several parts of the world also contributes to the growth of such crimes.

The battle against cybercrimes must be waged on many fronts. Both technological and non-technological measures can reduce the externality effects. At the micro level, technological and behavioral factors should be considered in design and implementation of computer networks to reduce the financial incentive for hackers and provide negative cognitive feedback to cybercriminals. Technological measures range from disconnecting databases containing sensitive information from the Internet to deploying sophisticated CS technologies. Similarly, behavioral measures such as CS training programs may reduce cybercrimes. The impact of such training programs would likely extend beyond the immediate purpose of creating CS awareness among target employees. Other employees and society as a whole may also learn by observing.

National and local governments need to take measures to develop technological and human resource capabilities. Stronger law and enforcement systems are also needed to discourage cybercrime activities.

National governments, possibly in collaboration with the private sector, should initiate efforts to educate businesses and consumers in CS-related issues from technical, behavioral, and cognitive perspectives. CS also needs to be integrated in the curricula of schools and colleges. National CS Awareness Month, which is observed in October in the US, focuses on strengthening the CS orientation of businesses and consumers.[73] In India, the National Association of Software and Services Companies (NASSCOM), Data Security Council of India (DSCI), and Mumbai Police announced a similar initiative in December 2015.[74]

Given cybercrimes' global nature, a high degree of cooperation with foreign law enforcement agencies is needed to reduce jurisdictional arbitrage. There are already some encouraging initiatives that provide a platform for such cooperation. In 2016, the No More Ransom[75] project was launched by the Dutch National High Tech Crime Unit, Europol, Kaspersky Lab, and Intel. As of early 2017, more than two dozen police agencies and many private firms had joined the initiative.[76]

4.5 DISCUSSION QUESTIONS

1. What positive feedback mechanisms encourage hackers to engage in cybercrimes?
2. What factors facilitate the diffusion of cybercrime expertise and technology among cybercriminals?
3. How do cryptocurrencies facilitate cybercrimes?
4. Why can law enforcement agencies not investigate all reported cybercrimes?

4.6 END-OF-CHAPTER CASE: RUSSIAN BUSINESS NETWORK'S UNDERGROUND CRIMINAL BUSINESS OFFERINGS

The Russian Business Network (Rossiiskaya Biznes Set) was among the most notorious bulletproof hosting (BPH) service providers. Note that BPH is an "underweb" or "deep web" term to refer to a service provider that hosts virtually any type of illicit content, such as those used in phishing and spamming, carding sites, botnet command centers, online gambling, illegal pornography, and browser exploit kits.[77]

Cybercriminals use so-called bulletproof service providers, which allow their customers to upload and distribute any type of material, including illegal content. The content hosted by bulletproof service providers is not indexed by search engines. As of 2017, there were 15 main BPH underground operations. Most of them were run from Moldova, Russia, or Ukraine.[78] Other popular BPH countries include Bulgaria, Iraq, Panama, Romania, and Yemen.[79]

Former spies[80] allegedly grouped together to establish the Russian Business Network (RBN), which was the first criminal outfit that spotted opportunities in offering digital hosting services to other cybercriminals. In this way, it was a pioneer in the underworld activities of this category and thus heavily influenced the trajectory and development of the entire BPH industry.[81]

In addition to providing website hosting services to criminals. RBN allegedly offered spyware, trojans, phishing, spam, malware distribution, botnet command and control systems, and even child abuse materials.[82] It also laundered money.[83]

A point to note is that the activities of BPH firms are not illegal in some countries.[84] An analyst with the antivirus firm Kaspersky Lab was quoted as saying, "They make money on the services they provide … the illegal activities are all carried out by groups that buy hosting services.… RBN … does not violate the law. From a legal point of view, they are clean."[85]

4.6.1 The Flagship Product Mpack

The virus creation tool Mpack was RBN's flagship product and was designed to extract data from infected PCs. It was packaged with personal tech support from RBN and cost US$500–US$1,000. Mpack exploited known software security holes in browsers.[86]

Cybercriminals launched attacks against websites and installed malicious programs created with Mpack. When someone visited such sites with a browser unequipped with software security updates, a password-stealing program was installed on the visitor's computer. It then scanned the computer for vulnerabilities.[87] The stolen data was forwarded to a "drop site" in RBN servers. Mpack monitored the success of its operation through various metrics on its online, password protected control, and management consoles.[88]

4.6.2 Political Protection

Observers noted that RBN seemed to have political protection.[89] The rumor for a long time was that its founder was the nephew of a high-ranking politician from St. Petersburg.[90] According to the Serious Organized Crime Agency, RBN also allegedly bribed local police, judges, and government officials.[91] An Economist.com article[92] noted:

> Despite the attention it is receiving from Western law enforcement agencies, RBN is not on the run. Its users are becoming more sophisticated, moving for example from simple

phishing (using fake emails) to malware known as "trojans" that sit inside a victim's computer collecting passwords and other sensitive information and sending them to their criminal masters.

4.6.3 Death and Rebirth of RBN

RBN stopped operations in November 2007. Some analysts suspected that "whatever protection RBN enjoyed was withdrawn because the group had overreached."[93] Some suggested that the group operating RBN may have shifted its operations to Asian countries.[94]

There was an accusation that, like most Russian cybercriminals, RBN worked closely with Russian intelligence agencies. RBN was believed to be a key player in cyberattacks on Estonia in 2007.[95] In July 2008, Russia's patriot hackers launched DDoS attacks against government agencies in Georgia. A CS researcher claimed that the servers used in the cyberattack on Georgia were connected to RBN.[96]

Some analysts suspected that RBN expanded its operations, which included the creation of a hacking software program called BlackEnergy, which was used to control a botnet.[97] According to a *Wall Street Journal* article published in December 2009, the FBI was looking at a hacking that attacked Citigroup and stole tens of millions of dollars.[98] The hacking had used BlackEnergy and was traced via traffic on ISPs that RBN had used previously.[99]

A decade later, in 2017, some cybercriminals believed to be the founders of RBN were operating from Vladivostok in Russia and from the Ukraine. One of the RBN employees, who lived in Beijing, China, in 2008, operated from Vladivostok and provided a cloud proxy network that abused the services of legitimate cloud service providers such as Amazon Web Services (AWS). Another individual affiliated with RBN operated from the Ukraine after 2010, providing fast flux proxy content delivery networks for criminal activities.[100] Note that fast flux is a Domain Name System technique that is used to mask botnets by rapidly shifting among a network of hosts that are compromised, which act as proxies. Doing this, cybercrime activities are less likely to be detected.[101] Fast flux proxy networks are thus an efficient form of BPH.[102] The individual mainly utilized compromised Iranian IP addresses for such purposes.

PART TWO

Macro-level Factors Affecting Cybercrime and Cybersecurity

CHAPTER FIVE

Political, Cultural, Organizational, and Economic Factors Affecting Cybercrime and Cybersecurity

CHAPTER PREVIEW

This chapter offers a detailed description of how political, social, cultural, organizational, and economic factors are linked to cybercrime and CS measures. It highlights how regulative, normative, and cultural-cognitive institutions are associated with cybercrime and CS activities in a nation. The chapter also takes a look at how cybercriminals consider these factors in the selection criteria for the target network. It gives special consideration to the effect of the stocks of hacking skills relative to the availability of economic opportunities on the degree of cybercrime activities originating from a country. This chapter argues that formal regulations and policies related to CS are among the most powerful forces that affect organizations' and individuals' CS measures. It also gives an overview of some key elements of culture from the CS standpoint. It highlights efforts being taken at the industry, trade association, and inter-organizational levels. A detailed description is provided of CS measures being taken by small and medium-sized enterprises in light of increasing cyberthreats facing them.

5.1 INTRODUCTION

Political, social, cultural, and economic factors are tightly linked to cybercrimes. For instance, in some countries, political and law enforcement institutions are weak, inefficient, corrupt, and indifferent to cybercrimes. There is thus a general neglect of fighting cybercrimes. To take an example, it is reported that government officials in Nigeria claimed that they were ignorant of cybercrimes originating from the country and some labeled it as Western propaganda.[1]

Next, culture influences how issues around cybercrimes are constructed and how they are defined. For instance, while viewing pornography is a crime in Saudi Arabia, viewing pornography is considered a personal freedom in some other countries. As such, cybercrimes and cyberoffenses have varying degrees of social and cultural acceptability across the world. They range from passive acceptance to outright celebration of "patriotic" hacker-heroes.[2] Indeed, important cross-cultural variation exists regarding what behavior is acceptable or unacceptable. As noted earlier, in some societies, rather than being stigmatized, some forms of cybercrimes are socially legitimized.

Cultural factors are also linked to how cybercriminals choose their victims. A related concept is cultural Others, which refers to "groups of people perceived to be outside defined boundaries, at the margins of acceptance."[3] Political, social, and cultural factors such as religious belief, ethnicity, and nationality form boundaries that separate in-groups and out-groups. Some cybercriminals tend to be more willing to victimize organizations or individuals that are viewed as cultural Others or as belonging to an out-group rather than their own group (in-group).

These macro factors also influence CS measures. For instance, in 2010, Saudi Arabia's Commission for the Promotion of Virtue and Prevention of Vice (known as the Haia) announced that it would establish a unit to fight cybercrimes. The Haia also noted that the cybercrime unit's initial focus would be on cases involving online blackmail of women.[4]

CS-related norms and practices are also evolving at the industry level. For instance, India's NASSCOM and DSCI have taken measures to strengthen CS standards.

5.2 INSTITUTIONS' EFFECTS ON CYBERCRIME AND CYBERSECURITY

Douglass North defines institutions as "macro-level rules of the game."[5] The rules can be formal as well as informal. In order to understand institutions' effects on cybercrime and CS, in this section we use three institutional pillars proposed by W.R. Scott: (i) regulative, (ii) normative, and (iii) cultural-cognitive.[6] These pillars relate to "legally sanctioned," "morally governed," and "recognizable, taken-for-granted" behaviours, respectively.[7]

5.2.1 Regulative Institutions

Regulative institutions consist of "explicit regulative processes: rule setting, monitoring, and sanctioning activities."[8] Some examples include regulatory bodies (such as the US DoJ) and existing laws and rules that influence cybercriminals to behave in certain ways.[9] In this section, we discuss rules and laws as key components of regulative institutions in influencing cybercrimes.

5.2.1.1 Effect on Cybercrime

The most relevant issue concerning regulative institutions is cybercrime laws, which deal with criminal activities in which computers or computer networks are the principal means of committing an offense. According to the United Nations Conference on Trade and Development (UNCTAD), as of April 2020, 79% of the countries in the world had enacted such laws. (See figure 5.1.) Some economies such as Cambodia had draft legislation.

In countries that have enacted cybercrime laws, online perpetrators can be tried and convicted explicitly based on such laws. In countries that lack cybercrime laws, provisions related to conventional crimes are stretched to accommodate cybercrimes and may cover only peripheral acts related to cybercrimes.[10] Cybercrime laws may also make it mandatory for victim organizations to report cybercrimes, which can help combat such crimes.

The existence of cybercrime law on the books, however, is not sufficient to control cybercrimes. In many countries, due to weak enforcement, such laws are ignored with impunity. Some countries, such as Russia, allegedly do not enforce cybercrime laws, thereby providing a fertile ground for cybercrimes.

Figure 5.1 Percentage of countries worldwide with different stages of cybercrime legislation

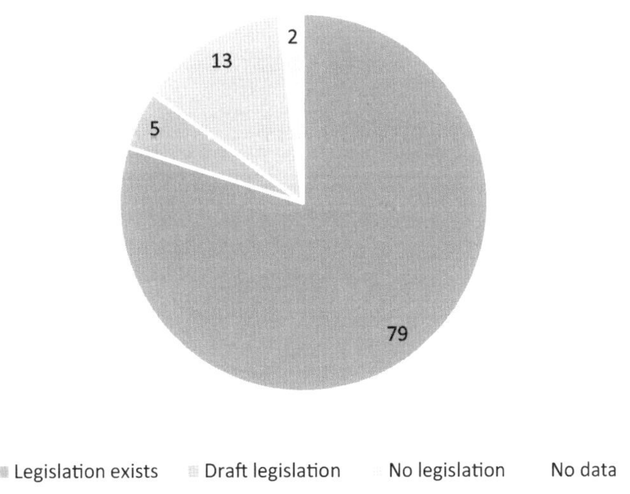

Data source: UNCTAD. (n.d.). *Cybercrime legislation worldwide.* Retrieved from http://unctad.org/en/Pages/DTL/STI_and_ICTs/ICT4D-Legislation/eCom-Cybercrime-Laws.aspx

Cyberattacks thus benefit from jurisdictional arbitrage. The lack of strong rule of law is associated with origination of more cyberattacks. Put differently, organized cybercrimes are initiated from countries that have few or no laws against cybercrimes and little capacity to enforce existing laws.

5.2.1.2 Effect on CS

Regulations and policies related to CS are a powerful force that affects organizations' and individuals' CS measures. For instance, countries worldwide vary greatly in terms of their formal regulatory environment related to CS. The most relevant regulations for this purpose are data protection and privacy laws, which deal with how personally identifiable information of individuals collected by governments, public, or private organizations is stored and used. According to the UNCTAD, as of April 2020, 64% of the countries in the world had enacted such laws. (See figure 5.2.) The UNCTAD lacked relevant data to determine the status of such legislation in another 11% of countries. All European countries for which data were available had enacted such laws. On the other hand, only 52% of countries in Asia-Pacific and Africa had data protection and privacy laws.

Countries also differ in the way their existing CS resources (e.g., workforce) are utilized. A major difference concerns whether a country's ruling elites and policy-makers view internal

Figure 5.2 Percentage of countries worldwide with different stages of data protection and privacy legislation

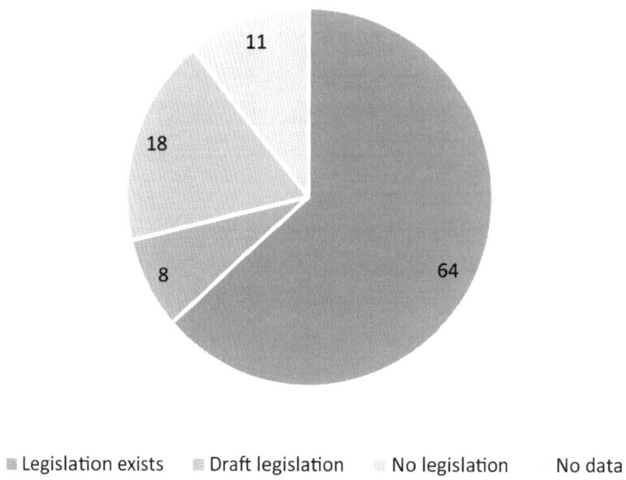

Data source: UNCTAD. (n.d.). *Data protection and privacy legislation worldwide.* Retrieved from http://unctad.org/en/Pages/DTL/STI_and_ICTs/ICT4D-Legislation/eCom-Data-Protection-Laws.aspx

or external threats as more severe. When internal security is perceived to pose a bigger threat, the defense against external attacks is a less important security goal and the government is likely to focus its CS efforts to deal with internal threats. Internal conflicts arise because social, economic, and political systems work to the national elites' advantage and they exploit the majority of the people in the country.[11] Internal security is often desirable for the ruling elites to remain in power, which does not necessarily involve making all citizens equally secure and safe.[12]

Some governments leverage their cyberpower to suppress domestic political dissent. To take an example, it was reported that the Government of Burma used its cyberwarfare department within the police force to track its online critics and send virus-attached emails to exiled activists. As another example, according to the Electronic Frontier Foundation, the Vietnamese government recruited local hackers to spy on journalists, dissidents, and activists.[13] Likewise, according to the 2012 annual report of Reporters Without Borders, Iran was ranked "the number one enemy of the Internet" due to its strict measures to restrict Internet access, filter content, and imprison bloggers. On these measures, the country was stricter than other authoritarian regimes such as Saudi Arabia, Bahrain, Syria, China, and Belarus.

In some other states, internal security threats are almost non-existent; instead, external threats are more prominent. One such example is Japan, which has an orderly and prosperous society with a homogeneous population characterized by a high-trust culture. India is another example of a country where external security threats, such as those associated with Pakistan's nuclear weapons and China, are viewed as more serious, despite significant internal threats associated with poverty, government corruption, and illiteracy.[14] South Korea is also an example in which the threats are primarily external, mainly associated with North Korea.

5.2.2 Normative Institutions

Normative components introduce "a prescriptive, evaluative, and obligatory dimension into social life."[15] As noted above, hackers in some countries are considered to be just living up to the moral standards of Robin Hood. Elements of normative institutions also include trade associations or professional associations that can use social obligation requirements or business ethical principles such as codes of conduct to influence the behaviors of individuals and organizations. For instance, the members of the Honker Union of China (also known as the Red Hackers) may be required to behave according to the guidelines set by that organization.

5.2.2.1 Effects on Cybercrimes

As discussed in chapter 3, hackers with a common moral framework sometimes launch cyberattacks to achieve their goals. An example is the hack by Anonymous at the beginning of the COVID-19 pandemic of about 25,000 email addresses and passwords allegedly belonging to employees at the Gates Foundation, the WHO, and National Institutes of Health. Hackers like Anonymous are sometimes considered vigilantes acting with moral, although not legal, justification.[16]

In other cases, national governments and ideological hacking groups hope to achieve the same goals. For instance, Hong Kong's organizations supporting the pro-democracy movement, also known as the Umbrella Revolution, which started in September 2014, faced a series of DDoS attacks that year. In October 2018, the pro-democracy newspaper Apple Daily and the online forum HkGolden used by the demonstrators to plan their activities both faced significant cyberattacks.[17] FireEye found some overlaps in the tools and infrastructure used in the DDoS attacks and custom malware used by China-based APT actors for espionage.[18] Specifically, the attacks were linked to a Chinese threat group that FireEye calls "Operation Poisoned Hurricane," which allegedly carried out cyberespionage against Internet infrastructure providers, at least one media company, financial services, and Asian government organizations.[19]

5.2.2.2 Effects on CS

Trade associations and industry bodies rely on business principles such as codes of ethics and codes of conduct to influence the behavior of organizations. One example is India's NASSCOM, which as of April 2020 had 2,800+ members and represented a US$191 billion industry employing 4.4 million professionals.[20] NASSCOM has established the DSCI as an independent and self-regulatory organization to focus on CS. DSCI monitors member companies to ensure they adhere to CS standards. For instance, it requires members to self-police and provide additional layers of security at the infrastructure, application, and other levels. Companies failing to secure their data may be fined as much as US$1 million. Noncompliant companies might also lose DSCI membership.

5.2.3 Cognitive Institutions

Cognitive institutions are associated with culture. These components represent culturally supported habits that influence hackers' behaviors or organizations' and individuals' orientation toward privacy and CS.

5.2.3.1 Effect on Cybercrimes

Cognitive institutions affect cybercrimes through various mechanisms and processes.

The orientation of a culture toward a cybercrime activity may differ depending on who the cybercriminals, victims, and potential beneficiaries are. The society may view a cybercrime differently when the victim is a foreign rather than domestic company. The idea of cultural Others may help us understand this phenomenon better. It is argued that cultural Others are defined according to "who they are not" rather than "who they are."[21] It is argued that "all humans may possess the seeds of pseudospeciation, of prejudice against dissimilar groups and values."[22]

5.2.3.2 Justifiability and Acceptability of Victimizing Cultural Others

Some examples of boundaries that separate perpetrators and potential victims and are likely to make cyberattacks more justifiable and acceptable from the perpetrators' perspective are presented in table 5.1.

5.2.3.2.1 Different Nationality Patriotism and nationalism legitimize some hackers' activities. We can find many instances of hackings linked to nationalism and patriotism. Consider the US–China cyberwars. In 1999, following the accidental bombing of the Chinese Embassy in Belgrade, a group of hackers that identified itself as Level Seven Crew defaced the website of the US embassy in China and replaced the home page with racist and antigovernment slogans.[23] Following the collision of a US surveillance plane and a Chinese fighter in 2001, a Chinese hacking group published its plans for a "Net War," which was planned to continue until the anniversary of the bombing in Belgrade. In response, hacking groups from the US, Brazil, and Europe attacked Chinese websites. Chinese hackers reportedly attacked about 1,100 US websites, including that of the White House, while US hackers broke into 1,600 Chinese websites.[24, 25]

5.2.3.2.2 Different Ideology As noted above, hackers' interests are also framed by ideologies. For instance, the Anonymous hacking group has launched several high-profile cyberattacks

Table 5.1 Some examples of Otherness that make cyberattacks justifiable

Source of Otherness	Explanation	Examples
Different nationality	• Launching cyberattacks against individuals and organizations in other nations	• Chinese hackers' cyberwars with Indonesian, Japanese, Taiwanese, and US hackers
Different ideology	• Using cyberattacks as a tool to attack others with different ideologies	• "Anonymous" hacking group's attacks against the websites of the Church of Scientology
Different religion	• Using cyberattacks against those who challenge one's religious beliefs and values	• India–Pakistan and Israel–Palestine cyberwars
Different economic privilege	• Launching financially motivated cyberattacks against rich people	• Indonesian hackers targeting rich Westerners

against political and religious organizations. In 2008, it attacked the websites of the Church of Scientology. The stated goal of the attack was to help "save people from Scientology by reversing the brainwashing."[26]

A report of the US DHS noted that non-state actors are rapidly emerging that possess the same level of capabilities as those of sophisticated nation-states.[27] One such example is the Anonymous hacking group. An Internet activist, who was formerly affiliated with Anonymous, was reported to say that the only guiding principle of the hacking group is "anti-oppression."[28] While it is difficult to determine the actual size of the group, it is a widely distributed global movement of ideological hackers.

5.2.3.2.3 Different Religion Hackings by Islamic activists are interesting examples of cyberattacks in this category. Except for occasional India–Pakistan and Israel–Palestine cyberwars, hacking by Islamist activists was insignificant before September 11, 2001. The mi2g Intelligence Unit reported increasing Islamist hacking, the targets being networks of Australia, the UK, the US, and other coalition partners as well as domestic networks of Indonesia, Kuwait, Morocco, Pakistan, Russia, Saudi Arabia, and Turkey.

5.2.3.2.4 Different Economic Privilege Cultural attitudes toward potential victims are also related to the perpetrators' propensity to engage in cybercrimes. It is, for instance, reported that many Indonesian hackers feel that cyberfraud is wrong but acceptable, especially if the credit-card owner is rich and not an Indonesian. An Indonesian hacker who used stolen credit cards to buy things online for himself (known as a carder) reportedly said: "Yes, it's wrong but it really only hurts other rich countries that were dumb enough to let us. Why should an Indonesian get arrested for damaging American business?"[29] Another carder was reported as saying, "I only choose those people who are truly rich. I'm not comfortable using the money of poor people. I also don't want to use credit cards belonging to Indonesians. Those are a carder's ethics."[30]

5.2.3.2 Effect on CS and Privacy

Cultural factors are linked to users' attitudes toward CS systems and measures. When CS measures are incompatible with culture, they may face resistance. For instance, Indian firms engaged in outsourcing have taken measures to prevent cybercrimes by current and former employees. Indian and foreign firms follow the security practices of Western firms, some of which are incompatible with local culture. For instance, call center employees must undergo security checks that are considered to be "undignified."[31] Firms have established biometric authentication controls for workers and banned cellphones, pens, paper, and Internet and email access for employees.[32] Computer terminals at outsourcing firms lack hard drives, email, CD-ROM drives, or other ways to store, copy, or forward data. Indian outsourcing firms also extensively monitor and analyze employee logs. The US computer firm Dell faced difficulties

Figure 5.3 Percentage of respondents willing to accept less privacy for faster Internet

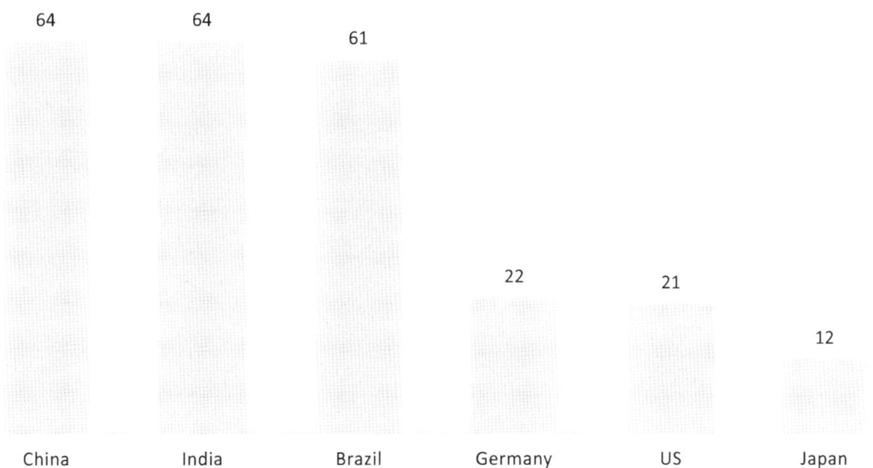

Data source: Politico/Qualcomm global survey as reported in Hendel, J. (2020, February 25). *The 5G world: What people care about.* Politico. Retrieved from https://www.politico.com/news/agenda/2020/02/25/poll-5g-what-do-people-really-want-110831

retaining its employees in its Indian call centers when the company attempted to monitor their activities. Such practice was viewed as demeaning to employees.[33]

Cultural factors affect how much people value CS and privacy and their willingness to trade privacy in exchange for economic benefits and convenience. As figure 5.3 shows, significantly higher proportions of consumers in emerging economies such as Brazil, China, and India than in industrialized countries are willing to trade their privacy for faster Internet.

The cultural and political acceptance of surveillance is arguably a major factor that contributed to South Korea's higher level of success compared to other major economies in controlling COVID-19. South Korea tracked credit- and debit-card transactions to know card users' movements. Cellphone customers are required to provide real names and national registry numbers to their phone companies. The locations of phones can be tracked. Closed-circuit television cameras also allowed government agencies to identify people who were in contact with COVID-19 patients.[34]

5.3 STOCK OF SKILLS

The enactment of data protection and privacy laws does not guarantee that they will be enforced and effective. Various barriers to enforcement exist. Many countries face increasingly severe personnel and skills shortages to enforce CS laws. For instance, while India has enacted data protection and privacy laws, the country's CS personnel is reported to be "grossly inadequate." As of 2013, all government organizations in India were reported to have a combined CS personnel strength of 556 compared with China's 125,000, the US's 91,080, and Russia's 7,300.[35]

The availability of skills, however, cuts both ways. The lack of economic opportunity can force IT personnel to engage in cybercrime. As explained above, cybercrime pays better than a legitimate job in many countries, such as Russia and Eastern European countries. The situation was exacerbated by a financial crash in 1998 that left many computer programmers unemployed.

A large number of extortion-related cyberattacks originate from Eastern Europe and Russia. These hackers are capable of very sophisticated attacks with limited computer power, which can be attributed to Russia's highly educated workforce and programming skills.[36] Russian hackers have a deep understanding of networks and know how to "get in and out without a trace."[37]

One of the most appropriate indicators to measure the stock of hacker skills is the number of science, technology, engineering, and math (STEM) students graduating from universities. Russia and the US produce about the same number of STEM graduates every year. There are technical universities in every major city in Russia. As shown in table 5.2, the Russian economy is less than 6% of the size of the US economy. The size of the Russian information and communications technology (ICT) market as a proportion of the US market is even smaller. For instance, total ICT spending per STEM graduate in the US is more than 28 times that of Russia. (See figure 5.4.)

Table 5.2 A comparison of China, India, Russia, and the US in terms of indicators related to the stock of skills and available opportunities

	China	India	Russia	The US
Number of STEM graduates (2016)[38]	4,700,000	2,600,000	561,000	568,000
CS market size (2019, US$ billions)	7.35[39]	1.97	1.06[40]	44.7[41]
Total ICT spending (2019, US$ billions)	488[42]	108.3[43]	47[44]	1,324[45]
ICT spending per STEM graduate (2019, US$ thousands)	103.8	41.7	83.8	2,331

Figure 5.4 ICT spending per STEM graduate in some major countries (US$, thousands)

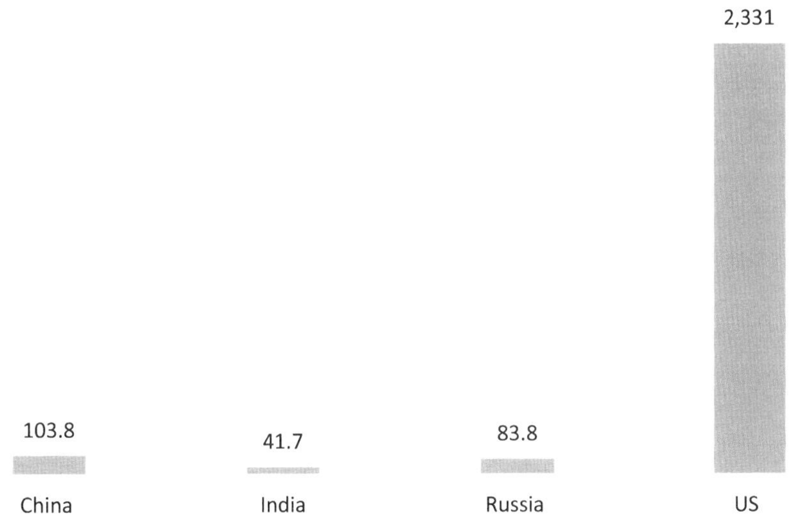

China	India	Russia	US
103.8	41.7	83.8	2,331

Data source: Calculated based on data from table 5.2.

5.4 INSTITUTIONAL AND ORGANIZATIONAL CHANGES

While institutions and organizations are characterized by inertia and resistance to change, they do nonetheless change. This means that we can observe changes in culture, mindsets, and policies related to cybercrimes and CS. (See In Focus 5.1.) Such changes may take place incrementally and gradually or discontinuously. We use cyberinsurance and changes in hackers' motivations to illustrate gradual and discontinuous changes in cybercrime and CS.

IN FOCUS 5.1: ROMANIA'S GROWING CS PROWESS

When communism ended, Romania had a lot of technically trained individuals. The market was too small to absorb them. Hacking became an attractive opportunity.[46]

Significant changes can be observed in Romania. While Romania has a "thriving" cybercriminal community, it is now also well-known for its CS prowess.

Romania's Bitdefender is a global CS leader. It has over 500 million customers in more than 150 countries[47] and 1,600 employees worldwide.[48] Its network of research labs in Romania is the largest in Europe. About 40% of antivirus and CS companies worldwide use at least one technology developed by Bitdefender.[49]

5.4.1 Gradual Changes

The types of risks that are covered by cyberinsurers vary significantly. Moreover, in many instances policyholders and insurers disagree about the coverage under cyberinsurance policies.[50]

Disagreements between insurers and policyholders have led to many lawsuits. (See In Focus 5.2.) To address this, cyberinsurers have started writing cyberinsurance contracts that more explicitly define what types of losses are covered. This trend is expected to reduce the number of lawsuits related to cybercoverage. Moreover, court decisions may lead to clearer language in cyberinsurance policies.[51]

IN FOCUS 5.2: MONDELEZ INTERNATIONAL'S LAWSUIT AGAINST CYBERINSURER ZURICH

In June 2017, the international food company Mondelez International was victimized by the NotPetya cyberattack. The attack affected its 1,700 servers and 24,000 laptops, which became permanently dysfunctional. Mondelez claimed that its loss due to property damage, disruption in commercial supply and distribution, inability to fulfill customer orders, reduced margins, and other losses exceeded US$100 million.[52] Mondelez filed a claim with its insurer - Zurich American Insurance Company - for these damages.

Zurich offered an initial payment of US$10 million but later rejected the claim altogether. It argued that the NotPetya ransomware attack was an act of "cyberwar" and thus was not covered by the policy. According to Zurich, the policy excludes "hostile or warlike action in time of peace or war" by a "government or sovereign power."[53] Mondelez sued Zurich for breach of contract.[54]

As of March 2021, the litigation was ongoing.[55]

5.4.2 Discontinuous Changes

While cyberinsurance has been available since the 1990s, it still has low penetration among organizations. An estimate indicated that while over 75% of large businesses in certain categories had cyberinsurance, less than 5% of SMEs carried cyberinsurance.[56] In the US, around 19% of SMEs had obtained cyberinsurance coverage. In OECD countries, the proportion of SMEs with cyberinsurance is reported to be in the single digits.[57] Many organizations have not realized the need for cyberinsurance. Recent high-profile data breaches such as WannaCry, NotPetya, and ransomware attacks in 2017 changed managers' and companies' mindsets about cyberinsurance. New regulations are also an important source of discontinuous change. For instance, the US exhibited higher growth in the cyberinsurance industry compared to other economies due to data protection regulations and fines for data breaches. Reports of such experiences have convinced companies to buy insurance. A similar trend has been observed in the EU economy following the enactment of the GDPR in 2018 and several big fines following this regulation.[58] As discussed in chapter 1, adoption and functioning of cyberinsurance are becoming more similar to other established insurance products.

Finally, institutional changes can also be observed in cybercriminals' and hackers' behaviors. Increasing opportunities in the CS industry have led to less ideological hacking activity. Many members of China's first hacktivist groups, such as the Green Army, China Eagle Union, and Honker Union, joined the country's rapidly growing CS industry.[59]

5.5 CYBERSECURITY AND SMES

Many SMEs do not allocate sufficient time, resources, and effort to secure their IT systems. They tend to think that CS investments involve high costs and low benefits. However, as figure 5.5 makes clear, a significant proportion of SMEs are victimized by cyberattacks.

5.5.1 Poor CS Orientation among SMEs

Despite the problems shown in figure 5.5, increasing the level of preparedness to defend themselves against cyberattacks has not been a major priority for most SMEs. A Dell Security survey conducted in 2015 found that only 3% of SMEs in the UK were "adequately prepared" for a cyberattack.[60] A survey of over 1,000 SMEs in the UK indicated that about half planned to spend less than £1,000 (about US$1,320) on CS annually.[61] One study found that 87% of small businesses in the US had no formal CS plan.[62]

Many SMEs tend to expect that investments in employees' CS training and awareness activities will have a low return on investment (ROI). Surveys and anecdotal evidence indicate that SMEs lack initiatives to provide adequate training and support to enhance their

Figure 5.5 Percentage of SMEs facing different types of attacks (2018)

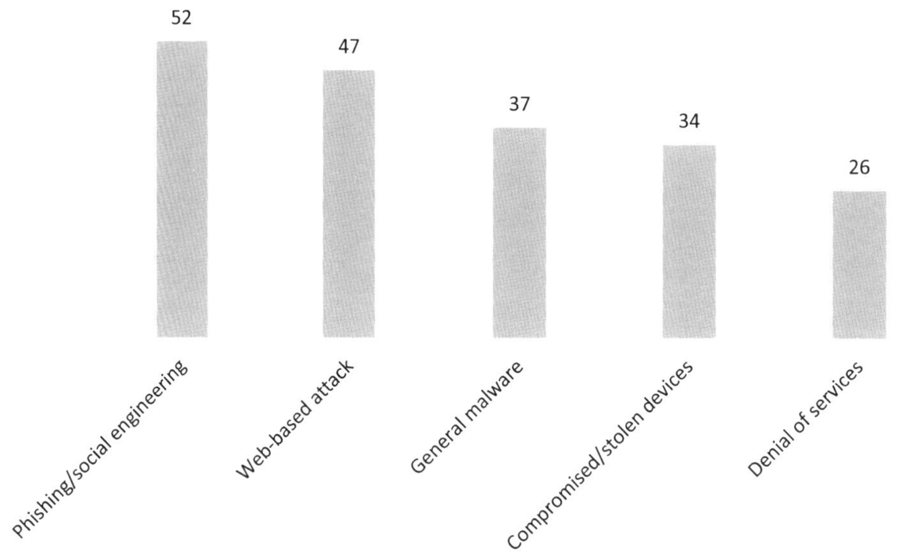

Data source: Ponemon Institute. (2018). *2018 state of cybersecurity in small & medium size businesses*. Retrieved from https://www.keepersecurity.com/assets/pdf/Keeper-2018-Ponemon-Report.pdf

employees' CS competence. According to a 2016 survey conducted among SMEs in the manufacturing sector in the UK by the manufacturers' organization Engineering Employers' Federation, 40% had not increased CS spending in the past two years, and one-fifth failed to make their employees aware of cyberrisks. Only 56% of the respondents said that CS was given serious attention by the board.

The above findings are of concern and have important implications for SMEs' cyberdefense capabilities. This is because recent studies suggest that 40% of SMEs have experienced a security breach resulting from employees' visits to malware-hosted websites.[63]

A poor CS orientation among SMEs can be attributed primarily to high perceived costs of cyberdefense measures and relative newness of cyberthreats. According to the UK government's "Cyber Streetwise" campaign, which aims to change how CS is viewed, a quarter of UK SMEs were reported to think that CS is too expensive to implement. The survey conducted among 1,000 SMEs in the UK also found that a fifth of the respondents did not know "where to start."[64]

5.5.2 A Changing Cost-benefit Structure of CS for SMEs

SMEs face a broad range of environmental pressures and threats. Consequently, the cost-benefit structure of CS is changing for SMEs. SMEs face pressure from regulators, customers, and organizations in their value delivery networks (VDNs) to strengthen CS measures.

First, firms face greater regulatory pressures from governments to strengthen CS, which affects the cost-benefit calculus of CS investments. A notable example is the Department of Financial Services (DFS) in New York, which regulates banks and insurance companies in the state. A 2014 guidance issued by the DFS specified stricter rules in corporate governance, login security, management of third-party vendors, CS insurance, and other areas. The DFS asked financial-sector firms to explain the processes and mechanisms used to track potential vulnerabilities at their third-party vendors and suggested they develop more CS expertise on their boards.[65] The DFS head also urged financial companies to invest in cyberinsurance.

Regulators are using not only sticks but carrots. To take an example, in 2015, the UK government announced a £1 million (about US$1.3 million) CS innovation voucher scheme, which offers micro, small, and medium-sized businesses up to £5,000 (about US$7,000) for specialist advice to strengthen CS.[66]

Some governments, singly as well as in collaboration with the private sector, are taking measures to strengthen SMEs' CS. For instance, in the UK, Cyber Essentials is a government-backed information assurance scheme that requires an organization to complete a self-assessment questionnaire; the responses are reviewed independently by an external certifying body. The scheme was developed as part of the UK's National Cyber Security Program in close consultation with industry. It is reported that most viruses, spyware, or malware found in the most commonly detected cyberattacks can be prevented if SMEs are certified by Cyber Essentials. Since October 2014, any UK government tenders are required to hold Cyber Essentials accreditation.[67]

Second, there is an increasing tendency among organizations to evaluate the CS practices of the members of the VDNs, such as distribution channels and supply chain partners. The goal is to make sure that companies require supply-chain partners to comply with at least the same CS standard that companies set for themselves.[68] In a survey conducted by KPMG among the UK's procurement managers at large organizations across several sectors, 94% of respondents said that suppliers' CS standards were important when awarding contracts to SMEs, 70% of the respondents were of the view that SMEs could do more to protect valuable data, and 86% noted that SME suppliers that suffered a data breach would be removed. Two-thirds of the respondents asked their suppliers to demonstrate cyberaccreditations such as the UK government's Cyber Essentials or the Payment Card Industry Data Security Standard scheme.

Third, SMEs face more demanding customers that place a high importance on the CS measures of companies they do business with. About a quarter of medium-sized businesses in the UK were reported to be asked by a current or prospective customer about their CS measures.[69]

Finally, the deployment of cyberdefense mechanisms diffuses from large to small organizations. As large companies implement stronger defense mechanisms against cyberattacks, SMEs are more likely to become cyberattack targets.

5.6 CHAPTER SUMMARY AND CONCLUSION

Political, social, cultural, and organizational factors play key roles in cybercrimes as well as the CS landscape. This chapter provided details of regulative, normative, and cognitive institutions from cybercrime and CS standpoints and explained how some disruptive events can bring institutional changes.

Countries worldwide differ in relative emphasis placed on fighting different types of cybercrimes. For instance, compared to most other parts of the world, Saudi Arabia and other Arab countries seem to be more serious about controlling cybercrimes that are considered to be against the religion.

When selecting a target, an attacking unit may consider whether the victim belongs to an in-group or an out-group. Put differently, cultural Otherness is positively related to the probability of being victimized.

Countries vary widely in terms of the development of legal and political institutions related to CS. They also demonstrate various levels of capacity to enforce CS laws.

Cyberthreats are not limited to large organizations and government agencies. SMEs are being victimized by cyberattacks at an increasingly higher rate, yet many SMEs are insufficiently prepared to defend themselves from cyberthreats. Thus, it is increasingly important for SMEs to increase their capability to deal with such threats.

5.7 DISCUSSION QUESTIONS

1. How are institutions related to cybercrimes and CS?
2. Who are cultural Others? How is this concept related to cybercrimes?
3. How do SMEs perform in terms of CS?
4. What is the current status of data protection and privacy legislation worldwide?
5. How are ICT skills of the population in a country related to cybercrimes and CS?

5.8 END-OF-CHAPTER CASE: CYBERSECURITY IN BRAZIL

Cyberattacks have become a serious problem in Brazil. It was reported that in October 2019, personal information of 92 million Brazilians was auctioned in an underground forum. The criminals were suspected to have illegally accessed the data from a stolen government database.[70] According to Symantec, in 2017, over 60% of Brazilian Internet users (i.e., 62 million people) were victimized by cybercriminals, which led to a loss of US$22 billion.[71] In 2018, such attacks affected more than 70 million people in Brazil.[72] The CS firm Trend Micro's study indicated that Brazil was the world's second largest victim of ransomware attacks in the first quarter of 2019. Brazil's ransomware attacks accounted for 10.6% of global ransomware attacks, which is close to the US's share of 11.1%.[73] A 2019 report of the International Telecommunication Union (ITU) noted that Brazil's cyberattack-related economic losses were second highest in the world.[74]

In 2018, Brazilian companies lost more than US$20 billion to cyberattacks. According to a study by the global professional services company Accenture, the average annual cost of cybercrime for a Brazilian company was US$7.24 million in 2018.[75]

CS initiatives have not been well-developed in Brazil to deal with these growing threats. While Brazil has developed guidelines and best practices to deal with cyberthreats, there have not been sufficient actions to prevent such threats.[76]

5.8.1 Security

In terms of strategic planning to deal with cyberthreats, Brazil has lagged behind other regional economies such as Argentina, Chile, and Mexico. For instance, in 2017, Chile and Mexico published their CS strategies that outlined what, how, and when to address various types and categories of CS risks. Mexico's focus has been to boost its economy and innovation, strengthen civil society and public institutions, and improve public and national security. Chile's CS strategy aims to improve its digital infrastructure and people's rights in cyberspace, develop a CS culture, and promote the growth of its CS industry. Likewise, Argentina's Programa Nacional de Infraestructuras Críticas de Información y Ciberseguridad seeks to improve the regulatory framework for identifying and protecting its digital infrastructures.[77]

However, Brazil has realized the urgency of addressing CS threats. In February 2020, the Brazilian government released its national cybersecurity strategy (NCS), which has the goal of increasing the country's digital trustworthiness and resilience against cyberthreats.[78] In January 2021, the country's data protection authority Agência Nacional de Proteção de Dados (ANPD) published a regulatory strategy for 2021–2023 and a work plan for 2021–2022. Key components of the ANPD's regulatory strategy include strengthening the country's data protection culture, improving the data protection regulatory environment, and enhancing the agency's ability to enforce the Lei Geral de Proteção de Dados (LGPD) rules.

Brazil has launched additional high-profile initiatives to address cyberthreats. In November 2019, the Brazilian government announced the creation of a network of eight R&D labs.[79] One of them will focus on the use of AI in CS and will involve the Brazilian army.[80]

5.8.2 LGPD: An Attempt to Replace Disparate Data-Protection Norms and Legal Regimes

On the privacy front, the LGPD was ratified in Brazil's Congress in mid-2018, which became effective on September 18, 2020.[81] In the pre-LGPD era, Brazil had over 40 legal norms at the federal level that directly and indirectly dealt with the protection of privacy and personal data.

These norms often functioned in a sector-based system and sometimes conflicted. The lack of clear legal rules hindered legal certainty and predictability, which also reduced the country's competitiveness.[82] The LGPD is expected to address these concerns.[83]

The LGPD is described as Brazil's first significant attempt to deal with digital privacy.[84] Whereas pre-LGPD laws were primarily directed toward regulating ISPs and required them to store records and make them available for law enforcement and government agencies, the LGPD addresses Brazilian citizens' right to data privacy and security. It replaces the existing patchwork of legislation governing CS issues such as the Civil Rights Framework for the Internet (Internet Act), the Civil Code, and the Consumer Protection Code.

The LGPD affects all industries and economic sectors. For instance, Article 18 of the LGPD gives consumers rights for their personal data. Organizations are required to ensure personal data is "anonymized, redacted, or eliminated."[85]

The LGPD is inspired by and modelled on the EU's GDPR.[86] However, Brazil lacks a national data protection authority to enforce basic provisions of the LGPD. The country's Congress passed the creation of a federal agency (ANPD) to enforce data protection rules,[87] but then President Michel Temer vetoed the legislation, arguing that such initiatives should be undertaken by the executive branch rather than the Congress. Before leaving office, his government had formed an interim agency with a two-year mandate; however, after President Jair Bolsonaro took office in January 2019, the ability of the new agency to impose penalties for the violation of the LGPD was restricted.[88]

Like the GDPR, the LGPD outlines new rules regarding how personal data can be collected, used, processed, and stored.[89] Just like the GDPR and the CCPA, the LGPD does not explicitly require organizations to encrypt data. Nonetheless, organizations need to provide reasonable CS when dealing with a consumer's personal information. Encryption is one way to satisfy the requirement.[90]

5.8.3 Wide Jurisdictional Reach

The LGPD applies to domestic as well as foreign entities that collect or process personal data in Brazil or provide goods or services to individuals in Brazil. For instance, a business that collects or processes personal data of individuals in Brazil is required to follow the LGPD even if it does not have a physical presence in Brazil. The fines for violating the LGPD can be up to 2% of the company's gross revenues derived from Brazil, or 50 million reais (about US$13 million).[91]

5.8.4 Summary

Brazil is exposed to significant cyberthreats thanks to the country's financially and politically motivated cybercriminal gangs. Detection of threats is difficult because many of the domestically originated attacks do not pursue foreign targets and hence are unknown to international security researchers. Brazil's CS initiatives have been criticized in the past for the lack of a vision, a cohesive strategy, or a common set of priorities. While different government departments developed initiatives and projects, their activities lacked coordination. They were not able to devote significant resources to implementation and enforcement. Many of these issues have been or are being addressed with recent efforts such as the enactment of the LGPD and the ANPD's regulatory strategy and work plan.

CHAPTER SIX

Cybersecurity Policies and Strategies of Major Countries

CHAPTER PREVIEW

Many governments recognize that it has become imperative to give CS high national priority. This chapter gives an overview of the general features of CS policies and strategies of major countries in the world. Specifically, it highlights the evolving regulatory frameworks in China, Japan, the EU, and the US and compares these four regions in terms of some key aspects of CS-related regulatory frameworks. Finally, the chapter evaluates the effects of the evolving CS-related regulatory frameworks on businesses and consumers.

6.1 INTRODUCTION

CS is currently a high national priority worldwide. Most countries in the world have adopted CS policies and regulation. (See figure 6.1.)

In the US, CS has been described as one of the most serious economic and national security challenges facing the country.[1] In 2018, the US White House released its national cyber strategy.[2] The strategy aims to use CS as a tool to protect people, the country, and its value system; foster the country's prosperity; establish and promote peace; and strengthen US influence.

The UK's 2010 National Security Strategy rated cyberattacks as a "Tier 1" threat and one of the four highest-priority risks facing the UK. Likewise, the document *National Security*

Figure 6.1 Number of ITU member countries with cybercrime legislation and an NCS (2018)

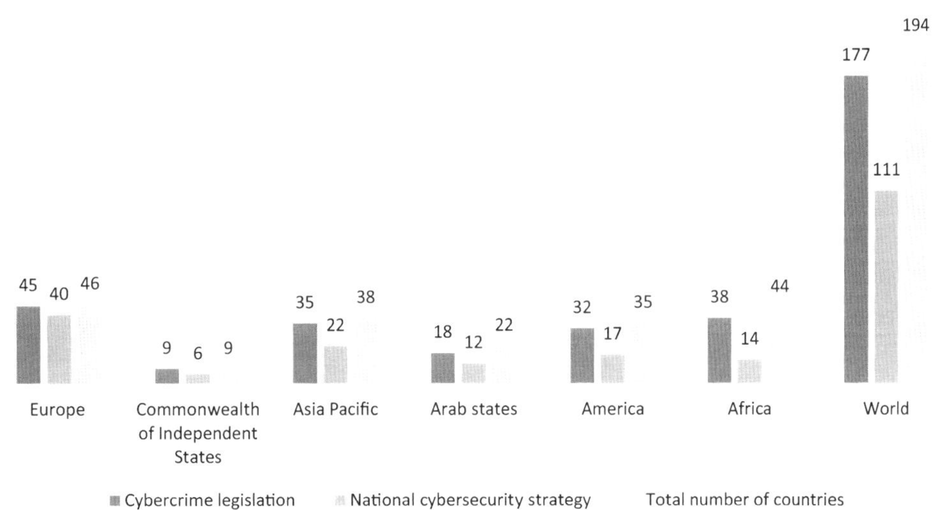

Data source: International Telecommunication Union (ITU). (2018). *Global cybersecurity index 2018*. ITU Publications. Retrieved from https://www.itu.int/dms_pub/itu-d/opb/str/D-STR-GCI.01-2018-PDF-E.pdf

Strategy and Strategic Defence and Security Review 2015 noted that cyberthreats are one of the four challenges driving the UK's security priorities and affirmed that it will treat a cyberattack as seriously as a conventional attack. The National Cyber Security Strategy 2016–2021 was launched in 2016, the third and final iteration of which is expected to be completed in 2021. A key goal of the strategy is to make the UK one of the most secure countries in the world to do business in cyberspace.

In China, CS is embraced as a key national priority demonstrated by the fact that President Xi Jinping headed the Central Internet Security and Informatization Leading Group. Premier Li Keqiang and the First Secretary of the Central Secretariat of the Chinese Communist Party (CCP) Liu Yunshan were named as the group's deputy heads.

As late as 2012, Japan had not officially acknowledged cyberattacks as a security threat;[3] however, CS issues have become a prominent part of the national discourse in recent years. Due primarily to high-profile cyberattacks facing the country as well as internal and external pressures, Japan has revised CS strategies and implemented new CS practices. As of December 2020, Japan was the only Asian country that the EC recognized as providing adequate data protection.

The formulation of CS strategy is not limited to rich countries or big economies. CS strategies and policies are diffusing rapidly across the world. However, compared to regulatory frameworks to fight cybercrimes, fewer countries have an NCS. (See figure 6.1.) Many policy-makers consider CS a peripheral issue, either too specialized or a technical rather than a strategic problem.[4] We can also find significant interregional variation in the proportion of countries that have an NCS. For instance, of International Telecommunication Union (ITU) member countries, only 32% in Africa had an NCS in 2018 compared to 87% in Europe. (See figure 6.1.)

There is also wide variation in the focus of CS strategies of major countries. For instance, significant differences can be observed in the CS strategies of the UK and the Netherlands. The UK efforts have mainly focused on technical resilience measures. Among the key elements of the UK cyberstrategy is establishment of the National Cyber Security Centre.[5] The goal is to ensure the availability of acceptable service levels even if ICT infrastructures face severe disruptions. The strategy also notes the lack of effectiveness of international efforts in the UN Government Group of Experts on CS. The Dutch National Cyber Security Agenda, on the other hand, emphasizes a value-driven strategy. It has listed one of its priorities as "contributing to international peace and stability in the digital domain."[6]

6.2 KEY COMPONENTS OF NATIONAL CYBERSECURITY STRATEGY

The International Telecommunication Union has noted that an NCS is tightly linked to a country's socioeconomic security.[7] The above examples point out the importance of CS strategies to achieve various political and economic goals.

The ITU's Global Cybersecurity Index (GCI) provides a useful index to understand the five key components of an NCS.

(i) *Legal*: legal institutions and frameworks to deal with CS and cybercrime (e.g., cybercrime laws on unauthorized access of computers, systems, and data; laws dealing with data protection, breach notification, CS certification/standardization requirements, and implementation of CS measures)
(ii) *Technical*: CS-related technical institutions and frameworks (e.g., CS standards implementation framework, use of cloud in CS, child online protection)
(iii) *Organizational*: institutions to coordinate CS-related policy and strategies at the national level (e.g., an agency responsible for implementing the NCS)
(iv) *Capacity building*: R&D, education/training programs (e.g., public awareness campaigns), certified professionals, and existence of public-sector agencies facilitating these activities
(v) *Cooperation*: engagement in CS-related partnerships, cooperative frameworks, and knowledge/information-sharing networks (e.g., bilateral and multilateral agreements, membership in international forums, etc.)[8]

The GCI measures a nation's CS readiness. For 2018, the GCI varied from the lowest of Maldives' 0.004 to the highest of the UK's 0.931.

Countries such as Maldives need to deploy resources to improve on any of the dimensions in order to improve their readiness according to the index. CS is not one of the priority areas where developing countries allocate their scarce resources and time. For instance, they lack resources to invest in R&D and education/training programs to build capacity (component [iv] in the GCI) in CS. Likewise, to improve in cooperation (component [v] in the GCI) they need to engage in multilateral agreements and gain membership in international forums related to CS. Many developing countries lack the capacity to commit to international norms and confidence-building measures (CBMs).[9]

6.3 CYBERSECURITY POLICIES AND STRATEGIES OF MAJOR COUNTRIES

The above examples point out the importance of CS-related regulations for achieving various political and economic goals. This section provides an overview of the general features of CS policies and strategies of the US, the EU, China, and Japan.

6.3.1 The US

CS concerns have elicited extensive regulatory responses in recent years. Federal regulatory agencies such as the Federal Deposit Insurance Corporation, the FDA, the FTC, the DoJ, the Securities and Exchange Commission (SEC), and state agencies such as New York's Department of Financial Services (DFS) are weighing in on the CS debate.[10]

6.3.1.1 The 2013 CS Executive Order

Among major policy documents, Executive Order (EO) 13636 "Improving Critical Infrastructure Cybersecurity" was issued in 2013 under the Obama administration. The EO was aimed at increasing CS for critical infrastructure vital to the economy, security, and daily life such as financial, transportation, communication systems, and power plants.

The National Institute of Standards and Technology (NIST) and the Pentagon are among the key agencies that implement the strategy. The role of the NIST is to develop a cybersecurity framework, which involves finalizing voluntary standards and procedures to help companies address CS risks. (See In Focus 6.1.) Likewise, the Pentagon was asked to recommend whether CS standards should be considered in contracting decisions.

IN FOCUS 6.1: NIST CYBERSECURITY FRAMEWORK

EO 13636 "Improving Critical Infrastructure Cybersecurity" called for the development of a voluntary cybersecurity framework (CSF) specifically targeted at critical infrastructure services. It aimed to provide a "prioritized, flexible and performance-based" approach that can serve as a basis for managing and minimizing cyberrisk for organizations in the critical infrastructure sector. However, the approach has also become popular among companies in other sectors.

The EO gave the NIST the responsibility to develop a framework.[11] The NIST worked with academia and the CS industry to do so, releasing the final version of the framework in 2014.

The NIST CSF's major goal is to improve security of organizations in the critical infrastructure sector. The CSF is made up of standards, guidelines, and practices that are intended to prevent, detect, and respond to cyberattacks.

Small or less regulated entities are most likely to benefit from the NIST's CSF. Larger organizations may already have a CS strategy in place. The framework provides guidance and does not focus on compliance. The idea is to encourage organizations to prioritize CS risks, just as organizations consider other risks, such as financial, personnel, safety, and operational. Another goal of the framework is to make CS a part of organizations' day-to-day discussions and decision-making.

Parts of the NIST CSF

The framework has three main parts: the framework core, framework implementation, and framework profile.

The Framework Core

The framework core provides a strategic view of the CS risks organizations face and consists of the set of CS activities and outcomes that are desired. It includes five functions: identify, protect, detect, respond, and recover.[12]

Identify: Organizations should develop an understanding of the ways they can manage CS risks to their systems, assets, and data.

Protect: Organizations should put in place safeguards to ensure that critical infrastructure services are delivered.

Detect: Organizations should have ways to identify a CS event.

Respond: Organizations should define actions to be taken when a CS event is detected.

Recover: Organizations should identify services that focus on resilience. They should also outline how to restore services that have been affected by a CS event.

The Framework Implementation

The framework implementation segments customers into four different tiers on the basis of CS risks and processes that have been put in place to manage the risks. Organizations that belong to Tier 1 have ineffective risk management methods. Their risk management processes are unsystematic and risk management programs lack reliability. Moreover, risk management participation is unresponsive, which means that the organization lacks an understanding of its own ecosystem and how it fits into the external ecosystem. The organization also does not collaborate with external entities.

Tier 2 organizations are characterized by informal risk management methods. Their risk management processes are unfinished and risk management programs are underdeveloped. They also exhibit incomplete risk management participation. For instance, while the organization may understand its role in the larger ecosystem, it does not understand how its dependents fit into the larger ecosystem and how other external entities depend on it.

In Tier 3 organizations, risk management methods are structured. They display orderly risk management processes as well as robust risk management programs. The risk management processes are approved by management and are expressed as a formally approved policy. CS risk management practices are regularly updated based on internal and/or external changes. These organizations also have routine risk management participation, which means that the organization understands its role in the larger ecosystem, collaborates and discusses CS issues with external entities, develops familiarity with CS risks in supply chains, and takes action to address them.

The highest level of preparedness is in Tier 4 organizations, which have adaptive risk management methods. Their risk management processes are dynamic: they can rapidly adapt to a changing CS environment, continuously improve CS risk management methods, and effectively and quickly respond to sophisticated threats. Their risk management programs are responsive: CS and other risks are monitored and treated by senior executives with the same priority. CS risk management is a key part of the organization's culture. They have interactive risk management participation.[13] Such participation is more valuable and effective in fending off cyberattacks than the routine risk management participation of Tier 3 organizations. For instance, Tier 4 organizations exhibit higher levels of width, breadth, depth, and scope in sharing information and using such information compared to Tier 3 organizations.

Tier 4 organizations share information that is prioritized based on profiles of information needs. Such information is shared with internal personnel as well as external collaborators. Such organizations use (near) real-time information to address cyberrisks. Timely information is used to understand and act upon cyberrisks in diverse settings such as products and services, a supply chain, and suppliers' products and services.

The Framework Profile

The framework profile describes the gap between the current ("as is") state and the desired ("to be") state of an organization's CS program. An understanding of the profile would help organizations identify opportunities for improving their CS position. Organizations can establish roadmaps, strategies, plans, and metrics to reach the desired state in a particular implementation scenario. Profiles are important in conducting self-assessments and communicating within an organization as well as among organizations.[14] The information related to the current profile can be a useful tool for the prioritization of resources and measurement of progress toward the target profile in light of other business needs such as cost-effectiveness and degree of innovation.

6.3.1.2 The 2017 CS Executive Order

In May 2017, an EO titled "Strengthening the Cybersecurity of Federal Networks and Critical Infrastructure" was issued. Section 1 of the EO deals with CS of federal networks. It requires federal agencies to be guided by the NIST Framework for Improving Critical Infrastructure

Cybersecurity. All federal agencies were required to develop plans to implement the NIST Framework within 90 days of the release of the EO.

Another key section of the EO provides federal support for the owners and operators of critical infrastructure. It builds on the 2013 EO to identify the entities that are supported, which are known as section 9 entities because they are defined in section 9 of the 2013 EO. These entities face the greatest cyberattack risks, which could "reasonably result in catastrophic regional or national effects on public health or safety, economic security, or national security."

Another substantive section of the EO focuses on consumer CS by fostering the development of a secure Internet and helping the growth of a CS-trained workforce. In order to achieve this goal, federal agencies were mandated to identify international CS priorities. The agencies were also required to assess the current state of the development of a CS-capable workforce and compare it to the current "workforce development efforts of potential foreign cyber peers."[15]

6.3.1.3 CS Regulations to Address Threats Facing Critical Sectors and Important Industries

6.3.1.3.1 Power and Electricity In 2014, US federal regulators adopted new CS standards for the electricity industry. The DoE released guidance to help the industry develop risk management plans shaped by the NIST's framework of CS standards. The ES-C2M2 provides guidance for utility companies to assess CS. The model was developed in consultation with the White House, the DoE, the DHS, and power companies.[16]

6.3.1.3.2 Healthcare Some regulatory agencies and public gatekeepers use their regulatory powers to ensure that CS becomes a priority instead of an afterthought. For instance, in 2014, the FDA, which oversees and regulates drug companies directly through its powers to control or approve products' market release, published recommendations for protecting medical devices from cyberattackers. The guidance suggests medical device makers should take CS risks into account early in the design and development process. They are required to show documentation to the FDA about the risks identified and the steps taken to protect devices from such risks. In addition, the FDA asks device manufacturers to submit plans to provide appropriate patches and updates to operating systems and software for newly emerging risks.[17] The FDA has also collaborated with the DHS to create public awareness of CS for medical devices.[18]

6.3.1.3.3 Banks and Financial Institutions Banks and financial institutions face unprecedented regulatory pressures to enhance CS measures. Raising concerns about high-profile cyberattacks facing the US financial system, Senate Banking Committee leaders pressed financial regulators and the Treasury Department to take regulative measures to strengthen CS.[19] Senate

Banking Committee leaders asked concerned regulators about the details of gathering cyberattack-related information, coordination of efforts across agencies, and involvement of the Financial Stability Oversight Council in monitoring risks. The Senate Banking Committee leaders asked about the details of the Federal Financial Institutions Examination Council's plan to help SMEs address CS gaps.[20]

The SEC, which was established to protect investors, is another federal agency active in the CS arena. The SEC has been sending strong and clear signals regarding the expectation of firms to put in place CS policies and procedures.[21] In 2011, the SEC issued guidance that recommended that companies should inform investors of "material" cyberrisks and attacks.

Regulatory measures are also being introduced to strengthen the CS of financial intermediaries and market participants such as brokerage firms. In 2019, the Financial Industry Regulatory Authority, which is a watchdog agency funded by the industry, published a cybersecurity checklist. The checklist's goal is to help establish data protection policies, which include identifying threats, implementing protections, and developing recovery plans.[22]

State-level regulatory and administrative agencies are also becoming active. A notable example is New York's DFS, which regulates banks and insurance companies. A new guidance issued by the DFS in 2014 specifies stricter rules in corporate governance, login security, management of third-party vendors, CS insurance, and other areas.

6.3.1.4 The 2018 National Cyber Strategy

In 2018, the US White House released the National Cyber Strategy. The new strategy defines four main priority areas known as pillars. The strategy also provides a "roadmap" of key actions that need to be taken to support these pillars. "Protect the American People, the Homeland, and the American Way of Life" is the first pillar of the strategy. It identifies key tasks for achieving this goal: securing critical infrastructure, federal networks, and information; fighting cybercrime; and improving incident reporting systems. The second pillar is "Promote American Prosperity." The important mechanisms of this pillar include stimulating the growth of a vibrant and resilient digital economy and developing a high-quality workforce. The plan is to work with technology companies to promote CS testing in new products and improve recruitment and retention of high-quality CS professionals.

"Preserve Peace through Strength" constitutes the third pillar. Its key focus areas are the achievement of cyberstability through norms of responsible state behavior and attribution and deterrence of unacceptable cyberspace behaviors. The strategy makes it clear that in order to deter cyberattacks, the US will use "all instruments of national power" and impose "swift and transparent consequences" against malicious and hostile actors. In addition, the strategy emphasizes a "Cyber Deterrence Initiative," working with foreign allies to support each other's responses to significant cyberattacks. An increased focus on deterrence is among the key aspects of the strategy, which showed a notable shift toward a more offensive CS posture.[23]

Then National Security Advisor Bolton noted that the US had authorized offensive cyberoperations mainly "to create the structures of deterrence" in order to discourage adversaries from launching cyberattacks against the US.[24]

"Advance American Influence" is the fourth pillar of the strategy. Activities that support this pillar include the promotion of an open, interoperable, reliable, and secure Internet. It also emphasizes international cybercapacity, which entails building cybercapabilities of US allies in order to address "threats that target mutual interests."[25]

6.3.2 The EU

In 2013, the EC released its CS strategy document *An Open, Safe and Secure Cyberspace*.[26] This document along with its directive on Network and Information Security[27] represent the EU's comprehensive vision on how best to prevent and respond to cyberdisruptions and attacks. The European Network and Information Security Agency (ENISA), which is a center of network and information security expertise for the EU, is the key agency for implementing the strategy.

Some EU member states expressed concerns about compliance costs associated with the EU CS strategy. Newer EU members, due to lack of national administrative, economic, and technical capacity, are especially likely to face higher burdens to comply with the EU regulations. For instance, in a 2012 questionnaire addressed to the national Parliaments of the EU, Romania's Committee for Information Technologies and Communications expressed concerns about financial burdens on private data controllers. The committee also argued that some of the proposed obligations should be analyzed further to look for the possibility of reducing additional burdens.[28]

The requirement for compliance with strict regulations and the lack of economies of scale and efficiency due to market fragmentation act as barriers for the development and diffusion of cloud and other technologies in EU countries. The GDPR is expected to address this challenge.

6.3.2.1 The EU Cybersecurity Act

As mentioned in chapter 2, the EU penalizes companies with a substantial fine for violating the GDPR. There has been, however, a lack of objective standards to determine the CS level of ICT products, services, or processes.[29] The EU Cybersecurity Act, which became effective on June 27, 2019, aims to address this challenge. While the GDPR's focus has been on data protection by design and involves both privacy and security components, the Cybersecurity Act focuses mainly on security.

The certification scheme specifies security assurance levels such as basic, substantial, or high. ICT manufacturers or service providers may carry out the assessment required for the basic level. For substantial or high levels, the national CS certification authorities perform the assessment.[30] CS certification is voluntary. The CS certification framework also supports the EU's CS and the Commission's Digital Agenda. A benefit is that it harmonizes the EU's digital ecosystem, which will result in better exploitation of ICT.

Among the leading provisions of the EU Cybersecurity Act is that it has granted a permanent mandate to the ENISA.[31] Before this Act was passed, the ENISA's mandate needed to be renewed every seven years.

The ENISA plays a key role in establishing and maintaining the European CS certification framework. It has a dedicated website to inform the public about schemes and certificates that have been issued.[32] The ENISA develops CS-related guidelines and best practices.[33]

The ENISA also increases CS-related operational cooperation at the EU level. It helps individual EU member states to handle CS incidents. In cases of large-scale cross-border cyberattacks, it facilitates coordination at the EU level.[34]

In January 2020, the commission developed a set of recommendations in order to help EU member states address risks associated with fifth generation (5G) technology, referred to as a "toolbox" of security standards for 5G.[35] The EU member states will put measures in place to respond to the risks that have been identified as well as to possible risks that will arise in the future. They also agreed to apply a risk-based approach to ensure that they will assess the risk profile of 5G telecommunications providers. They may also impose specific requirements and conditions for such 5G equipment.[36]

In February 2021, the European Commission announced a plan to launch a cybersecurity certification scheme for 5G networks across the EU in order to mitigate technical vulnerabilities in such networks. The ENISA has been designated as the agency responsible for coming up with the standards that are required to qualify for the certification.

6.3.3 China

One complaint about China's CS regulation is its piecemeal approach that does not adequately provide systematic and comprehensive personal data protection.[37] CS issues in China have been governed by many regulations, laws, and guidelines.

6.3.3.1 The 2017 Cybersecurity Law

The Cybersecurity Law of the People's Republic of China, which was passed in November 2016 and became effective June 2017, provides guidelines for maintaining network security and protecting individuals' and organizations' rights and interests.[38]

In November 2018, China added new provisions named "Regulations on Internet Security Supervision and Inspection by Public Security Organs" to the 2017 Cybersecurity Law. The provisions gave the Ministry of Public Security additional powers, such as conducting inspections (in-person or remote) of companies' network security defenses, looking for "prohibited content," copying user information during the inspection, performing penetration tests for vulnerabilities, conducting remote inspections without informing companies, and sharing collected data with other state agencies.[39]

Clarifications of standards related to the 2017 regulations were introduced in May 2019.

Among key aspects of the new law are changes in the Multi-Level Protection Scheme (MLPS) law that governed IT security standards for critical information infrastructure operators for many years. The new version of MLPS, known as MLPS 2.0, which introduced tougher standards, became effective on December 1, 2019. It covers all corporations operating in China.

Before the new law was enacted, the Chinese government needed to approach foreign companies in order to obtain access to their data. MLPS 2.0 allows the government to access the data of all companies operating in China without permission.[40] Under MLPS 2.0, the networks that fall under critical information infrastructure and hence are subject to scrutiny have been dramatically expanded to include cloud computing, mobile networks, Internet of Things, and other modern technologies.[41] MLPS 2.0 divides networks into five levels from Level 1 (least sensitive) to Level 5 (most sensitive). Networks in levels 2 to 5 are required to open their networks for onsite evaluation of security control measures performed by a government-designated "security auditing" firm.[42]

International companies have been concerned about the invasive nature of this approach. This concern has been amplified by China's alleged use of its hackers to steal intellectual property from foreign companies. Some analysts argue that China uses all possible means to help its domestic businesses.[43] The opposing argument is that China can employ many other ways to steal intellectual property if it wants to do so. The government thus does not need to use formal implementation of the law as a disguise for policies that aim to gain access to foreign intellectual property. The Cyberspace Administration of China (CAC) made it clear that it will not use the law to steal private information.[44]

The law requires a "network operator" to have privacy policies, designate responsible personnel, and take technical measures to ensure security. Article 22 requires network operators to notify users and relevant authorities immediately when they discover data breaches, destruction, or loss. According to Article 28, network operators are obligated to provide technical support and assistance to the police and national security agencies for criminal investigations.[45]

Article 31 requires network operators in critical industries, such as those related to telecommunications, energy, transportation, information services, and finance, to have higher CS standards. Article 35 states that if a proposed purchase of IT products and services may affect national security, network operators in such industries must pass security inspections by government agencies for cyberspace and the State Council.

Under Article 37, network operators in the critical industries need to comply with data localization requirements. "Critical and personal information" that is collected and produced during their operations in China needs to be stored within Chinese territory.

6.3.3.2 Regulations on the Management of Blockchain Information Services

More general regulations have been laid down to control blockchains. The Regulations on the Management of Blockchain Information Services became effective on February 15, 2019.[46] The regulations require users to provide real names as well as national ID card numbers, mobile phones, or company registration to use blockchain services. User anonymity is thus not allowed. Blockchain services are required to remove "illegal information" quickly in order to stop it from spreading among users. Providers of blockchain services are also required to retain backups of user data for six months. Moreover, law enforcement must be able to get access to data whenever it is necessary. Blockchain service providers are required to keep relevant records of transactions and report to authorities in cases of illegal use. They are also obliged to prevent the production, duplication, publication, and dissemination of content that is banned by Chinese laws. In April 2019, the CAC released the first list of 197 companies that were approved to conduct business using blockchain.[47]

6.3.3.3 Tackling External Threats

China has pursued an integrated political, military, and economic strategy to deal with external threats. (See In Focus 6.2.) It is believed to have an extensive network of hacking groups that consist of a mix of independent criminals, patriotic hackers who focus their attacks on political targets, intelligence-oriented hackers inside the People's Liberation Army (PLA), as well as other groups that are believed to work with the government.[48] The PLA's doctrine of cyberwarfare is to "knock out the enemy's information infrastructure" and its doctrine of CS is to "go on the offensive to defend itself against attacks."[49]

IN FOCUS 6.2: CHINA'S MEASURES TO INCREASE GLOBAL INFLUENCE

China seeks to enhance its global influence in CS-related norms. It has engaged in UN efforts to adopt resolutions regarding CS. It has participated in the processes of the UN Group of Governmental Experts (GGE) on cyberspace issues. China is one of the UN GGE members for 2019–2021. It has promoted cooperation and coordination with member states in the Shanghai Cooperation Organisation (SCO), BRICS, and the ASEAN Regional Forum (ARF).

China has established bilateral partnerships on CS with a number of countries. For instance, following a consensus with the US in 2015,

China signed CS agreements with several countries, including Australia, Brazil, Canada, Russia, and the UK.[50]

China has also played an active role in multilateral groups of rising power countries in which it is a member. At the Russia-India-China (RIC) trilateral meeting that took place at the G20 summit in Osaka, President Xi emphasized that the three countries should "expand cooperation in 5G network, high technology, connectivity, energy and other areas."[51]

6.3.3.4 Defensive and Offensive Motives

States can have both defensive and offensive motives in cyberspace. Defensive motives may include a desire to deny a rival state the opportunity to increase its power, prevent the escalation of a possible conflict, increase stability, or maintain the state's international influence.[52] Offensive motives are concerned with a desire to extend the state's political, ideological, and economic interests.[53] China views the US approach as "capitalistic hypocrisy" and has complained about the system of rules designed by the US to define "legal" and illegal spying activities.[54] China's approach shifted toward more offensive strategies following Edward Snowden's revelation of the US PRISM surveillance program.

The Chinese government's response to internal security threats can also be interpreted in terms of these motives and attitudes. China's approach to CS has defensive motives, such as influencing consumers to use the Internet in a way that minimizes the threat to the CCP. An example is the Green Dam Youth Escort firewall software program, which was launched in 2008. The Chinese government had announced a plan to make it mandatory to have the Green Dam installed on all new PCs in the country. The stated goal of the mandate was to protect children from violent and pornographic content. However, it was mainly used to block politically sensitive content.

Chinese regulators have also responded to internal threats with retaliation. For instance, it has been reported that the Chinese government sent a virus to attack banned websites.[55]

China has one of the world's most sophisticated and advanced censorship and surveillance systems. There are reports that Internet content providers (ICPs) are fined or even shut down if they fail to comply with censorship guidelines. Some ICPs are reported to employ thousands of censors.[56] Among the most impressive of all measures is China's plan to assign a credit code, in addition to a government-issued identity card, to every adult. Factors such as financial standing, criminal record, and social media behavior will be used to assign every citizen a numerical rating.[57] It is argued that information included in the rating may also include what books people read. It is referred to as "Amazon's consumer tracking with an Orwellian political twist."[58] Individuals with bad credit are likely to be subject to sanctions, such as financial penalties that might include no housing, no credit to start a company, and ineligibility for certain jobs. High-status positions, such as government official or chief executive officer (CEO) of a company, may require a certain minimum score.[59]

6.3.3.5 Cybercontrol as a Key Element

China's relatively stable internal security does not mean that there is no internal security threat. Chinese policy-makers have given a high priority to cybercontrol measures – that is, administrative, legislative, and technical measures as well as procedures and resources to monitor, control, and regulate users' access to and activities in cyberspace. It is argued that only North Korea has stricter Internet censorship than China.[60]

The stringent ISP recordkeeping requirements and the requirement to provide technical support to government authorities and prosecutorial authorities' power to access private information under various regulations reflect a strong emphasis on cybercontrol measures. The National People's Congress succinctly stated the rationale of the 2012 Decision of the Standing Committee of the National People's Congress to Strengthen the Protection of Internet Data (2012 Decision): "to protect network information security, protect the lawful interests of citizens, legal persons and other organizations, [and] safeguard national security and social order."[61]

6.3.4 Japan

The Japanese government lacks the public's trust in protecting the country's networks from cyberattacks. According to a 2018 Pew Research Center survey of Japanese public opinion, 52% of the respondents thought that Japan was ill-prepared to handle a major cyberattack.[62] The public's concern about the government's lack of preparedness to fight cyberattacks was further amplified by the fact that in 2018 the country's then CS minister admitted that he had never used a computer in his professional life. It was reported that he was confused and did not understand a question that was asked to him about whether USB drives had been used at Japanese nuclear facilities.[63] An opposition lawmaker commented: "It's unbelievable that someone who has not touched computers is responsible for cybersecurity policies."[64]

Attempts have been made to develop an extensive network of CS-related rules, regulatory bodies, and enforcement agencies. In 2014, Japan enacted the Basic Act on Cybersecurity. In 2015, the Cybersecurity Strategic Headquarters and the National Center of Incident Readiness and Strategy for Cybersecurity (NISC) became parts of the Cabinet and the Cabinet Secretariat, respectively. These organizations are responsible for basic CS policy and for medium- and long-term CS measures.[65]

Japan's national CS strategies released in 2017 were reported to have two key changes. First, the new strategy documents stress the importance of business executives' proactive involvement in CS and management of cyberrisks by treating CS as a key part of their corporate strategy. This new emphasis is especially urgent for businesses in the critical infrastructure sectors. Second, the Japanese government now emphasizes the importance of cross-sector collaboration and information sharing.[66]

Three major categories of responses and actors in the government sector deserve mention in Japan's efforts to fight against cyberthreats and enhance economic competitiveness: (i) cybercrimes that fall under the purview of the national police; (ii) industrial policies, such as the standardization and codes and other protection measures, that are mainly handled by the Ministry of Economy, Trade and Industry (METI) and the Ministry of Internal Affairs and Communications (MIAC); and (iii) national security in which the Ministry of Defense (MoD) is the key player.[67] In addition, an emerging network of academic institutions and the private sector interested in CS has begun to develop.

6.3.4.1 Anti-cybercrime Initiatives

In 2004, the National Police Agency (NPA) installed the Cybercrime Division as well as the High-Tech Crime Technology Division in each Prefectural Info-Communications Department. In March 2013, the NPA announced the launch of a nationwide police task force. In addition, the police have played a key role in public education about CS.

6.3.4.2 Industrial Policies and Other Protection Measures

The METI established the Personal Data Working Group under the IT Integration Forum in 2012. In its report released in May 2013, the working group recommended utilizing intermediary organizations that provide consumers with information regarding businesses' credibility and help businesses to handle personal information in an appropriate manner. The working group also pointed out a deficiency of the current framework, which requires consumers to disclose personal information required by the businesses according to the latter's terms in order to use the services. The working group recommended a framework that involves businesses providing different levels of services based on the type of information consumers want to disclose. That is, consumers can select the "disclosable information." Likewise, the Technical Review Working Group of the MIAC's Research Society for Use and Circulation of Personal Data had several meetings in order to gather perspectives on the issue of handling user data. The Research Society released its official report in June 2013, which favored transparency, user participation, and proper means of data collection and management of user information, among other things.[68]

6.3.4.3 National Security

The government has realized that ensuring the stability of cyberspace is critical for the Japan Self-Defense Forces to achieve their missions. In 2014, the Cyber Defense Unit (CDU) was established, whose task is to monitor and respond to attacks on the MoD's networks. Some of the key goals that are expected to be achieved through the CDU include protecting its own information systems and contributing to the government's response to cyberthreats facing the country by advancing relevant knowledge and skills.

The MoD's plan was to use 25.6 billion yen (about US$236 million) for developing cyber-defense capabilities in 2020. The MoD would increase the personnel in the Cyber Defense Group from 220 to 290. It also wanted to establish a Cyber Protection Unit in the Ground Self-Defense Force. Other plans included utilizing cutting-edge technologies, such as a Cyber Information Gathering System; designing an "AI-enabled system to respond against cyberattack"; and researching the security of network devices in 5G networks.[69]

Particularly relevant in view of current Japan–US relations on the cyberfront is Article V of the Japan–US Security Treaty, which was signed on January 9, 1960. The article recognizes the danger of an armed attack against Japan or the US in the territories of Japan and requires the two countries to act to "meet the common danger." In April 2019, the Japan–US Security Consultative Committee (SCC) agreed that international law applies in cyberspace and the 1960 mutual security treaty will also apply in the case of serious cyberattacks against either country. The SCC noted that under certain circumstances a cyberattack could constitute an armed attack for the purposes of Article V of the Japan–US Security Treaty.[70]

6.4 COMPARISON OF PRIVACY AND CYBERSECURITY REGULATIONS OF MAJOR COUNTRIES

Table 6.1 compares privacy and CS regulations in China, the EU, Japan, and the US. China has stricter data privacy regulations for sensitive personal information compared to other economic sectors. In the banking sector, the People's Bank of China issued a Notice to Urge Banking Financial Institutions to Protect Personal Financial Information (Notice) in 2011. The notice requires banks and other financial institutions operating in the country to follow these rules when they collect, process, and store personal financial information (PFI). The notice prohibits banks from storing, processing, or analyzing any PFI collected in China outside the country. It also forbids providing PFI collected in China to an offshore entity.[71] Branches of foreign banks, however, can send such data outside China with written consent from customers.[72]

Likewise, in 2014, the National Health and Family Planning Commission issued the Management Measures for Population Health Information. According to this regulation, an

Table 6.1 A comparison of China, Japan, the EU, and the US

Dimension	China	Japan	The EU	The US
Salient feature	▪ Encourages economic use of ICTs and maintains strict cybercontrol measures	▪ Does not have a privacy commission	▪ Strictly enforces privacy rights through the GDPR	▪ Prefers to rely mostly on voluntary self-regulation but has sector-specific regulations for sensitive data
Key driving factors	▪ Balances economic modernization and maintenance of unity and stability through political control	▪ Has felt compelled to respond to high-profile cyberattacks facing the country	▪ History of misuse of information: citizens fearful of the prospect of information abuse	▪ Encourages marketing and innovations
Effects on IT providers	▪ Many provisions are unclear ▪ Selective enforcement	▪ Some private-sector self-regulation	▪ Stricter than US regulations: more wide-ranging impact on all businesses	▪ Fear among foreign consumers that US CSPs disclose data to the US government without the owner's knowledge
Effects on users	▪ Unavailability of some services ▪ Some foreign firms have located their servers in neighboring countries ▪ Impact on service quality	▪ Insufficient privacy protection ▪ Businesses have no general obligation to delete personal data after use	▪ Users enjoy a high level of privacy	▪ Some concerns related to the government's monitoring and companies' misuse of PII

entity "in charge of the collection, utilization, management, security and privacy protection of population health information" cannot "store population health information in overseas servers, [or] host or rent overseas servers." In addition, such an entity is required to "establish a tracing management system under which any user who creates, modifies and accesses population health information shall be subject to stringent real-name identity authentication and authorization control."[73] Likewise, it is a crime for employees of government institutions and organizations in the financial, telecommunications, transportation, education, and medical sectors to unlawfully provide personal information to third parties.[74]

Another important difference between China and the US is that China lacks a self-regulatory component in data privacy laws, which is likely due to its strong state and weak civil society. China's regulations on data transfer to foreign countries also differ from that of the US, which has no general prohibition against transferring data outside its borders. In the US, sensitive data, such as healthcare and financial information, are regulated and companies dealing with such data are expected to protect personal information irrespective of location.

There are some similarities between Japan's CS landscape and that of the EU. Japan has followed some elements of the EU directive mode. For instance, just like the EU, it has established a detailed regulatory framework and requires its ministries to implement it for all types of personal data. Another similarity with the EU is that due to concerns about privacy and data protection, most Japanese companies' use of big data and cloud computing has been limited.[75]

6.5 CHAPTER SUMMARY AND CONCLUSION

The rapid escalation of cyberattacks has been a major force influencing the development of CS regulatory frameworks across the world. These frameworks are geared toward responding to different combinations of external and internal threats.

Countries have adopted diverse regulatory approaches to deal with growing cyberthreats. Two major approaches have been used to characterize data privacy regulations: the EU model and the US model. The EU set a baseline common level of privacy to protect its citizens' rights, irrespective of data location. The US, on the other hand, relies more on voluntary self-regulation in an attempt to encourage firms' marketing and innovation. However, it has sector-specific strict regulations for sensitive data.

Although policy-makers claim that they are acting in the best interests of their countries, one can ask whose interests are being served by these regulatory frameworks. A look at new regulations in countries such as China indicates that a main goal is to advance the interests of the ruling regime.

Regulatory frameworks in most countries affect different economic sectors and industries differently. The costs of new regulations are especially high for firms in economic sectors that are considered sensitive and related to national security.

6.6 DISCUSSION QUESTIONS

1. What is the current situation regarding the adoption of CS policies and regulations worldwide?
2. What are the key components of an NCS?
3. How are the framework core, the framework implementation, and the framework profile related in the NIST Cybersecurity Framework?
4. How do China, the EU, Japan, and the US differ in terms of the key drivers of CS regulations?

6.7 END-OF-CHAPTER CASE: ISRAEL'S NATIONAL CYBERSECURITY STRATEGY

Israel is a small country in the Middle East with an area the size of the US state of New Jersey. Its population in 2019 was 8.9 million or 0.12% of the world population. The country is rapidly developing as a major global CS hub. For instance, in 2018, worldwide venture capital (VC) funding in CS was US$5.3 billion,[76] of which Israeli CS companies received US$1.03 billion.[77] (See figure 6.2.) Israel thus accounted for 19.4% of global CS VC funding in 2018.

Figure 6.2 VC deals and investments in Israeli CS startups

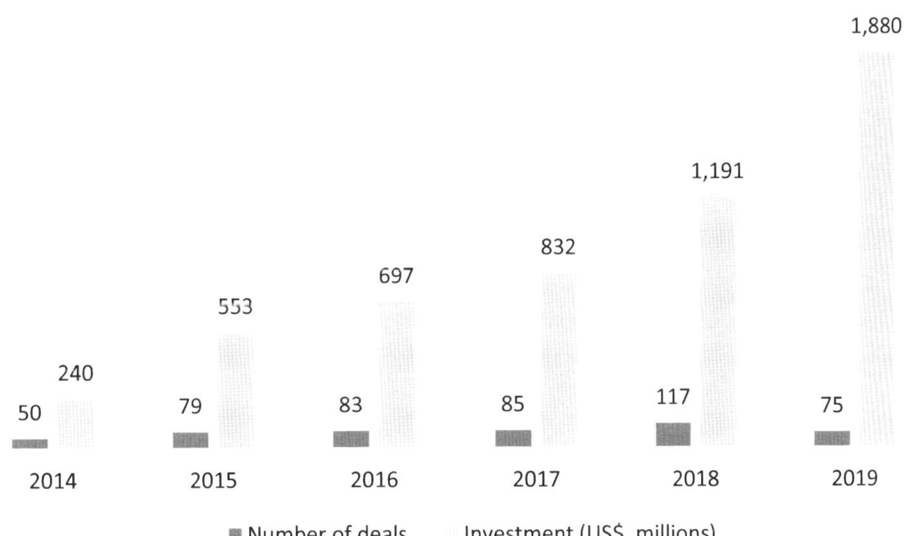

Data sources: Data for 2014 to 2018 as reported in Press, G. (2019, February 26). Israeli startups shine in the $92 billion cybersecurity market. *Forbes*. Retrieved from https://www.forbes.com/sites/gilpress/2019/02/26/israeli-startups-shine-in-the-92-billion-cybersecurity-market/#5c6b289b451d; data for 2019 as reported in *Cybersecurity startup exits in Israel total to US$11.3 billion: IVC report*. (2020, January 21). CISOmag. Retrieved from https://www.cisomag.com/cybersecurity-startup-exits-in-israel-total-to-us11-3-billion-ivc-report/

By 2019, Israeli CS startups attracted funding of US$6.32 billion[78] from investors across 594 deals.[79] (See figure 6.2.)

6.7.1 Comprehensive and Proactive National CS Policy

Israel has developed a comprehensive and proactive national CS policy. As an illustration, Israel was the first country to adopt the NIST's 2014 Cybersecurity Framework. The framework was translated into Hebrew, shared with the Israeli private sector,[80] and tailored to fit the needs of the Israeli business environment.[81]

6.7.2 Key Agencies Responsible for CS

In an effort to foster cooperation and coordination in CS-related issues among the private sector, academia, CS experts, and government agencies and to promote long-term research and analysis of cyberthreats, in 2011, Israel created the National Cyber Bureau (NCB). The NCB's tasks included coordinating national CS efforts and policy. It reported directly to the prime minister.[82]

In 2015, Israel's Cabinet approved a plan to form the National Cyber Defense Authority, whose goal was to strengthen cyberprotection for institutes, defense agencies, and citizens. The new authority was described as an "Air Force" against new cyberthreats.[83] In 2018, the National Cyber Directorate (NCD) was formed by merging the NCB and the National Cyber Defense Authority.

6.7.3 A Three-layer Model

Israel's NCS consists of a three-layer model. The first layer focuses on "Aggregate Cyber Robustness" and organizations act as the first line of defense. The idea is that organizations and processes have the ability to prevent and defend against most cyberattacks. They should be able to continue operating despite facing cyberthreats. The country believes that such a strategy can counter 95% of the threats.[84]

The NCD plays a key role in increasing the effectiveness of this layer by influencing the demand as well as the supply side of CS. On the demand side, it provides guidance for the private sector such as specific CS-related advice to critical infrastructures, mandatory CS standards in critical areas, and promotion of knowledge and awareness. On the supply side, it facilitates the availability of CS professionals, products, and services as well as related technological service.

The second layer involves "Systemic Cyber Resilience." This layer is related to the systemic ability to confront and deal with cyberattacks before, during, and after the incidents have occurred. A goal of this layer is to prevent cyberattacks from spreading so that the cumulative impact to the nation can be minimized. While the first layer is preventive and aims to reduce attacks on networks, this second layer helps achieve an effective response after cyberattacks. In order to enhance systemic resilience, the state facilitates and encourages information sharing. It also generates and disseminates valuable information. Additionally, it assists organizations if they face cyberattacks.

The third layer is "National Cyber Defense." The goal of this layer is to develop a national campaign against severe, high-profile threats by resource-rich attackers such as terrorist organizations and adversary states. Such attackers are often determined and pose the greatest danger to the nation. Defensive efforts are made to contain such attacks. These are combined with efforts to deal with and confront the sources of the threats. The NCD manages the defensive campaign within the civilian sector and coordinates the efforts of relevant agencies. It also performs national situation assessment.[85]

6.7.4 Firms with Expertise in Diverse Areas

In 2018, Israel had 421 CS companies.[86] (See figure 6.3.) They include NASDAQ-listed companies such as Check Point Software and CyberArk Software.[87] These firms exported CS solutions worth US$6.5 billion in 2018.[88]

Figure 6.3 Number of CS companies in Israel at the end of 2018

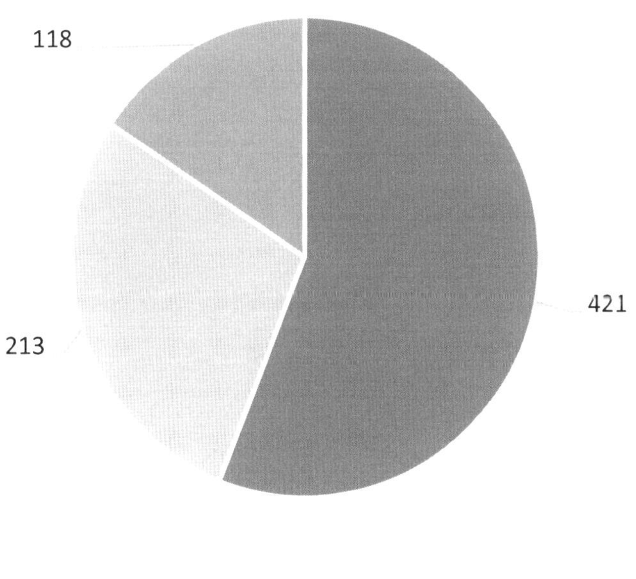

Data source: Solomon, S. (2019). *Israel cybersecurity sector hamstrung by shortage of labor, report says*. The Times of Israel. Retrieved from https://www.timesofisrael.com/israel-cybersecurity-sector-hamstrung-by-shortage-of-labor-report-says/

The data firm CB Insights' *Cyber Defenders 2018* report selected 29 CS startups. Cyberdefenders are defined as early- to mid-stage "high-momentum companies" with pioneering technologies that have the potential to transform CS. With six firms in the CB Insights' 2018 list, Israel ranked second after the US in the concentration of cyberdefenders.[89]

Israel-based firms have achieved the depth needed in cyberprotection expertise. Some high-profile CS firms from the country include the financial data security firm Trusteer, CyberArk (specialized in securing and managing privileged passwords/identities), NativeFlow (specialized to manage the Bring Your Own Device [BYOD] problem), Check Point Software Technologies, Actimize (provider of solutions to financial institutions to detect fraud, prevent money laundering, and manage risk), and Aorato (which learns and graphs people's network behavior and identifies suspicious activity).[90]

Israel is making attempts to build an image of the country as the center of global CS excellence. CyberArk's CEO noted: "Everybody understands that you buy Swiss watches from Switzerland and information security from Israel."[91]

6.7.5 CS Human Resources

In 2018, the Israeli CS industry employed 20,500 people, excluding those working in defense firms. Half of the national CS workforce worked for private-sector startups. R&D centers set up by foreign firms employed about 4,500 people and about 5,900 worked in the public sector.[92]

Israel starts CS education during middle school. It is also the only country that provides CS as an elective in high school. Several universities offer undergraduate specializations in CS.[93]

A strong relationship between military intelligence and private industry has been a key force in the country's prominence in the world CS market. The Israeli army's military intelligence and technological units, such as Unit 8200, provide much of the training and experience needed for the country's success on the CS front.[94]

Israel's strong entrepreneurial spirit and top technical universities have been key ingredients of its global leadership in CS. The government has provided favorable tax incentives and subsidies, which have amounted to as much as 40% of some companies' staff salaries.[95]

6.7.6 Case Summary

Israel has been successful in cashing in on the trend of rapidly rising cybercrime, which has spawned numerous globally competitive CS firms. The country's success on the cyberentrepreneurship front can be attributed to a combination of several factors, including the emergence of the CS sector as a magnet for VC investment, high R&D intensity, and the government's initiatives to strengthen the country's CS.

CHAPTER SEVEN

Cybersecurity Issues in International Relations and the Global Political System

CHAPTER PREVIEW

CS is emerging as a key issue in international relations and national security. This chapter discusses key challenges associated with international legal regimes and institutional frameworks related to CS. It also takes a look at critical issues and current sources of disagreement among nations. It provides a framework that can be used by nations in strategic policy choices for cyberconflicts. As well, the chapter presents detailed examples of cyberwarfare between major countries. Finally, it evaluates the impacts of these issues on businesses.

7.1 INTRODUCTION

The significance of CS needs to be considered in the context of the global political system and international relations issues facing nations. CS policies of many countries emphasize the need for international cooperation.[1] The EU has identified "a need for closer cooperation at a global level to improve security standards, improve information, and promote a common global approach to network and information security issues."[2] The US Cybersecurity Strategy has emphasized the need to strengthen "the capacity and interoperability of those allies and partners to improve our ability to optimize our combined skills, resources, capabilities, and perspectives against shared threats."[3]

Due to the growing insecurity of cyberspace, states and other stakeholders have created numerous and diverse initiatives focused on increasing its security. Discussions in diverse forums have led to the creation of "a new ecosystem of 'cyber norm' processes."[4]

These initiatives and measures have been introduced by international and regional organizations and relevant bodies and forums for reducing cyberthreats and promoting a greater cyberpeace. Despite these encouraging attempts, widespread allegations and counter-allegations involving cyberspace among major countries and some nations' engagement in the cyberwar arms race point to the fact that current legal regimes, institutional frameworks, and instruments are insufficient, inappropriate, and ineffective to address cyberthreats.

Nations increasingly engage in cyberwarfare and cyber cold war, and state actors launch cyberattacks against other states. (See In Focus 7.1.) A report by the think tank Center for Strategic and International Studies (CSIS) in Washington, DC, provides a glimpse of this phenomenon. CSIS's analysis of publicly available information on cyberespionage and warfare, which excludes cybercrimes, indicated that China and Russia were the largest source of cyberattacks that resulted in losses of more than US$1 million each from 2006 to 2018. (See figure 7.1.)[5]

Organizations and individuals are affected in many different ways by such phenomena. For instance, a number of cyberattacks allegedly carried out by North Korean hackers on South Korean banks, insurance companies, and other targets from 2009 to 2013 led to significant economic losses. The hackers erased computer hard drives of one of South Korea's largest banks, which left 30 million customers without ATM services for many days.[6]

Figure 7.1 Number of cyberattacks with losses of more than US$1 million each between 2006 and 2018 attributed to major cyberpowers

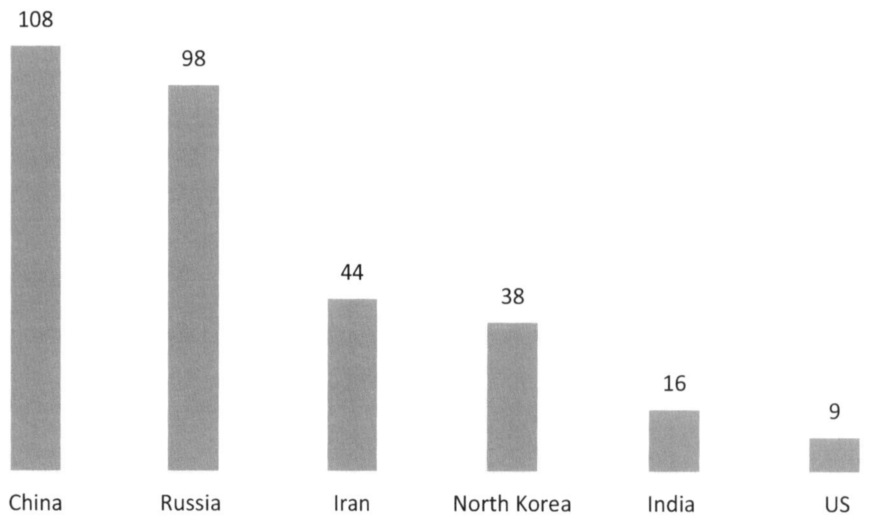

Data source: *Significant cyber incidents*. (n.d.). Center for Strategic & International Studies. Retrieved from https://www.csis.org/programs/technology-policy-program/significant-cyber-incidents

IN FOCUS 7.1: MAJOR ADVERSARIES LAUNCHED CYBERATTACKS AND DISINFORMATION CAMPAIGNS AGAINST THE US AND ITS ALLIES DURING THE COVID-19 PANDEMIC

During the COVID-19 pandemic, China, Iran, North Korea, and Russia launched cyberattacks and disinformation campaigns against the US and its allies. For instance, a spokesperson for the Chinese Ministry of Foreign Affairs claimed the US was responsible for creating the virus.[7]

According to the New York-based startup Graphika, which analyzes social networks, the International Union of Virtual Media, a group linked to the Iranian government, used Facebook and Twitter to engage in disinformation campaigns related to COVID-19. Among the narratives it promoted was that "COVID-19 was a biological weapon created by the US."[8] Likewise, a website run by the Russian Defense Ministry promoted the conspiracy theory that Bill Gates was behind the creation of the virus.[9]

Hackers belonging to the Lazarus Group, which is believed to have links with the North Korean government, launched cyberattacks against South Korean consumers. They sent emails that purported to come from a local government agency fighting COVID-19.[10] While the cyberattack campaigns from China, Iran, and Russia were politically motivated, attacks associated with North Korea were believed to have economic motivations.

North Korea has been of special interest. Cybercrime organizations associated with the country such as the Lazarus Group have been victimizing individuals and organizations worldwide for several years. For instance, in May 2017, the Lazarus Group launched WannaCry ransomware attacks on computer networks worldwide. The victims included the UK's National Health Service. The attacks resulted in 6% fewer admissions than normal in the NHS.[11] The country has thus received attention due to the regime's isolation from the rest of the world and the suspicion that its weapons program is being financed by sophisticated cyberattacks that target banks and cryptocurrency exchanges. One estimate has suggested that by 2019, cyberattacks associated with the North Korean government had generated US$2 billion for the country's weapons programs.[12]

This chapter explores the nature, sources, and consequences of international cyberthreats. It also specifies possible actions that nations can take to deal with various deficiencies of the current regimes, frameworks, and instruments.

7.2 INTERNATIONAL LEGAL REGIMES AND INSTITUTIONAL FRAMEWORKS

Some argue that a key problem of the Council of Europe Convention on Cybercrime (CoE-CoC), the only multilateral treaty focused purely on cybercrimes, is its adoption of vague definitions of cybercrime and related concepts that are subject to different interpretations by different states. Many nations that have ratified the CoECoC have done so under several reservations (e.g., "more than a half dozen" for the US).[13] These interpretations and reservations have reduced the scope of cybercrimes covered by the treaty and led to obligations that are less demanding and that lack uniformity across countries. A National Research Council study concluded, "a signatory nation may decline to cooperate with its obligations under the convention on fairly broad grounds, and the convention lacks an enforcement mechanism to assure that signatories will indeed cooperate in accordance with their obligations."[14]

Nations have also relied on global intergovernmental organizations such as the UN to address cybercrime issues. In the first meeting of the Intergovernmental Group of Experts of the UN Crime Prevention and Criminal Justice Program held in January 2011, the Chinese delegation complained that the country was suffering from foreign-originated cyberattacks. The delegation noted that in 2010, servers for over 90% of network sites that were used to commit cyberfrauds such as phishing, pornography, and Internet gambling against Chinese targets were located outside China. The delegation also stated that over 70% of botnet control sites were in foreign countries.[15]

Formal standards-setting international institutions such as the International Telecommunication Union (ITU) have also become a venue where these issues are being discussed and debated. For instance, while governments of the US and EU countries have argued that the Internet Corporation for Assigned Names and Numbers (ICANN) should continue to be the central organization, governments of some other major countries such as Brazil, China, India, South Africa, and Middle Eastern countries including Iran and Saudi Arabia want to move the Internet management system under the ITU. The countries in the latter group also want to define Internet governance more broadly to include issues such as spam and illegal content as opposed to the ICANN's narrow technical mandate, management of the Domain Name System. Since the ICANN is US-based, many governments think that ICANN's central role in governance would put the US in a position of power. These governments also think that the US may have exploited its advantage to create malware such as Flame and Stuxnet, which attacked sovereign nations.

New regional multilateral exclusive groupings established for politico-security arrangements such as the Shanghai Cooperation Organisation have also dealt with CS. Before the

2018 SCO Summit, the secretary-general of China-based SCO Research noted CS issues are a threat to the SCO region's stability.[16] The SCO countries' approach to CS differs in several important and fundamental ways from the CoECoC signatory countries. The two groups differ in the definition and assessment of the scope of the problem. One such difference is that SCO countries consider it important to focus on the broader problem of information security rather than narrowly on CS. In 2008, the SCO Agreement in the field of International Information Security emphasized and expressed concerns about the "digital gap" between the West and the East. These countries have been particularly concerned about the West's monopolization of ICT products and less developed countries' dependence on the West.

Multilateral groups of rising power countries, such as BRICS (Brazil, Russia, India, China, and South Africa), have also taken initiatives to develop international cybernorms. One such example is the CyberBRICS project. The objectives of CyberBRICS include (i) evaluating the features of existing regulations, (ii) identifying best practices, and (iii) developing CS governance policy suggestions.[17] Some key focuses of the BRICS countries have been to refine and retool digital policies and data privacy regulations in order to reduce the power of foreign technology companies and strengthen and affirm their sovereignty and independence.[18]

Finally, military, political, and economic organizations such as Asia Pacific Economic Cooperation, Association of Southeast Asian Nations (ASEAN), the EU, North Atlantic Treaty Organization (NATO), and OECD have also addressed CS issues. For instance, in an attempt to enhance NATO's cyberdefense capability, the NATO Cooperative Cyber Defense Centre of Excellence was established in 2008.[19]

The US and EU countries have also established deep and strong collaborations and partnerships. For instance, the Italy-based European Electronic Crimes Task Force, which has dedicated personnel from the countries involved to investigate and prosecute cybercrimes, provides a forum for law enforcement agencies, the private sector, and academia from the US and EU nations. In the same vein, a virtual forum for ASEAN CS is being formed to develop a common framework to coordinate exchange of information, establishment of standards, and cooperation among enforcement agencies.

7.3 CYBERSECURITY-RELATED INITIATIVES AT THE UN

Several recent UN initiatives on the cyberfront deserve mention. In 2004, the UN General Assembly established the GGE to examine the impact of developments in ICT on national security and military affairs. The GGE works by consensus-based decision-making. By 2020, six GGEs had been convened. The 2016/2017 GGE concluded the last round of deliberations in June 2016 but failed to produce a consensus outcome report because the GGE members expressed significantly divergent views.

The US wanted to include "clear and direct statements" regarding the use of ICTs in international humanitarian issues, the right to self-defense, and international law regarding state responsibility and countermeasures.[20] States such as China, Cuba, and Russia opposed the inclusion of such provisions.[21] For instance, Cuba argued that the inclusion of provisions suggested by the US may allow states to take "unilateral punitive force actions" and "sanctions and even military action" against other states by claiming that they have become "victims of illicit uses of ICTs."[22] It was argued that all of these would militarize cyberspace.

In 2018, another working group – the Open-Ended Working Group (OEWG) – was established by the UN.[23] Whereas the GGE comprises about 25 states, the OEWG accepts participation of any interested state. Just like the GGE, the OEWG's work is based on consensus. In order to produce an outcome report, it requires agreement among all the participants.[24] The OEWG started its work in the summer of 2019 and issued a consensus report in March 2021.[25] All 193 countries that participated, including China and Russia, agreed to follow voluntary and non-binding norms for responsible cyberspace behavior.

7.3.1 Russia-led Resolution of December 2019

Another notable event was that in December 2019, the UN passed a resolution countering the use of information and communications technologies for criminal purposes. The resolution was led by Russia and backed by China. Russia's preference has been to develop cyberspace norms based on UN Charter principles that include respect for national sovereignty and no interference in internal affairs.[26]

Several major Western powers opposed the resolution.[27] UN members voting in favor of the resolution included many smaller countries in the Caribbean, the Pacific, and Africa. Critics noted that the governments of the countries voting for the resolution did not talk to their cybercrime experts.

In an open letter, the nonprofit organization Association for Progressive Communications (APC) had raised concerns about the vagueness of the Russian draft convention. The APC especially argued that the resolution's language regarding what constitutes the use of ICTs for criminal purposes was not defined in the draft.[28] The resolution's critics voiced the concern that its vagueness can be used by authoritarian governments to violate Internet users' human rights.[29] For instance, the Russian government has been cracking down on the Internet in order to intimidate and suppress its opposition.[30]

7.4 CRITICAL ISSUES AND CURRENT SOURCES OF DISAGREEMENT AMONG NATIONS

There has been a relatively low level of acceptance of the CoECoC. For instance, none of the BRIC states (Brazil, Russia, India, and China) is currently a signatory to the CoECoC. Its failure to attract the support of a large number of countries can be attributed to the general Western focus. Many countries have objected to a number of clauses of the treaty, including definitions of cybercrimes, criminalization of intellectual property rights violations, and mechanisms that involve intrusion of national sovereignty.[31] Russia's SCO National Coordinator Ambassador Barsky described the Convention as less than satisfactory.[32]

A state victimized by cyberattacks is likely to encounter several problems in applying formal and informal sanctions. As discussed in chapter 1, one problem centers around the technical difficulty in attributing the sources of cyberattacks. For instance, there have been talks in Washington about fines and trade penalties against a country guilty of cyberattacks.[33] The near impossibility of identifying the actual source means that economic and trade sanctions as an instrument of diplomacy in response to cyberattacks would generally be an unworkable idea. As a result, it is impossible to regulate cyberweapons in the same manner as nuclear, chemical, and biological weapons. In this section, we outline key unaddressed issues.

7.4.1 Outdated Legislative Framework and the Lack of Law Enforcement System Capacity

Most international cybercriminals can "jurisdictionally shield" themselves just by operating from countries that lack law enforcement system capacity. The primary reason they are not prosecuted is due to outdated regulative institutions and the unwillingness of law enforcement agencies to pursue cyberfraud cases because the criminals mainly victimize foreigners. In Ukraine's case, for instance, despite international cooperation, crime-fighting efforts are hindered by an underdeveloped and outdated institutional framework. While the Security Service of Ukraine (Sluzhba Bespeky Ukrayiny) cooperated with the West in 2010 and arrested five alleged kingpins of a criminal group that stole US$70 million from US bank accounts, the kingpins were freed immediately without a court trial. It is also argued that foreigners also account for international cybercrimes originating from Ukraine.[34] Corruption has arguably enabled and generally encouraged them to obtain the right to reside in and operate criminal activities in the country.

7.4.2 Concerns regarding the Fairness of the Procedures and Outcomes of Formal Frameworks

The problem of system capacity applies equally to other developing countries that are not signatories to the CoECoC. An additional significant issue is that many are neither ready nor willing to join the CoECoC. Some countries are concerned with issues of equity, voice, and just representation and have questioned the fairness of the procedures and outcomes of international regimes. In the first UN forum on Internet governance, countries such as Iran and South Africa complained that they had not been given an opportunity to adequately express their views.[35]

7.4.3 Disagreement regarding the Nature and Dimensions of Cyberthreats

In particular, the West's relationships with some major world powers such as China and Russia are confronted by complex issues. Some argue that it is insufficient simply to consider political and military issues as the only dimensions of national security. The issues of national values (e.g., national unity, social harmony, state sovereignty, nuclear non-proliferation, freedom of expression, tolerance) and identity have also been identified as sources of insecurity.[36] An article in China's People's Liberation Army Daily warned that "information colonialism" is a real threat to national security and "will be a major cause of future wars."[37]

The views and positions of China and Russia are reflected in the approach of the SCO. The 2008 SCO Agreement in the field of International Information Security expressed concerns about the "digital gap" between the West and the East. The SCO countries are interested in controlling information that is likely to provoke what they call the three "evils" (terrorism, extremism, and separatism). They consider it important to prevent other nations from using their technologies to disrupt economic, social, and political stability and national security. Western countries, on the other hand, maintain that too much government regulation and control may harm cyberspace security. They emphasize the importance of the private sector in the formulation of international norms.[38]

States face a unique governance challenge associated with the Internet. Their degree of trumping power over internal illicit actors is declining and power over their citizens has been reduced due to broader and more diverse flows of information. Some regimes have come under cyberattacks from their own citizens. For instance, in 2014, in order to protest corruption, activists hacked the Twitter accounts of the Kenyan defense forces and its spokesperson.[39] Hacking groups such as Anonymous have attacked the government websites of a number of countries such as Brazil, Colombia, Ecuador, and Venezuela.

Where internal security poses a bigger threat, the defense against external attacks is likely to be a less important security goal. Instead, the government is likely to focus its CS efforts on internal threats. Internal security is often desirable for the ruling elites to remain in power, which does not necessarily involve making all citizens equally secure and safe.[40]

7.4.4 Isolation from Most of the Economies of the World

First, it is important to note that, despite their capabilities, countries such as Russia and China have little incentive to launch cyberattacks against the West that may cause widespread economic harm. For instance, if Russia and China launch economic cyberattacks against the West, due to economic global integration, the aftershocks from economic disaster in the West are also likely to harm them.[41] This is not the case, however, for countries such as North Korea.

Some regimes' political and economic isolation from most of the world's economies is clearly the dominant challenge facing the West to integrate them into formal institutional frameworks. Iran and North Korea present challenges that are unique and different from those presented by China and Russia. For instance, the US and other Western countries have a clearer understanding of the points of view of China and Russia. China and Russia also maintain diplomatic and economic ties with most countries. The general lack of such understanding and the lack of diplomatic and economic ties with North Korea increase complexity. Due to the regime's unpredictability and desperation to survive, North Korea arguably poses high threats to its adversaries.[42] Moreover, unlike China's PLA, which regularly publishes academic journals and policy documents, North Korea's Korean People's Army (KPA) does not publish any documents. Assessing the KPA's advancement in cyberweapons has thus been extremely difficult.[43]

As noted earlier, the asymmetric nature of cyberattacks means that entities with limited financial and technical resources or militarily weaker actors can compromise high-value targets.[44] This means that so-called rogue and economically destitute countries have a higher propensity to engage in cyberattacks. For instance, North Korea may launch cyberattacks and avoid sanctions and retaliatory attacks. In this way, the development of cyberwarfare capability would give countries such as North Korea the ability to harm adversaries without potential negative consequences.[45]

7.5 NATIONS' STRATEGIC POLICY CHOICES TO DEAL WITH CYBERCONFLICTS

The international institutional framework for CS is structurally imperfect. Nations hold diverse interests, viewpoints, and perspectives on CS. Nations also have varied levels of resources and capabilities available for developing CS-related capabilities. These differences make it necessary to develop strategic responses to possible cyberthreats associated with a nation based on its individual characteristics as well as the nature of its relationship with others. In this section, we discuss some policy choices in dealing with other countries on the cyberfront.

7.5.1 Local Capacity Building in Law Enforcement and Institutional Development

Some emerging countries' limited law enforcement system capacity to deal with cybercrime has become the dominant challenge that inhibits their ability to fight cybercrimes. Local capacity building and institutional development can produce effective results in strengthening global CS. As an example, consider Romania. In response to the rise of Romania-originated cybercrimes, the Council of Europe selected the capital city Bucharest for its latest cybercrime program office. As of 2013, the FBI had trained about 600 Romanian law enforcement personnel in fighting cybercrime.[46]

Some private-sector players also contribute to institutional and law enforcement capacity development. To take an example, eBay has been educating Romanian prosecutors about cybercrimes, including how to explain cybercrimes to a judge using layperson's language.[47]

7.5.2 Creation of Informal Networks and Agreements

Informal networks and agreements among states and transnational actors are becoming an important feature of world politics.[48] Such networks are found in areas such as financial markets, aviation, antitrust, data privacy, pharmaceuticals, and the environment. State and sub-state officials from a number of countries work together to share information with each other, develop harmonized guidelines and best practices, and reduce friction associated with globalization.[49]

As of July 2018, the FBI had 63 legal attaché offices and 27 sub-offices around the world, covering more than 200 countries and territories.[50] Many of these offices focus on cybercrimes. It has permanently based a cybercrime expert in Estonia.[51] The FBI worked closely with the Estonian Police and Border Guard, the Dutch National Police, and other relevant agencies, which led to the arrest of six Estonians who allegedly used Domain Name System Changer malware to hijack more than 4 million computers in over 100 countries, which generated at least US$14 million in profits.[52]

7.5.3 Providing Opportunities for Developing Countries' Voices and Participation

Making efforts to understand developing countries' problems from their point of view and providing opportunities for their voices and participation would help explore mutual advantage and encourage their participation in formal international frameworks. Attempting to achieve too much too rapidly is often counterproductive. One way to gain cooperation would be to exclude issues from formal treaties that are objected to by developing countries and that are only tangentially related to cybercrime, such as software piracy.

7.5.4 Helping, Encouraging, and Providing Incentives to Integrate with the West

As noted earlier, cybercriminals can take advantage of jurisdictional arbitrage by operating from countries with outdated legislative frameworks and little law enforcement system capacity. Emerging countries' international cooperation and integration would help them modernize CS-related legislative frameworks and enhance system capacity and law enforcement performance.[53] Cyberoffenders would then be less likely to be jurisdictionally shielded. For instance, North Macedonia became a NATO member in March 2020. NATO is expected to provide assistance to North Macedonia in fighting fake news.[54]

7.5.5 Harnessing the Power of Successful Regional Organizations

Given the lack of membership of most developing countries in the CoECoC, one way to compensate for some of the deficiencies of existing CS-related international institutions is to harness the power of regional organizations, especially those of developing nations that are not signatories to the CoECoC. Successful regional organizations that are internally cohesive and have security as a key focus are logical candidates for developing meaningful CS relationships. The ASEAN is one example that fits these criteria and has been effective in managing its internal security relations. Some major countries' engagement with the ASEAN has focused on deepening a CS relationship. For instance, since 2009, the ASEAN and Japan have been working to promote CS-related cooperation and collaboration. Relevant ministries and agencies such as computer security incident response teams of ASEAN member states and Japan are working toward addressing common concerns through initiatives such as the Internet Traffic Monitoring Data Sharing Project (TSUBAME Project).[55]

Especially in light of the inability of the global forums such as the UN GGE to build consensus, policy-makers consider multilateral efforts and regional activities' roles in strengthening CS. In the Asia-Pacific region, the ASEAN and ARF are considered key venues to implement cybernorms and develop CBMs.[56]

The ARF was established in 1994 as a platform for security dialogue in the Indo-Pacific region. It comprises 27 members, which include the 10 ASEAN member states (Brunei, Cambodia, Indonesia, Laos, Malaysia, Myanmar, Philippines, Singapore, Thailand, and Vietnam) as well as key global and regional powers such as Canada, China, the EU, India, Japan, New Zealand, South Korea, Russia, and the US.[57]

7.5.6 Offensive and Defensive Capabilities Tailored to Specific Threats

The responses discussed thus far are appropriate only if the source countries of the cyberattacks have formal diplomatic and economic ties. In the absence of such ties, CS may require an appropriate balance of offensive and defensive capabilities. It is argued that many countries, including the US, have a strong offensive capability but are weak in defensive power. Some analysts have called for more proactive defense involving the design and implementation of secure software rather than the current focus on reactive defense.[58] South Korea's cyberdefense measures against cyberthreats originating from North Korea illustrate this point. In response to GPS attacks against commercial flights and maritime navigational units that originated in North Korea in 2012 and 2013, South Korea enhanced GPS system capability. In 2013, the country's Ministry of Science and Future Planning announced detailed plans to strengthen the surveillance system against North Korea's electronic jamming signals as well as develop systems that are able to locate the attack point and assess the impact of jamming attempts.

7.6 CHAPTER SUMMARY AND CONCLUSION

Views on CS of nations across the world are far from uniform. CS is thus becoming a key source of tension and dispute among major nations. We see a wide range of cybersecurity concerns in international relations and the global political system, such as inappropriate regulatory frameworks and poor law enforcement in some countries, that fail to deter cyberperpetrators. There is also the perception of unfairness in the global norms surrounding mechanisms to fight cyberthreats. Some major source countries of cyberthreats are isolated from most world economies, which poses another challenge to controlling cyberthreats.

Countries worldwide have recognized the need to use diverse measures and perspectives for fighting international cybercrimes. For instance, informal networks have become an increasingly common mechanism for solving trans-border cybercrimes. Overall, despite some progress in recent years, building international consensus and norms has been a challenging task.

7.7 DISCUSSION QUESTIONS

1. What are some of the key limitations of the CoECoC in fighting international cybercrimes?
2. How successful are CS-related initiatives that are being taken at the UN?
3. What roles are played by regional multilateral groups to promote a greater cyberpeace?
4. What are some of the challenges in fighting cyberattacks that originate from Iran and North Korea?

7.8 END-OF-CHAPTER CASE: US–CHINA RELATIONS IN CYBERSECURITY

Cybercrime is arguably the key challenge that the relationship between China and the US has been facing for some time.[59] Allegations and counter-allegations have been widespread in the US–China discourse on the governance of cyberspace. There has been too little cooperation and too much conflict in this area. The cyber cold war between the two countries has been underway for a long time.

7.8.1 Espionage and Cyberattack Allegations against China

There has been a lot of controversy about allegations of cyberattacks and cyberespionage by Chinese firms and actors sponsored by the Chinese government. According to a previous US counterintelligence chief, during the decades of the 2000s, over 2,000 companies, universities, and government agencies in the US experienced cyberattacks originating from China.[60]

The US government in May 2014 indicted five high-ranking members of the PLA for committing cybercrimes against US companies. The Chinese government questioned evidence used in the indictment and, in response, suspended its participation in a CS working group with the US.[61] China blamed the US for disrupted CS cooperation between the two countries.

In June 2015, it was widely believed that hackers allegedly working for the Chinese government stole the personal information, including financial and biometric data, of 22 million current and former federal employees from the US Office of Personnel Management (OPM).[62]

7.8.2 China's Response to US Allegations and Edward Snowden's Revelation

In September 2012, Huawei's corporate senior vice president and ZTE's senior vice president for North America and Europe testified at the US House committee about the allegation that the two companies' operations had posed CS threats to the US. In an interview after the US House committee hearing of September 2012, ZTE's senior vice president for North America and Europe noted that "many Congress members still harbor a Cold War mentality and know little about China's development."[63]

Edward Snowden's revelation of the US PRISM surveillance program constituted an important turning point in China–US relations involving cyberspace. A *China Daily* article on June 5, 2013, noted that "the cyber-attacks from the US [on China] have been as grave as the ones the US claims China has conducted."[64]

Commenting on the US intelligence agencies' alleged hacking of China's major mobile companies and universities, an editorial in China's state-run Xinhua News Agency noted, "These, along with previous allegations, are clearly troubling signs. They demonstrate that the United States, which has long been trying to play innocent as a victim of cyber-attacks, has turned out to be the biggest villain in our age."[65]

In October 2014, a Chinese Foreign Ministry spokesperson expressed dissatisfaction regarding "the United States' unjustified fabrication of facts in an attempt to smear China's name" and demanded that the US "cease this type of action."[66] In the same month, China's state councillor overseeing foreign affairs, Yang Jiechi, told then US secretary of state John Kerry that resuming CS cooperation between the two countries would be difficult because of "mistaken US practices." He emphasized that the US "should take positive action to create necessary conditions for bilateral cyber security dialogue and cooperation to resume."[67]

China also announced that it will perform national security inspections for imported technology products. China's nationalist bloggers described the announced inspections as a "hard blow to anti-China forces" and suggested that US companies such as Cisco, IBM, and Microsoft would be negatively affected.[68] In 2014, anti-monopoly investigators raided four Microsoft offices and allegedly copied large amounts of the company's data.[69]

7.8.3 Major Breakthrough in 2015

A major breakthrough in the negotiation process was seen in September 2015, when Chinese president Xi Jinping visited the US. During the visit, then US president Barack Obama and Xi reached a "common understanding" that aimed at stopping cyberespionage activities. They agreed that their governments would not knowingly support cyberespionage activities aimed at stealing corporate secrets and business information. This agreement became a major milestone in China–US cyberrelations.

The US side in the negotiation process wanted Xi to publicly acknowledge that the Chinese government would not help its companies to use cyberattacks to steal intellectual property from US companies. This understandably made the Chinese side uncomfortable because China had always denied that its companies had stolen US secrets. The request upset the Chinese side and almost caused the deal to collapse.[70]

This agreement produced some very good results. According to the CS firm CrowdStrike's co-founder Dmitri Alperovitch, while Chinese espionage activities against US government targets such as the Pentagon continued, China-originated hacks against US companies declined by over 90% in the 18 months following the agreement.[71]

According to some experts, a main reason Xi agreed was because he wanted to take action against Chinese military elites who allegedly used state-sponsored hacking to benefit themselves and their cronies. Another major motivation was arguably to bring China's cyberforce networks under tighter control of the CCP.

7.8.4 Little Long-term Impact

The 2015 agreement has failed to overcome the mutual mistrust between the two countries. Each party has accused the other of violating the agreement. In an unclassified document, the US Department of Defense published its cyberstrategy in September 2018.[72] The strategy document noted that China is among the four "states that can pose strategic threats to US prosperity and security" (the others being Russia, North Korea, and Iran). A Chinese CS analyst, Lyu Jinghua, noted that the US approach to view China as an adversary was troubling and problematic in the context of the deteriorating US–China relationship on the cyberfront.[73] Lyu observes that such a viewpoint would send a negative signal to the Chinese government.[74]

In November 2018, a senior US intelligence official was quoted as saying that China had been violating the 2015 agreement between Obama and Xi.[75] The Chinese Foreign Ministry spokeswoman Hua Chunying rejected the US allegations, saying, "The US accusations lack factual basis. China firmly opposes them."[76]

In a blog post published in March 2020, the Chinese antivirus firm Qihoo 360 alleged that the Central Intelligence Agency (CIA) had launched hacking activities against the Chinese airline industry and other targets for more than a decade. Qihoo noted that it reached its conclusion by comparing the samples of malware found in the networks of the victim companies against the CIA's spy tools released by WikiLeaks in 2017.[77]

For the governments of some of the major developed countries, concerns associated with Chinese 5G players such as ZTE and Huawei are more visible and prevalent. The US as well as its allies, such as Australia and New Zealand, have banned Huawei's 5G equipment.[78] In May 2019, the US Department of Commerce further blacklisted Huawei by adding it to the "entity list."[79] This means that US firms cannot do business with Huawei without obtaining a government license.[80]

In an attempt to place more restrictions on Huawei and ZTE, in October 2019, the US Federal Communications Commission (FCC) put forward a proposal to ban companies receiving government money from purchasing equipment or services from these firms. The FCC argued that if Huawei's 5G network equipment operates in sensitive locations, such as near a US military base, China could ask the company to install a secret back door or malware in a US network and the US officials might not be in a position to know it.[81]

Since Huawei is on the Department of Commerce's "entity list," Google cannot supply Android services, updates, and apps for Huawei 5G handsets.[82] Huawei decided to delay sales of its 5G-enabled Mate 30 smartphone series in Europe, which is its biggest market outside China. The company noted that the value and usability of these 5G smartphones would be reduced without Google apps.[83]

The alarms raised about Chinese firms by countries such as Australia, Japan, New Zealand, and the US have also triggered security fears among EU countries.[84] EU officials have expressed concerns and worry related to the use of China-originated 5G technologies in critical infrastructures and systems ranging from road and rail management to household devices. Dusan Navratil, the director of the Czech National Cyber and Information Security Agency, argued that Chinese laws require companies such as Huawei to cooperate with intelligence services. He went on to say that using Huawei products in critical systems might pose national security threats for the Czech Republic.[85]

7.8.5 Summary

Despite some progress in reducing the cyberconflict between the world's two largest countries, mutual mistrust has continued. While the 2015 agreement was a major milestone, the agreement has not stopped the two countries from attacking each other's networks.

PART THREE

Strategic and Organizational Issues Associated with Cybersecurity

CHAPTER EIGHT

Corporate Cybersecurity Strategy

CHAPTER PREVIEW

This chapter presents a framework delineating the drivers and outcomes of CS strategy. It provides information on which to build a comprehensive CS strategy to minimize and prevent cyberattack risks facing a company as it examines the corporate CS environment within the larger context of data-driven decision-making and increasing cybersecurity risks. It classifies cyberthreats in terms of various dimensions: capabilities and motivations of the attacker(s), potential consequences to the victim organization(s), relationship with actions and positions of the target, and the time-variance relationship. The chapter also examines the relevance and limitations of various types of resources in the context of CS and articulates processes by which internal and market-based resources can be improved and redeployed in order to strengthen CS and convert it into a competitive advantage in the big data era.

8.1 INTRODUCTION

CS is among the most critical issues facing organizations today. In order to understand the critical significance of CS, consider the impact of the December 2013 high-profile cyberattacks on the US retailer Target, which compromised 40 million credit- and debit-card accounts and 70 million people's personal data.[1] In May 2014, Target's then CEO Gregg Steinhafel "held himself personally accountable" and resigned.[2] In the same month, the proxy advisory firm Institutional Shareholder Services suggested that seven of the ten directors at Target be removed from office because they did not take sufficient actions to prevent the data breach.[3] As of July 2014, Target's expenses related to the breach had reached US$146 million. More than 100 lawsuits related to the breach and Target's breach response had been filed against the company. Target blamed the attack for its sales decline in the fourth quarter of 2013.[4]

In another high-profile example of data breach, 56 million credit cards of Home Depot's customers were compromised in 2014. After the breach, the data were available for sale on black markets. One estimate suggested that they could be used to make US$3 billion in illegal purchases before the credit cards would be deactivated by the issuing banks.[5] The US Secret Service estimated that about 1,000 companies in the US faced the same types of cyberattacks that compromised Target and Home Depot.[6]

The above examples illustrate the increasing importance of CS in determining the strength of a company. CS has become a central component of corporate strategy due to its increasingly stronger links with firm performance. It is thus a critical part of the overall vision of a firm's value creation and key to achieving competitiveness.

Many organizations, however, have not realized the importance of developing CS strategies. A survey conducted by executive coaching organization Vistage found that 62% of SMEs lacked an up-to-date CS strategy or any CS strategy at all.[7]

Organizations face pressures from diverse stakeholder groups to strengthen CS. In addition to rapidly escalating cyberthreats, regulatory pressures are also increasing. For instance, a guidance issued by New York's Department of Financial Services in December 2014 specifies stricter rules in corporate governance, login security, management of third-party vendors, CS insurance, and more. Likewise, the US SEC is becoming stricter with regard to CS-related follow-up. It increased the amount of information any public company must include in its 10-K report about financial performance, which is filed annually, to include security risks that could negatively impact it. To take an example, Amazon.com's 2012 annual report filed with the SEC did not mention the online theft in January 2012 of customer data that were held by its subsidiary, Zappos. After the SEC's request, the company agreed to modify the report.[8] To take another example, in December 2014, a judge in the US state of Minnesota rejected Target's motion to dismiss the lawsuit brought against the company by several banks. The judge argued that the retailer's actions were negligent because it failed to give careful attention to warnings from a security alerting system and disabled some security features.[9]

Organizations may, however, look at the pressures to strengthen CS not only as a threat but also as an opportunity. For instance, higher CS spending and a better CS model and performance will likely contribute to bottom-line results and competitive advantage.

As a key component of CS strategy, it is imperative that organizations have a well-developed plan for post-breach resilience in order to quickly return to normal business operations in case

Figure 8.1 Percentage of organizations with different stages of cyberattack and breach response plans

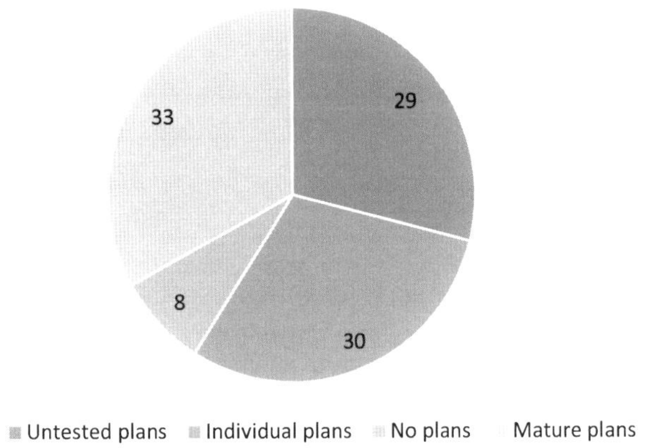

Data source: FireEye. (2019, November 11). *FireEye's cyber trendscape 2020.* Retrieved from https://www.fireeye.com/content/dam/fireeye-www/summit/cds-2019/presentations/cds19-executive-s05-fireeye-cyber-trendscape-2020.pdf

of a cyberattack. A 2019 study conducted by FireEye and market research firm KANTAR in Asia, Europe, and North America found that only about a third of organizations had mature cyberattack and breach event response and communications plans. These organizations regularly reviewed and updated the plans. About 29% of organizations had cyberplans but they had not been tested or updated for at least one year. Cyberattack and breach event response plans in 30% of organizations were owned and maintained by individual businesses or applications. They were not incorporated within an organization-wide plan. About 8% of organizations had no such plans at all. (See figure 8.1.)

8.2 GOALS, PERFORMANCE, AND CONTROL MEASURES IN CYBERSECURITY STRATEGY

It is important to define CS goals and measure them regularly. Some suggest that CS goals should be measured on a quarterly basis or even more frequently. In order to formulate goals, it is critical to understand the latest CS threats and identify the tasks that need to be done. Goals must have measurable metrics.[10] An example of a short-term CS goal could be to conduct application vulnerability assessments and penetration tests for a company's Internet-facing applications that collect or process sensitive information.[11] Long-term CS goals may include enhancing operational CS by improving company-wide processes, guidelines, and decisions related to handling and protecting data and information assets.

For instance, rules and guidelines need to be developed and communicated regarding the permissions users are required to have when accessing a network. Likewise, detailed procedures must be developed for determining how and where different types of data may be stored and shared.[12]

Organizations should consider performance measures and metrics in order to determine whether their CS strategy has translated into good results. Some examples of CS-related performance measures and metrics include (i) *mean time to detect*, which measures the average time taken to become aware of a CS incident; (ii) *mean time to resolve*, which measures the average time taken to rectify a CS incident or threat after it is been discovered; and (iii) *phishing test success rate*, which is an indicator of the proportion of employees that are tricked by phishing.[13] Especially important is (iv) *mean time to remediate* urgent and critical vulnerabilities. In order to give context to these metrics, organizations can compare their CS ratings with third parties as well as with the average CS ratings for their industries. The CS ratings company BitSight provides these indicators.

In addition to the CS-specific metrics discussed above, management control practices such as budgeting systems (see In Focus 8.1) and traditional accounting measures such as ROI (see In Focus 8.2) provide important insights into organizations' CS performance.

IN FOCUS 8.1: CYBERSECURITY BUDGETING

Creating effective CS budgeting requires an understanding of CS risks faced by the company and knowledge of activities and resources required to respond to such risks and recover from CS events if they occur.[14] It is important to keep in mind that cyberpreparedness is not a one-time event. It is a continuous and ongoing process that requires flexibility in budgeting in order to address new and emerging vulnerabilities and threats.[15]

CS budgets are expressed as a percentage of the overall IT budget, revenue, or costs per full-time employee. In 2016, Gartner found companies' CS budgets as a percentage of the overall IT budget ranged from 1% to 13%.[16] Others have found an average of 5.6% of the overall IT budget. A 2018 survey of financial institutions by Deloitte and the Financial Services Information Sharing and Analysis Center found that companies spent about US$2,300 per full-time employee. The CS spending as a percentage of the overall IT budget was in the range of 6% to 14% with an average of 10%. As a proportion of the organization's revenue, the CS budget varied from 0.2% to 0.9%.[17]

IN FOCUS 8.2: RETURN ON CYBERSECURITY INVESTMENT (ROCSI)

Investment in CS can contribute to ROI through various mechanisms. Between 2008 and 2010, Booz Allen Hamilton invested US$15 million to enhance employees' skill through CS training and education. The company argued that the ROCSI can be realized in a variety of ways. For instance, it can meet client needs and government requirements related to technically certified staff. The company can position itself as a thought leader. It can also increase the firm's ability to recruit, train, and retain a high-quality cyberworkforce.[18]

Measuring ROCSI precisely, however, is not easy. The European Union Agency for Cybersecurity has suggested the following formula to calculate ROCSI:[19]

$$\text{ROCSI} = \frac{\text{Reduction in monetary loss} - \text{Cost of the solution}}{\text{Cost of the solution}}.$$

Annual expected loss (AEL) is the annual monetary loss associated with a specific cyberrisk for a given information asset.

$$\text{AEL} = \text{ARO} \times \text{SAL}$$

where
Annual Rate of Occurrence (ARO) is the number of times an information asset may face cyberattacks in a year; and
Single Attack Loss (SAL) is the expected monetary loss in a single cyberattack.

When an effective CS solution is implemented, it is expected to reduce the AEL. That is, the new value of monetary loss or new AEL (nAEL) is smaller than AEL.

$$\text{Reduction in monetary loss} = \text{AEL} - \text{nAEL} = (1 - n)\,\text{AEL}$$

$1 - n$ is also referred to as the mitigation ratio (m). For instance, if an antivirus solution is expected to block 90% of attacks, the mitigation ratio (m) is 90%.

$$\text{Reduction in monetary loss} = m\text{AEL}$$

$$\text{Thus, ROCSI} = \frac{\text{mitigation ratio} \times \text{AEL} - \text{Cost of the solution}}{\text{Cost of the solution}}.$$

Example

ABC Corp. needs to decide whether to acquire a new antivirus solution. Each year, the corporation faces three virus attacks and the loss associated with each attack is estimated at US$25,000. The deployment of the antivirus solution is expected to cost US$30,000 and is expected to reduce the attacks by 75%. Should the company buy the solution?

$$\text{ARO} = 5$$

$$\text{SAL} = \text{US\$25,000}$$

$$\text{Mitigation ratio } (m) = 75\% = 0.75$$

$$\text{Cost of the solution} = \text{US\$30,000}$$

$$\text{AEL} = \text{ARO} \times \text{SAL} = \text{US\$125,000}$$

$$\text{ROCSI} = \frac{0.75 \times 125{,}000 - 30{,}000}{30{,}000} = 2.125 \text{ or } 212.5\%$$

To make the correct decision, it would be necessary to take into account the time value of money: $1 today is worth more than $1 after a year. The CS investment needs to be made up front; however, the benefits of reduced cyberattacks will accrue over time.

In order to further illustrate some CS-related performance measures and metrics, we consider Accenture's *Third Annual State of Cyber Resilience* study.[20] In its survey of enterprise security practitioners worldwide, 17% exhibited superior CS performance. Accenture refers to this elite group as the "leaders," which showed significantly better results from investments in CS compared to others. Specifically, the leaders exhibited highest performance in at least three of the following four areas: (i) stopping more attacks, (ii) finding breaches faster, (iii) fixing breaches faster, and (iv) reducing the impact of the breach. The study also identified a second group, which is referred as "non-leaders" or "average performers." This group consisted of 74% of the respondents. Figure 8.2 presents how leaders and average performers differ in CS strategy and performance.

8.3 THE FIRST PRINCIPLE OF CYBERSECURITY

The first principle of CS is to reduce cybervulnerability and minimize potential negative consequences should the firm face cyberattacks. In a process-based model, the formula is to use a

Figure 8.2 How leaders and average performers differ in key aspects of CS strategy and performance

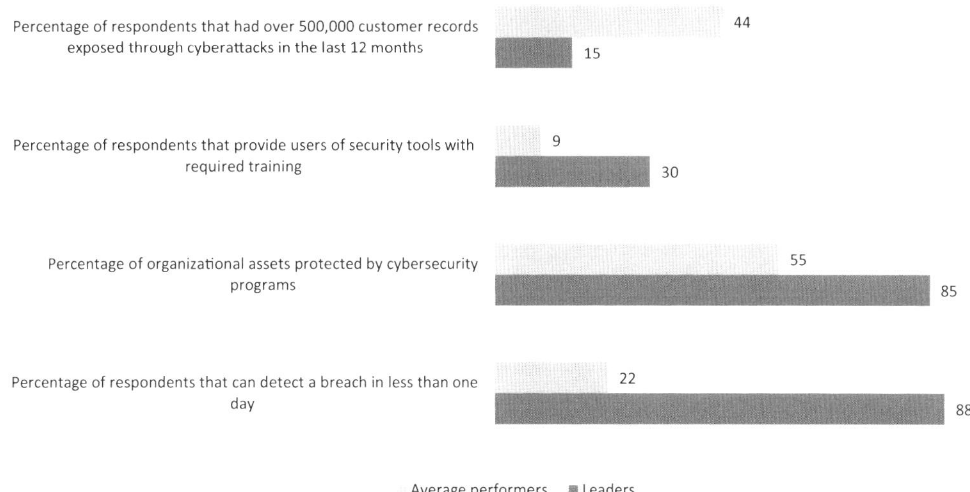

Data source: Geib, A., & Berard, D. (2020, February 3). *Only 17% of organizations globally considered "leaders" in cyber resilience, according to Accenture study.* Accenture. Retrieved from https://newsroom.accenture.com/news/only-17-percent-of-organizations-globally-considered-leaders-in-cyber-resilience-according-to-accenture-study.htm

risk-based strategy, where risk equals "threat plus vulnerability plus consequences."[21] A threat is a danger related to a cyberattack that has the potential to harm an organization. For instance, a number of factors, such as a firm's jurisdiction, physical location, nature of business, political orientation, and symbolic significance, are related to the degree of cyberthreats a firm faces.

Cybervulnerability refers to the degree to which an organization is susceptible to harm. A firm with poor CS practice is more likely to be harmed by cybercriminals. Note that vulnerability has two dimensions: objective and subjective.[22] Objective vulnerability is related to political, social, economic, and demographic characteristics of an entity that determine its vulnerability to cyberattacks. Vulnerability testing – automated testing or human "penetration testing" – can be performed to determine objective vulnerability. Automated testing includes static application security testing (SAST) and dynamic application security testing (DAST).

The SAST analysis involves analyzing an application's source code. The idea is to look for coding and design vulnerabilities that make the application susceptible to cyberattack. The DAST analysis is performed by attacking the concerned applications from outside in order to find vulnerabilities before malicious actors exploit them.

The subjective vulnerability refers to an entity's self-perception related to the risk of becoming a cyberattack victim. It is important to increase the overall accuracy of the assessment of objective vulnerability. A vulnerability analysis entails the systematic examination of factors such as security mechanisms of third-party vendors as well as software and hardware used in the company. For instance, a third-party service provider's weak cyberdefense mechanism is likely to increase vulnerability by exposing personally identifiable information.

Organizations that deploy technologies and systems used by other more attractive cyberattack targets are also exposed to higher degrees of vulnerability. To take an example, most of the computing platforms used in natural gas and petrochemical industries in Arab countries run on common operating systems that businesses and consumers widely use in other parts of the world. This means that any computing problem faced by the Arab gas and petrochemical industry is likely to spill over to organizations that use similar systems and software.[23]

Finally, consequences of possible cyberattacks need to be evaluated in terms of factors such as reputational damage, financial loss, and possible physical harm. The consequence of a cyberattack is determined by external and internal factors. A firm may suffer more negative consequences if it does not have cyberliability insurance. An even more severe consequence can arise if the jurisdiction of the firm's operations has strict laws against companies' failure to protect PII. To minimize the impacts, firms with higher probabilities of potential negative consequences must take measures such as internal and external information sharing to gain intelligence on fast-evolving cyberthreats, development of threat-specific policies and controls, and enhanced training and workforce messaging to boost CS awareness.

8.4 VARIOUS TYPES OF RESOURCES IN THE CONTEXT OF CYBERSECURITY STRATEGIES

Internal resources and capabilities can provide sources of competitive advantage. Moreover, the meaning and significance of resources may change due to a change in the environment. As noted earlier, executives consider cyberinfrastructure to be an extremely important factor to an organization's growth. In order to develop superior information capability, organizations may treat information as a strategic resource. The increasing importance of IT infrastructure requires enhanced CS measures to protect valuable information. Table 8.1 presents contrasting perspectives regarding the development of resources in the context of CS.

8.4.1 Human Resources

An organization's human capital resources include the CS-related experience, judgment, and intelligence of the managers and workers. Development of CS-related skills should be undertaken as a company-wide initiative that involves all departments. Most importantly, it is necessary to develop a CS mindset among employees.[24]

In some cases, organizations' employees tend to be a source of weakness rather than strength from a CS standpoint. According to the US FBI, insider cyberattacks from current and former

Table 8.1 Resources in the context of CS: Contrasting practices and perspectives

Resources	Encouraging Examples	Concerning Examples
Human	• 2019: JPMorgan had 3,000 dedicated workers for CS.	• Home Depot's top management's response to its CS team's demand to strengthen CS measures: "We sell hammers."
Organizational	• The Retail Industry Leaders Association, a US trade group that promotes public policy advocacy, announced the appointment of a new head of CS.[25]	• Most organizations lack mechanisms to ensure that permanent and temporary employees and third parties have appropriate access to data and that they comply with data-protection regulations.[27]
Market-based	• Following the enactment of the CCPA, which became effective July 1, 2020, an increasing number of companies were reported to be buying cyberinsurance.[26]	• The Danish shipping company Maersk and the US multinational courier delivery services company FedEx did not have cyberinsurance when they faced NotPetya malware attacks in 2017.[28]

employees are among the biggest threats organizations face and have risen in recent years. Such attacks have led to "several significant FBI investigations." Some current and former employees are found to use their access privileges to destroy and steal data, obtain confidential information about customers, and engage in frauds using customer accounts.[29]

8.4.2 Organizational Resources

Organizational resources involve structure, systems, and the relations among groups within the firm. This issue is extremely relevant because organizational rather than technology issues are becoming increasingly important in successfully protecting and defending digital assets. CS practitioners have emphasized the importance of building a "human firewall," which entails clearly defined CS responsibilities for every employee through awareness and training.[30]

8.4.3 Market-based Resources

Two types of market-based resources have been identified: relational and intellectual.[31] Relational assets are associated with external stakeholders, such as perceptions held by customers (e.g., brands, reputational capital, and customer loyalty), channel members, and other external organizations. Intellectual assets are related to a firm's knowledge about its competitive environment.

8.5 MAPPING CYBERSECURITY RESPONSES TO POTENTIAL CYBERTHREATS

Organizations are vulnerable to a wide range of cyberthreats. It is important to understand the cyberterrain, which is "the sum of all of operational assets, security controls, data assets, and overall decision-making within an organization."[32] Organizations must develop a high degree of visibility of the entire cyberterrain in order to defend it properly. For instance, high-value and critical information assets should be protected by higher levels of CS efforts.[33] Likewise, special efforts must be made to protect information assets that have a high degree of vulnerability because such vulnerability must be remediated. Similarly, certain parts of the cyberterrain may need higher CS capabilities.

In making strategic CS decisions, knowing the nature of threats facing the organization is critical. States are among the most prominent cyberoffenders. For most firms, it may make no sense to take preventive security measures against cyberthreats coming from nation-states or groups that have access to tools with the same level of sophistication as tools used by countries. For instance, according to the FBI, almost no security measure would have stopped the type of cyberattack that was launched against Sony Pictures in November 2014.[34] The cyberthreats associated with amateur criminal gangs, who use readily available well-known tools, or insiders focused on selling corporate intellectual property rights, on the other hand, can be minimized.

An understanding of the adversary's likely motivations, tools, tactics, and practices can help design measures to boost cyberdefense and resilience. We divide the threat sources into two categories: sources with a high degree of capability, commitment, and competence (e.g., state actors) and sources with a low degree of capability, commitment, and competence. Table 8.2 proposes strategies for managing vulnerability for various categories of threats and consequences. The four cells in table 8.2 indicate the levels of resources to be devoted. Regarding the basic CS measure (cell I), CS experts emphasize the importance of having a CS framework,

Table 8.2 Strategies for managing vulnerability for various categories of threats and potential consequences

	High Degree of Capability, Commitment, and Competence	Low Degree of Capability, Commitment, and Competence
High degree of severity of consequences	[IV] Level 4: Strongest possible CS measures (China's PLA, Russia's Federal Guard Service, Lockheed Martin's most sensitive data)	[II] Level 3: Very strong CS measures (JPMorgan, Citibank, Petrobras)
Low degree of severity of consequences	[III] Level 2: Strong CS measures (WeChat users who do not face significant threat)	[I] Level 1: Basic CS measures (most ICT users)

identifying external and internal cyberthreats, making employees aware of their CS role, managing CS oversight, and developing a response protocol in case of a breach.[35]

For firms dealing with sensitive data, basic CS measures are not sufficient. Since a data breach can lead to severe consequences (cell II), strong technological and organizational measures need to be in place. Cyberthreats facing the financial sector are an issue of critical importance. As an example, in his annual letter to shareholders in 2014, JPMorgan's CEO said that hackers' efforts to attack the bank's computers were becoming "more frequent, sophisticated and dangerous." In a worst-case scenario, in which cyberattacks delete records, drain account balances, and freeze networks, the resulting shocks will have a severe impact on the economy that will be on the scale of the attacks of September 11, 2001.[36]

In order to deal with an enemy with a high degree of capability, commitment, and competence, when the consequences are less severe (cell III), non-technological rather than technological factors should be the priority of CS measures. This is because it is very difficult or impossible to match the adversary's technical resources. Consider WeChat, which is a free messaging and calling app developed by China's Tencent. Some refer to WeChat as "China's equivalent of PRISM."[37] While WeChat use represents a severe threat for some activists, there are no severe threats to most users.

Governments and organizations with highly sensitive information are more likely to be targeted by attackers with a high degree of capability, commitment, and competence. Moreover, data breaches are likely to lead to severe consequences (cell IV). Following the first principle of CS, if an organization is likely to suffer extremely high consequences due to some types of data breach, extreme caution should be exercised when protecting such data. The most obvious mechanism for reducing cybervulnerability is to use non-ICT communications or use systems that are not connected to the Internet. To take an example, in 2013, Russia's Federal Guard Service, the equivalent of the US Secret Service, reportedly ordered typewriters due to CS concerns. The typewriters are used to produce hard-copy secret documents. This was reportedly in response to leaks of secret information by the National Security Agency (NSA) whistleblower Edward Snowden. A tender to purchase 20 electric typewriters for about US$15,000 was published on the government's procurement website zakupki.gov on July 3, 2013. The Federal Guard Service is responsible for ensuring the protection of top officials, including the president of Russia.

CS breaches associated with a given threat source may lead to different consequences across various targets. As noted earlier, while consequences associated with WeChat use are not severe for most users, some activists faced severe consequences. It was reported that two monks in China's Tibetan Autonomous Region were arrested in 2013 for posting pictures of self-immolation protests to WeChat. One of them was sentenced to a six-year jail term and the other was likely to get life imprisonment. A monk reported that he taped over his Mac's webcam, took the battery out of his cellphone when he did not want to be tracked, and avoided WeChat.[38]

Likewise, organizations possess various categories of data and information that differ in terms of sensitivity and therefore need to be protected by different CS measures. For instance, in Petrobras, scientists and others with access to data are prohibited from transferring critical data, such as seismic studies of the company's oil reserves, through the Internet.[39]

Obtaining an understanding of why and when organizations face a cyberattack is an obvious first step in taking precautions. For this purpose, we further categorize the cyberthreats associated with offenders with different levels of capability, commitment, and competence in terms of time-variance and in terms of the offender's perception of the nature, position, or actions of the target organization. Regarding the question of when, various categories of cyberattacks have been divided into two categories based on their time variance: time-invariant and time-variant. (See table 8.3.) An organization or an individual acting against the interest of a potential offender may become a target of revenge cyberattacks from the offender. For instance, as noted earlier, the Ethiopian government (offender) used spyware to launch cyberattacks against dissidents (target) living abroad.[40]

From a cybercriminal's standpoint, it is also important to emphasize the time-variant nature of attractiveness evaluations. (See table 8.3.) For instance, the 2013 cyberattacks on Target occurred during the peak holiday shopping season and during peak customer visit times. As noted earlier, Target had disabled some security features during the period to deal with high traffic flow.

Some cyberattacks and actions of some categories of cyberoffenders do not depend on time. For instance, many national governments and other highly organized groups constantly engage in cyberespionage activities against sensitive information networks. A *National Post*

Table 8.3 Classification of cyberattacks based on the nature and time variance of threats

	State Actors	Non-state Actors
Time-variant threats (change over time)	▪ State actors with a goal to spread disinformation become more active during high-profile events such as the US presidential election.	▪ Cyberattacks on Target occurred during the peak holiday shopping season and peak customer visit times during that season.
Time-invariant threats	▪ This includes national governments and other highly organized groups' cyberespionage activities. An Iranian hacking network created a fake news organization with fabricated journalists and attempted to interact with military, government, and diplomatic officials over Facebook and other social media sites.	▪ Most cybercriminals constantly look and wait for the right opportunity to attack a target.

article, citing iSight, reported that a three-year campaign of an Iranian hacking network, which was believed to be associated with the government, used social networks to develop friendships with US lawmakers, defense contractors, and a general in an attempt to extract data from them. The espionage group allegedly created a fake news organization with fabricated journalists and attempted to interact with some 2,000 military, government, and diplomatic officials over Facebook and other social media sites.[41] The hackers used 14 personas to make connections, 6 of which appeared to work for a fake news site, NewsOnAir.org. Eight personas purportedly worked for defense contractors and other organizations.[42] Likewise, many cybercriminals constantly look and wait for the right opportunity to attack a target. According to a report from Verizon, 83% of victims were targets of opportunity.[43]

As to the question of why certain targets are selected by cyberoffenders, we divide the causes of attacks into actions taken by organizations and their positions. (See table 8.4.) The cyberattack on Sony Pictures in November 2014 is a high-profile example of a cyberattack associated with actions (making *The Interview*) taken by an organization (Sony Pictures). CS experts linked the hacking group involved in the attacks, Guardians of Peace (GOP), to the North Korean government.[44] The malware reportedly contained Korean language code.[45] Analysts believed that the attacks could have been in response to Sony's movie *The Interview*, the storyline of which involved a plot to assassinate North Korea's leader, Kim Jong-Un.

State actors often pursue specific organizations due to the position or information held by the targets. For instance, hackers allegedly working for the Chinese government were believed to be behind the 2015 hack of the US Office of Personnel Management. The attack exposed personal information of 22 million current and former federal employees.[46] The breached information was not sold on the dark web, which indicated that financially motivated cybercriminals were unlikely to have been the attackers.

Some employees, viewing actions taken by their organization as unfair, may respond by engaging in cyberattacks against the organization. Some organizations have realized that they

Table 8.4 Classification of cyberattacks based on the nature of threats and related to actions/positions of the target

	State Actors	Non-state Actors
Due to actions	▪ The GOP's cyberattacks on Sony Pictures by the North Korean government.	▪ Insider cyberattacks (disgruntled former employees)
Due to the position or information held	▪ Organizations with high symbolic significance are more likely to face cyberattacks. Hackers allegedly working for the Chinese government were behind the 2015 hack of the US Office of Personnel Management.	▪ Financially motivated criminals interested in attractive information

require special processes and practices in order to protect themselves against potential cyber-attacks from disgruntled employees.

In other cases, an organization's position is likely to attract the attention of attackers. Financially motivated criminals, for instance, are more interested in attractive information such as banking credentials or sensitive data.[47]

8.6 CHAPTER SUMMARY AND CONCLUSION

Organizations have been undergoing significant transformation due to internal and external pressures. The discussion in this chapter suggests that there may be a need to revisit and update the way data are produced and collected. Some types of data and information may constitute a new type of liability for organizations.

There is no such thing as foolproof CS and providing 100% security is neither realistic nor feasible. What is important is having all possible safeguards in place and following appropriate processes to strengthen CS. It is important to document and justify CS measures and steps taken to choose the measures. A high priority must be given to protecting the most valuable assets.

Cyberthreats facing a firm are far from static. This means that periods of high threat would require the deployment of more resources to keep the system safe. Organizations may also need to utilize market-based resources such as VDNs during such periods. For instance, Target could have minimized the chance of successful attacks on its system by utilizing cloud services providers (CSPs) more effectively instead of disabling some security features.

It can be taken as axiomatic that organizations should consider the CS-related consequences and implications of actions in their decision-making. But in practice it may not always be possible to predict such consequences and implications. Nonetheless, a framework for cybervulnerability analysis based on systematic examination of political, social, economic, and demographic factors facing the firm; internal factors such as software and hardware used in the company; as well as factors such as security mechanisms of third-party vendors would help managers and practitioners take appropriate CS measures. Overall, while complete prediction and prevention may not be achievable, organizations can benefit from some degree of proactivity.

Industries and organizations differ widely in terms of the threats and vulnerabilities faced and their degree of risk tolerance. If vulnerable organizations take protective action, organizations in (formerly) less vulnerable positions will likely be targeted. That is, the relative position of an organization in terms of its attractiveness as a cybercrime target is also a function of CS measures taken by other organizations.

Organizations are expected to establish effective data privacy and consumer information security systems and practices. Nonetheless, in most countries, no specific CS laws provide clarity and definitive guidance on identifying the steps and procedures needed to establish such systems and practices. Moreover, the cyberattacks and cyberspying activities of governments in China, the US, and other countries illustrate that measures taken by governments to strengthen national CS are complex and frequently have conflicting effects and unanticipated consequences on businesses' CS initiatives.

A related point is that, most often, firms need to exceed the regulatory requirements or auditing standards that are generally accepted. For instance, the process by which banks audit their IT systems is an area that needs improvement. Often it is not possible for auditors to cover an entire banking application in a single audit. They thus rely on random sampling to assess a bank's CS profile. A problem is that auditors do not vary their sampling techniques to ensure a wider CS test. For this reason, it is important for banks to exceed rather than just comply with an IT audit.[48]

While CS measures taken by organizations discussed in this chapter are good signs, more remains to be done. Many organizations' CS strategies (e.g., Home Depot) reflect the misguided belief that strong CS measures are needed only for businesses that exhibit a high degree of digitization. It is clear that no business sector or industry is immune from cyberthreats. While many companies have become "unwilling pawns" in cyberwarfare, the case of the hack on Sony Pictures illustrates that companies can also become the direct and main target of cyberwarfare.

8.7 DISCUSSION QUESTIONS

1. Why is CS becoming an increasingly important part of corporate strategy?
2. According to Accenture, how do "leaders" differ from other organizations in terms of CS performance?
3. Why do companies face more pressure from regulatory agencies to strengthen CS practices?
4. How do SMEs perform in the development of CS strategies?

8.8 END-OF-CHAPTER CASE: CYBERSECURITY STRATEGY OF PETROBRAS

Petróleo Brasileiro SA (Petrobras) is Brazil's state-controlled oil company. It is the largest company in Brazil. In the 2019 *Forbes* global list of the 2,000 biggest public companies, Petrobras ranked fiftieth overall and as the ninth biggest company in the oil and gas operations industry.[49] Due primarily to some high-profile cyberattacks in recent years, it has taken several initiatives to strengthen and improve CS.

Thanks to intensive research and investment of billions of dollars, Petrobras has been a global leader in deepwater drilling and oilfield development technologies. The company's key strength is especially in exploiting pre-salt reservoirs. In addition, Petrobras leads the development of many major projects that are expected to contribute to a new production of about 2 million barrels per day of oil to the global market over the next decade. A key challenge has been to ensure that its intellectual property and other business secrets are secure.

8.8.1 Cyberattacks against Petrobras and Other Major Brazilian Companies

There have been several major cyberattacks against Petrobras. In 2011, LulzSecBrazil, a Brazilian component of a black hat computer hacking group LulzSec (Lulz Security), released personal information including identification numbers and bank details of Petrobras employees.[50]

In 2013, Petrobras and other major Brazilian companies felt a big jolt following the former CIA employee and whistleblower Edward Snowden's revelations of the details of the US NSA surveillance programs. According to Snowden, the NSA hacked into Petrobras's network.

Then president Dilma Rousseff argued that if the allegation of the NSA's breaking into Petrobras computers were true, then gathering economic information would be the motive.[51] The networks of the Ministry of Mines and Energy were also hacked, which, together with Petrobras, was involved in the auction of oil fields.[52] Specifically, many Brazilians thought that the company's data on Brazil's offshore oil reserves and plans for allocating licenses for exploration to foreign companies were the intended targets.[53]

In September 2013, Brazil's Congress opened an investigation and questioned oil industry regulator Magda Chambriard on whether the alleged spying by the NSA could have given US companies an edge in bidding for offshore production rights to be auctioned in October 2013. In 2012, the Dell SecureWorks Counter Threat Unit (CTU) research team tracked a cyberespionage campaign that targeted high-profile oil companies in a number of countries, including those in Brazil.[54]

In general, major Brazilian companies face high-profile cyberattacks. In one instance, the supervisory control and data acquisition (SCADA) systems of power grids were hacked, which resulted in a loss of access to power for about a quarter-million people for almost twenty-four hours. Note that SCADA networks use hardware, software, and applications to keep operations of critical infrastructures running. The Brazilian government has taken some initiatives to minimize the risk to the country's SCADA systems. Between 2008 and 2014, the Brazilian government set up several critical infrastructure technical groups to review the SCADA systems of Petrobras and other major organizations.[55]

Major cyberattacks against global oil and gas companies also served as eye-openers for Petrobras and other oil companies. Among the most notable were the 2012 cyberattacks against Saudi national oil company Aramco and Qatar's liquid natural gas company RasGas.

8.8.2 Security Systems and Policies

Petrobras has exemplar security systems and policies in place. Among the company's more than 86,000 employees, 3,000 worked in CS in 2013.[56] The IT software solutions general manager of Petrobras, Marcel José Kaskus, noted that the company made some use of private clouds but not public clouds. He expressed concerns regarding the security of clouds. Due to CS concerns, Petrobras also prefers to deploy proprietary systems and Brazilian-developed software. The company's sensitive data are stored in a highly secure processing center, access to which is controlled with biometrics such as thumb readers, weight monitoring, and video surveillance. It was also reported to be studying and piloting holography and 3D motion capture.[57]

Following Snowden's revelation that the company had been targeted by the NSA, Petrobras announced heavy investment in CS. In 2013, the company planned to spend US$1.8 billion in CS. It further announced another US$9.5 billion in CS investment by 2017.[58]

Sensitive data and activities are kept offline. For instance, scientists and others are prohibited from transferring critical data, such as seismic studies of the company's oil reserves, through the Internet.[59]

8.8.3 Precautionary Measures

Moreover, a large number of potentially malicious emails are blocked. For instance, of the 195 million emails sent to the company's systems from August 9 to September 9, 2013, only 16.5 million reached their destination.[60] Likewise, following the WannaCry ransomware attack that started in May 2017, as a precautionary measure, Petrobras disconnected its computers from the Internet.[61]

8.8.4 Case Summary

Petrobras is an illuminating example of a firm with strong CS measures. It has mobilized human, technical, and financial resources in effective ways to strengthen CS. Brazilian policymakers have expressed fear that Petrobras has been an industrial espionage target of economically motivated foreign perpetrators. Ironically, the truth or falsity of such claims is less relevant than the fear itself, which led to a clear and dramatic response to improve CS. Many of its CS initiatives can be considered a direct response to cyberattacks targeting the company as well as those launched against Brazilian organizations and firms in the energy sector.

CHAPTER NINE

Cybersecurity and Marketing: Illustration of Advertising and Branding

CHAPTER PREVIEW

CS is emerging as an issue of central concern in marketing. This chapter provides a critical evaluation of how cyberattacks and CS are related to some key marketing issues. First, the chapter examines the mechanisms by which cyberattacks may affect a firm's brand equity. It also looks at the contexts and processes associated with advertising frauds, especially the click fraud industry. The chapter then analyzes some measures that firms facing hacks and click fraud can take to minimize negative effects on brand equity and reduce the waste of advertising budgets.

9.1 INTRODUCTION

CS is emerging as an issue of central concern in marketing. To understand the significance of this, we can consider the effect of cyberattacks on brands. Surveys show that companies facing or experiencing cyberattacks are likely to suffer a loss in brand equity. According to Ponemon Institute's study released in January 2012 that was sponsored by Experian's Data Breach Resolution, the average loss in the value of brand equity due to a CS breach ranged from US$184 million to more than US$330 million or 17% to 31% of the brand's value.[1] The average value of brand equity for the participating organizations was estimated at US$1.5 billion. Unsurprisingly, the main motivation behind organizations' IT security spending has been to protect reputation and brand. The *Vormetric Data Threat Report*, which was based on a survey of retailers, found that the highest proportion of respondents (55%) considered reputation and brand protection to be among the key priorities for this CS spending. (See In Focus 9.1.) According to the report, 89% of retailers felt vulnerable to data threats and 51% had already experienced a data breach.[2]

IN FOCUS 9.1: ZOOM'S SECURITY DEBACLES LEAD TO CUSTOMER DISTRUST AND REPUTATION LOSS

Security analysts had been talking for a long time about security issues with the video-conferencing service Zoom. For each call, Zoom randomly generates ID numbers of between 9 and 11 digits that are used by participants to join a meeting.[3] In January 2020, CS research company Check Point Research reported that it found ways to predict valid meetings about 4% of the time. It also reported that it was able to join some meetings.[4]

Prominence of CS Vulnerabilities during COVID-19

The issue came to prominence during COVID-19 when its users were growing rapidly. Many instances were reported in which Zoom's weak CS gave rise to a form of harassment, known as "Zoombombing," in which intruders could gain access to video calls. They were reported to harass participants and post hate.[5] Many Zoom users were furious to discover that the app's iOS version was sending analytics data to Facebook. This was the case even if Zoom users did not have a Facebook account. Worse still, the company and its privacy policy did not make it clear that this data sharing would happen.[6] A report by the University of Toronto's Citizen Lab detailed a number of shortcomings in Zoom's security; some of them made it particularly vulnerable to China. The report noted that among the weaknesses in Zoom's encryption scheme, it used Chinese servers to route some encryption keys. Additionally, the company is allegedly vulnerable to pressure from the Chinese government due to the company's ownership structure and its reliance on China for labor and other resources.[7]

Customer Mistrust and Dissatisfaction

Certain CS researchers called the Zoom app "malware" and the company "a privacy disaster" and "fundamentally corrupt."[8] The security flaws and negative media publicity led to customer mistrust and dissatisfaction. Companies such as Google[9] and Standard

Chartered banned Zoom from company use.[10] Analysts noted that more secure alternatives that are encrypted include TeamViewer, Cisco Webex, and GoToMeeting.[11] Zoom suffered significant drops in stock price as a result of the news about its security flaws.[12]

Target of Lawsuits and Investigations

Zoom became the target of several lawsuits and investigations. It drew scrutiny from New York's Attorney General's office.[13] A lawsuit filed in the Northern District of California accused Zoom of failing to disclose to its users that the company was sharing their information with third parties. An additional accusation was that its apps lacked proper security features.[14] Another class action lawsuit was filed by Zoom's shareholders accusing Zoom of failing to disclose that its services lacked end-to-end encryption.[15]

A California resident also sued Facebook and LinkedIn for allegedly "eavesdropping" on Zoom users' video conferences.[16] According to the lawsuit, Facebook deceptively created detailed profiles of users who installed the Zoom app. Facebook benefited by being able to effectively target advertising to the users. Zoom was also able to target users more accurately for additional services the company provided in addition to being able to convert users to paying customers. Likewise, LinkedIn used the Zoom profile to effectively target users in order to sell its LinkedIn Sales Navigator. The tool cost US$780 per year in April 2020.[17] It allowed users to integrate the LinkedIn Sales Navigator app with the Zoom platform.[18]

Zoom's Response

Zoom was forced to take several actions to protect its reputation. In a blog published on April 1, 2020, the company's chief executive apologized for the company's failure to provide a high level of privacy and security.[19] The company also announced that it had hired a former Facebook chief of security to provide consulting services, particularly those involving privacy and security issues related to sensitive data and information.[20] Additional measures included teaming up with Luta Security in a "bug bounty" program.[21] Such a program offers recognition and compensation for those who report security vulnerabilities.

In addition to the effect on brand equity, which is often long-lasting, other marketing activities are also affected by cybercrime through various mechanisms. Juniper Research estimated that ad frauds cost online advertisers US$19 billion worldwide in 2018. A survey found that 78% of respondents were concerned about ad fraud and bot traffic.

According to a survey conducted by advertising agency MyersBizNet, 78% of respondents were concerned with ad fraud and bot traffic.[22] Illegitimate clicks on pay per click advertisements have rekindled debate about the effectiveness of online advertising. Cybercriminals involved in diverse activities are expanding their operations into lucrative businesses in the search advertising industry. Search engine network partners, competitors, and unhappy employees are receiving financial and psychological benefits by generating illegitimate clicks.

Traditionally, managing cyberrisks has been the responsibility of chief information officers (CIOs), chief information security officers (CISOs), and chief technology officers (CTOs).[23] Compared to some other departments, marketing currently plays a less prominent role in managing CS-related risks. According to a survey by Deloitte, only 22% of consumer product companies included chief marketing officers (CMOs) on post-breach incident-response teams.[24] The above discussion indicates that the CMO's role is no less important in managing such risks. It is critical for marketing teams to collaborate with the IT department to ensure that marketing activities do not suffer from cybercrimes and do not create security risks and vulnerabilities. It is thus imperative for marketing teams to work with the CISO to audit solutions before using them.[25]

In this chapter, we offer a detailed description of the aforementioned problems. We examine the mechanisms by which cyberattacks may affect a firm's key marketing activities. We also look specifically at cyberattacks' effects on brand equity and the contexts and processes associated with the click fraud industry.

9.2 CYBERSECURITY AND BRAND EQUITY

We start this section with the definition and determinants of brand equity. *Brand equity* is a brand's commercial value that is attributable to consumer perception of the brand's products or services. Recent surveys show that consumers show more negative perception of brands that have had consumer data compromised due to having been hacked.

What mechanisms lead to adverse changes in consumer brand perception associated with poor CS performance or a CS attack? To answer this question, we consider two key determinants of brand equity: *perceived quality* and *brand loyalty*.[26]

Perceived quality measures how consumers judge the value and attractiveness of a product or service. Successful cyberattacks can redefine quality perceptions.

It is generally understood that when businesses sell a product, they actually sell trust. There has been increasing use of software as part of products that consumers buy.[27] For purely digital

Table 9.1 Findings of recent surveys regarding the importance placed by consumers on organizations' CS measures (2019)

Survey Conducted by	Major Findings
Ping Identity	• 78% of consumers said that they would not engage in online activities with a brand if it experienced a cyberattack.[28]
Gemalto	• 66% of consumers said that they were unlikely to do business with a firm that experienced a breach involving their sensitive information.[29]
PwC	• Over 90% of consumers believed that organizations need to be more proactive in protecting their data.[30]
Morphisec	• About half of the respondents noted that a hotel's cyberdefense measures influenced their decision to stay at the hotel.[31]

products as well as physical products that include software, trust is tightly linked to how secure the software is. Among *Fortune* 500 CEOs in 2017, 71% claimed that they were running technology companies.[32] Products containing insecure software can be perceived by consumers as untrustworthy. A well-known recent example of the increasingly prominent role of CS on brand equity is the 2015 Jeep Cherokee. In this example, two researchers remotely changed the setting of the climate control system, turned on the radio and windshield wipers, and stopped the accelerator. A *Wired* article asserted, "Automakers need to be held accountable for their vehicles' digital security." The lead researcher noted, "If consumers do not realize this is an issue, they should, and they should start complaining to carmakers. This might be the kind of software bug most likely to kill someone."[33] This example illustrates the basic idea: high quality can only be achieved if reasonable CS is built in.

Brand loyalty is related to consumer tendency to repeatedly choose one brand over alternatives. Firms experiencing CS breaches may observe a decrease in loyalty. As surveys reported in table 9.1 indicate, consumers are increasingly concerned about organizations' data protection measures. Moreover, they are less likely to do business with a company if it fails to protect their sensitive data.

In the hospitality industry, a survey conducted among 1,000 consumers by Israel-based CS solution provider Morphisec in 2019 found that 69% of travelers felt that the hotels they booked did not invest enough in CS. About half of the respondents noted that a hotel's cyberdefense measures influenced their decision to stay with the hotel.[34]

9.2.1 Suffering Brand Image Erosion due to Cyberattacks

Table 9.2 shows businesses that experienced negative brand effects following cyberattacks. One widely used indicator is the "Buzz score" developed by the polling site YouGov BrandIndex.[35] The BrandIndex Buzz score measures brand popularity, which ranges from 100 to –100.

Table 9.2 Some firms facing cyberattacks and the effects on brand reputation

Organization	Brief Explanation of Cyberattacks	Some Examples of Impacts on Brand Equity and Related Indicators
Sony PlayStation	▪ *2011*: Hackers extracted PII of over 77 million people.	▪ Buzz score fell from 5 the week before to -14 one day after the announcement.[36]
Anthem (Blue Cross Blue Shield)	▪ *2015*: 80 million records were breached, including customers' and employees' Social Security numbers, birthdays, addresses, email, employment information, and income data.[37]	▪ The proportion of consumers who thought that Anthem Blue Cross Blue Shield was a better brand than other insurers decreased from 51% before the breach to 45% afterward.[38]
Target	▪ *December 2013*: Target faced a high-profile security breach.	▪ Buzz score for the week before the announcement of the data breach was 26. The score dropped by 35 points to -9 on December 20, 2013. By December 23, the score reduced to -19 or a drop of 45 points.
TalkTalk	▪ *October 2015*: Hackers accessed the details of 150,000 customers and 15,000 bank accounts.[39] The hack took place on October 21.	▪ Reputation score declined from -10 to -34 by 26 October. Recommend score fell from -7 to -39. Buzz score in a two-week period following the hack fell from -1 to -50.[40]
Yahoo	▪ *September 2016*: Yahoo reported that hackers may have stolen information of at least 500 million users in 2014.[41]	▪ A day after the announcement, there was a 474% increase in online mentions of Yahoo compared to a week earlier. 70% of the mentions were negative.[42]

A company's score is calculated by subtracting negative feedback from positive feedback. Surveys have shown how companies who experienced high-profile cyberattacks in the past few years, such as Target, Sony PlayStation, Sony Pictures, Anthem, and TalkTalk, have had significant declines in their Buzz scores. The case of Sony Pictures illustrates that firms facing cyberattacks are likely to perform poorly in terms of their relative standing with respect to competitors and experience negative press.

Brand loyalty is an important aspect of brand equity. Firms suffering from cyberattacks will likely suffer a decline in loyalty. In the three months following a hack, TalkTalk lost 101,000 customers, and 95,000 of them were estimated to be a "direct result" of the cyberattacks.[43]

Reliability is also linked to perceived service quality. Consumers may question the reliability of services provided by a firm if breaches of sensitive information occur. About 20% of those leaving TalkTalk cited "poor reliability" as the reason compared to less than 1% in the previous quarter.

In some cases, the damage to the reputation of a company suffering from a cyberattack could last for a long time. A KPMG study conducted with 448 consumers indicated that 19%

would stop shopping with a retailer that had been a cyberattack victim even if the company took measures to fix the issues.[44]

9.3 PPC ADVERTISING AND CLICK FRAUD

Cybercriminals enjoy a higher ROI from ad frauds than from most other cybercrime schemes. Consider this example: CS research company White Ops reported that it uncovered a gang that arguably ran the "largest and most profitable ad fraud operation." The gang created a fictitious advertising firm and promised advertisers that their ads would be hosted on websites of reputable companies like Fox News, ESPN, and CBS Sports. However, the ads were delivered on fake websites they built that no real person would visit. A sophisticated botnet consisting of 500,000 different US IP addresses was deployed to view the ads. The bots were programmed so that they would be active only during the daytime. They appeared to be using Chrome on a Mac. They also had fake Facebook accounts. To advertisers checking their metrics, the clicks seemed to be coming from real people. The gang was estimated to make about US$180 million in two months.[45]

Table 9.3 presents some indicators related to click fraud. Fraudulent clicks as a proportion of total clicks as presented in the third column are substantial. Automated scripts or computer programs are used by click fraudsters to generate fake clicks that imitate legitimate users. Networks of human clickers that engage in click frauds are reported to operate from countries all over the world.

Table 9.3 Indicators related to click fraud

Time	Study Conducted by	Fraudulent Clicks as a Proportion of Total Clicks	Remarks
2014[46]	• CS firm White Ops for the ANA	• One-fourth of online video ads and 11% of display ads were fraudulent.	• 36 advertisers such as AIG and MillerCoors were studied.
2017[47]	• Data service provider AdMaster (conducted in China)	• One-third were generated by click farms.	• They used powerful computers to control a large number of cellphones. They also installed malware in other smartphones and built botnets to perform click frauds.
2018–2019	• Ad verification company Cheq	• 77% of US ad traffic is "sophisticated invalid traffic."[48]	• The traffic uses malicious methods that are advanced and sophisticated.
2020[49]	• Madison Media (conducted in India)	• About a quarter of clicks were fraudulent.	• 78% of marketers considered click fraud to be their top concern.

According to the Association of National Advertisers (ANA), which represents the marketing community in the US, about half of all click fraud activities were associated with publishers that pay for third-party traffic providers.[50] Some website owners have reportedly formed international networks to click on ads on each other's sites. A *Washington Post* article claimed that one such network, Mutualhits.com, had over 2,000 members.[51] Click fraud is also associated with and facilitated by parked sites, which have little or no content except for ads supplied by search providers.

In the beginning, click fraudsters mostly targeted businesses in the US and other industrialized countries. However, click fraud's footprints across the world are getting bigger. A study found that the Indian subcontinent's mobile ad fraud in 2017 was 31.9% compared to global ad fraud of about 15%.[52]

9.3.1 The Cost-benefit Calculus of Activities Related to Click Fraud

This section provides the cost-benefit calculus for a click fraudster, a PPC provider, and an advertiser.

9.3.1.1 The Click Fraudster

For a click fraudster, benefits may include monetary gains and returns as well as psychological benefits such as enjoyment. Similarly, there are monetary as well as psychological costs, which are functions of the probabilities associated with the detection of the fraud and those of fines, arrest, and conviction.

9.3.1.2 The PPC Provider

A PPC provider's revenue is directly related to the amount of PPC advertising sold. Online advertising providers do not charge advertisers for clicks that are identified as invalid by their detection system. They thus directly benefit from valid clicks and may have a vested interest in labeling more clicks as valid ones. Thus, in the short run, PPC providers such as Google and Yahoo benefit from frauds committed by their affiliates. However, if a PPC provider is associated with frauds, it may suffer a decline in reputation and bear a cost in the form of fewer opportunities to serve advertisers in the future.

9.3.1.3 The Advertiser

Advertisers like to receive high-quality clicks on their ads at the lowest cost. Advertisers' difficulties in measuring and benefiting from Internet ads can be explained in terms of behavioral and technological factors. The first problem is that of an imperfect relationship between click-through rates and conversion rates – that is, a visitor to a website who clicked on an ad does not necessarily buy the advertised product. The second problem concerns a lack of built-in mechanisms to protect advertisers against click fraud.

From the advertiser's perspective, the cost-benefit calculus associated with preventing click fraud involves determining the optimum investment and types of measures. For small companies, identifying fraudulent clicks is a challenge. According to an *Inc* magazine article, tools such as Click Lab, Click Defense, and Click Detective, which are available to identify fake clicks, cost up to several thousand dollars per month.[53] Click fraud is painful for small companies that have small marketing budgets and are forced to accept fraudulent clicks as a cost of doing business.

9.3.2 Hypermediation and Click Fraud

A central feature of the Internet economy is its near zero transaction cost. An emerging body of literature asserts that businesses are undergoing hypermediation as opposed to disintermediation. New intermediaries have emerged to provide various services. The roots of click fraud lie partly in hypermediation, which is an increase in the number of subdistributors. PPC providers do not normally disclose the chain of intermediaries. Identifying them from outside is difficult. To understand hypermediation-led click frauds, consider one detail: an ad of the business cloud communications provider Vonage passed through eight subdistributors and was "illegally" downloaded to users' PCs.[54]

The PPC syndication networks are intermediaries that match advertisers with relevant audiences. In some cases, such networks are a better match than search engine results pages and provide high conversion rates at a low cost. Hypermediation, however, has acted as a fraud generator by bringing potential click fraudsters into the value chain. PPC syndication networks consist of players with different sizes, reputations, and external visibility.

9.3.3 A Click Fraudster's Strategic Elements

9.3.3.1 The Economic Geography: Locations of Click Fraud Operations

It is tempting to employ low-wage workers from developing countries to generate clicks on ads and collect commissions from PPC programs. An indiatimes.com article reported that housewives, college graduates, and professionals in India make US$100–US$200 per month by clicking on Internet ads. The so-called click farms in developing countries also engage in closely related phenomena, such as generating fake likes and followers in social media. *Dispatches*, a program on Channel 4 in the UK, reported that the owner of a Bangladesh-based click farm, who described himself as the "king of Facebook," employed a large number of workers. His employees created fake accounts and used them to generate fake likes for his clients. The workers needed to generate 1,000 likes or follow 1,000 people on Twitter to earn US$1.[55]

In June 2017, Thai police arrested three Chinese nationals for allegedly operating a click farm. They used hundreds of cellphones and 347,200 SIM cards to generate "likes" and views on WeChat. Each person in the gang earned US$2,950–US$4,400/month.[56]

China is also well-known for click farms. AdMaster reported that in a single day it detected over 100,000 phones that accessed a website from the same location in Shanghai. Chinese click farms were reported to charge about US$7 to buy 1,000 clicks for articles published on WeChat.[57]

Deployment of invalid click detection methods by a PPC engine or an advertiser, however, changes the calculus of employing low-wage workers. The low-wage workers face an entry barrier if advertisers and PPC engines activate fraud detection tools such as IP address filtering or geo-targeting and monitoring traffic from unusual locations. For instance, according to Chosen.com, Overture South Korea's "continental cut-off" services block clicks from Africa.

9.3.3.2 Economics of Labor versus Technology and Technological Economies of Scope

Click fraudsters are confronted with the problem of whether to employ the seemingly bottomless source of human clickers in developing countries or technologies. Some analysts assert that click fraud–enabling technologies are developing more rapidly than anti–click fraud technologies developed according to advertisers' reactive decisions.

The use of bots to perform click fraud has become more common in recent years. In such schemes, a large number of computers are infected with malware. The bots use the infected machines to visit a website to click on a PPC ad or watch a video. Botnet-generated clicks come from a large number of geographically distributed computers with unique IP addresses. Algorithms used by PPC providers and third-party auditors, which look for unusual traffic patterns, thus may fail to identify such clicks.

Consumers are duped by free software, games, and pornography. Easycracks.net, the Armenian company, lures consumers by offering free downloads of unauthorized copies of Windows XP and games that also require the installation of ECS International's software, ActiveX controls. When users approve the installation, 16 pieces of adware are downloaded without permission to deliver five pop-up ads per minute.[58]

Economies of scope exist if a technology is used in several activities. A *New York Times* article explained how botnets are used for click fraud.[59] Click fraud has been a lucrative activity for botnet criminals and is used in other activities such as mass distribution of spam, key-logging, identity theft, denial-of-service attacks, phishing, and spyware. Cybercriminals have used botnets to fraudulently increase traffic to online ads and generate false clicks. An article published in *The Register* documented how the KMeth worm targeted Yahoo Messenger users.[60] The worm reportedly directed infected users to a website hosting Google AdSense ads about mesothelioma. Similarly, an *E-Commerce Times* article highlighted how online pornographers are turning to click fraud.[61] Internet users are lured to click on naked pictures, which take them to a website to register a click.

9.3.3.3 Wasting Competitors' Budgets

Economic and psychological benefits are associated with wasting a competitor's advertising budget. In April 2020, the online golf equipment retailer Motogolf.com sued its competitor Top Shelf Golf for click frauds. The plaintiff alleged that the defendant company repeatedly clicked on Motogolf's PPC ads.[62]

Businesses usually limit how much they spend on PPC advertising. Once they reach the limit, search engines stop displaying their ads. Pushing competitors' links off the search sites would help the fraudsters' ads receive a higher priority. Such frauds mainly victimize small businesses with limited budgets. Some users benefit psychologically from wasting a competitor's budget. Psychologists refer to this phenomenon as *enjoyment-based intrinsic motivation*. A chief executive of a marketing company was reported to enjoy clicking on competitors' ads. The executive said that clicking on competitors' ads was "an entertainment."

9.3.3.4 Target Attractiveness

Two observations are worth making regarding click fraud targets. The first observation is that return on click fraud is positively related to the search term price. Site owners of Google's AdSense, Yahoo's Publisher Network, or other contextual networks earn a percentage of the PPC charge for clicks on ads on their sites. While some search terms cost US$0.10–0.15 per click, others cost several hundred dollars. Search terms related to law, medicine, finance, and travel industries are expensive. For instance, the *Inc* magazine article mentioned above stated that for "D.C. Hair Laser Removal," the maximum cost per click was US$146 and the average was US$69 in 2005. According to a 2017 *Inc* magazine article, the six most expensive keywords in 2017 were business services, bail bonds, casino, lawyer, asset management, and insurance.[63] (See figure 9.1.) Companies that buy higher-priced search terms are more likely to fall victim to click fraud.

Second, to avoid detection, click fraudsters are more likely to target companies that buy more terms. As noted earlier, advertising networks and third-party auditors employ various methods to identify invalid clicks. If several of a competitor's different keywords are searched, instead of a single term, a detection method such as rules-based algorithms could label a fraudulent click as legitimate competitor analysis and research.

9.3.3.4.1 Weak Defense Mechanisms Click fraud's main victims are advertisers. It would be erroneous, however, to assume that advertisers are the only victims of click fraud. Note too that

Figure 9.1 Most expensive keywords (PPC, US$, 2017)

Data source: Gabbert, E. (2020, October 14). *The 25 most expensive keywords in Google ads.* WordStream. Retrieved from https://www.wordstream.com/blog/ws/2017/06/27/most-expensive-keywords

weakness of defense mechanism co-varies positively with the likelihood of becoming a crime victim. Consumers are both instruments and victims of click fraud schemes. Naive users' poorly protected devices are especially more susceptible to such schemes. In August 2019, Facebook filed a lawsuit against two Android app developers – Hong Kong-based LionMobi and Singapore-based JediMobi – for engaging in a so-called click injection fraud scheme against Facebook ads. The two defendants allegedly created apps that had malware-like features. The apps were available to download from the Google Play Store. Once a user installed the apps on their smartphones, malicious codes that were hidden in the apps generated fake clicks on Facebook ads.[64]

Internet users in developing countries are attractive targets for botnet-generated click frauds. In these countries, many Internet users are connecting to the Internet for the first time and are not security oriented. Moreover, security products are unavailable in their languages or are unaffordable.

9.3.4 Institutions' Effects on Cost-benefit Analyses of Activities Related to Click Fraud

As presented in table 9.4, institutions can add financial and psychological value or cost to various actors. The first column shows the types of institutions affecting actors' behaviors.

Table 9.4 Institutional mechanisms associated with criminalizing and stigmatizing click frauds

Level of Institution	Mechanisms	Examples
National/state	• Statutes/regulations addressing click frauds • Cybercrime-related rule of law	• The FBI took action to dismantle 3ve.
Industry group, trade/professional association	• Codes of ethics to maintain higher standards of conduct than required by law • Economic exchange-related responses	• Global Alliance for Responsible Media created a plan to fight ad frauds.
Organizational	• Organizational rules, norms, and culture	• Microsoft updated its adware policy, which aims to block unsafe adware.
Individual	• Feeling of guilt/remorse	• Many clickers in developing countries may not know that businesses are victimized.

The second column summarizes mechanisms associated with each source. The third column presents some examples.

9.3.4.1 National/State Level

In nascent sectors such as Internet advertising, there is no developed network of regulatory agencies. Governments, however, are adopting statutes and regulations to deal with click frauds, which would change the cost-benefit equation for fraudsters. Law enforcement agencies are beginning to take a closer look at click fraud, criminalize associated activities, and dismantle click fraudsters' networks. As an example, in 2018, the FBI dismantled the operation of a major online ad fraud outfit that was referred to as "3ve" by the US government. (See In Focus 9.2.)

IN FOCUS 9.2: THE FBI DISMANTLED CLICK FRAUD OPERATOR 3VE

3ve operated one of the most sophisticated click fraud operations. It detected whether computers had installed anti-malware or antivirus software. Such devices were not targeted. For instance, computers in San Francisco were less targeted because they had higher probability of detection due to the concentration of more technologically savvy users. When it found vulnerable computers, 3ve opened a hidden browser and loaded ads on over 5,000 fake websites that it had created.

A fake website would start a video, which would stop in the middle. However, ads delivered on the fake websites would be clicked by bots. The bots mimicked human clickers. A simulated mouse would scroll down the fake websites that were being viewed with hidden fake browsers. An analysis over a 10-day period indicated that 3ve had controlled over 1.7 million unique IP addresses globally.[65] 3ve was estimated to make over US$29 million in fraudulent ad revenue before its operations were stopped.[66]

There were eight alleged perpetrators involved in the operation. In late 2018, the FBI charged three of them, who were located in Bulgaria, Estonia, and Malaysia. Crimes they faced included wire fraud, ID theft, computer intrusion, and money laundering. The remaining five perpetrators were not found.[67]

9.3.4.2 Industry Group, Trade/Professional Association Level

Some coalitions that represent the interests of advertisers are organized around their shared interest in minimizing click fraud. According to the South Korean newspaper *Chosun*, South Korean small businesses established the Online Advertisers Association to voice concerns about click fraud. Such trade associations can direct efforts to mobilize discourses and pressure PPC providers for change in existing technological and institutional arrangements. US advertisers pressured Google and Yahoo to be more accountable and demanded audited numbers and common measurement standards. One way to analyze the actions of these coalitions is to consider them as economic arbiters making decisions about economic exchange. Prior research in economics indicates that institutional arrangements favor organized groups and actors over those that are unorganized. Organized groups thus can play the role of an economic arbitrator more effectively.

In 2019, 17 of the world's biggest advertisers, including Adidas, Mastercard, Procter & Gamble, and Unilever, launched the Global Alliance for Responsible Media (GARM). A key focus area for the GARM is protecting consumers and brands, including controlling ad frauds.[68]

9.3.4.3 Organizational Level

Some organizational-level rules aim to reduce click frauds. For instance, Microsoft updated its adware policy to fight unsafe adware.[69] It aims to block a variety of man-in-the-middle attackers, which capture traffic and then play it back triggering unauthorized actions. One such example is changing consumers' DNS settings. (See the end-of-chapter case: Rove Digital's Victimization of Advertisers.)

9.3.4.4 Individual Level

As noted above, many click farms operate from developing countries. The workers in such farms may just click on ads or like Facebook content to make money and may not know that they are victimizing businesses.

9.4 WHAT CAN MARKETERS DO?

Firms need to introduce adequate CS measures directed toward minimizing the adverse consequences noted above. First, a post-breach resilience plan would repair broken trust. Companies that experience hacks and fail to disclose details are viewed suspiciously as dishonest and untrustworthy. Consequently, they find it difficult to repair broken trust and to rebuild reputation. TalkTalk was accused of a cover-up over its data breach. The company reportedly kept the breach secret for at least a week before it admitted to the scale of the hack.[70]

However, it may be possible to partially repair broken trust.[71] The degree to which broken trust can be repaired depends on the response.[72] So, what actions can a company take to restore consumer trust? Having a well-developed CS communication plan is one. The components of the plan should include detailed information for handling CS breaches, an internal year-long security messaging campaign, a prepared corporate spokesperson response in the case of a breach, and text for website, customer, and social media communications. The website should also have built-in emergency messaging capabilities to alert customers.[73] Another is to be transparent: explain what went wrong. Consider offering mitigation services, e.g., free credit monitoring.

As an advertising medium, part of the fascinating character of the Internet stems from its measurability and instant feedback. In this regard, the basic idea behind the PPC model is simple: from a marketer's standpoint, a genuine click represents the clicker's choice, which provides an opportunity to create and deliver value. Businesses are understandably willing to invest in generating genuine clicks. However, the Internet's measurability is more complicated than first meets the eye.

Advertisers need to be vigilant about click fraudsters' creative ways. Small companies that cannot afford tools to identify fake clicks may look for unusual clicking behavior and regularly track conversion rates to see whether their PPC ads are working. Businesses can also direct efforts toward harnessing the power of consumer reviews, blogs, and other forms of endorsement as an alternative to PPC advertising. Most often, these are less costly, relatively fraud free, and becoming more effective.

9.5 CHAPTER SUMMARY AND CONCLUSION

CS is emerging as an issue of central concern in marketing. This chapter presented an analysis of two main issues that are related to marketing activities.

The observations in this chapter suggest that a firm's CS measures are tightly linked to brand equity. The degree to which a firm's brand equity will suffer following an attack is a function of the level of sensitivity of the consumer information that is compromised, consumers' perception of the organization's CS practices, and a post-breach resilience plan and implementation. Firms facing hacks can take measures to minimize adverse effects on brand equity. As indicated by the KPMG survey, while reputation damage may not be completely repairable, measures can be taken to minimize the damage. Broken trust can be managed and repaired if a company responds quickly and correctly.

Click fraud has been a challenging reality facing the search advertising industry and has threatened the growth of this industry. The presumed infallibility of click measurement is eroding because of massive click fraud schemes. The roots of click frauds lie partly in asymmetric hypermediation consisting of a dense network of organizations on the supply side (such as PPC providers, subdistributors, and affiliates), and thin and dysfunctional institutions to perform trust-producing roles. Regarding the latter point, regulatory institutions such as networks of regulatory agencies to deal with Internet advertising and click fraud and codes of ethics to maintain higher standards of conduct in the Internet advertising industry are at a nascent stage. Various arbiters exert legal, ethical, and economic pressure on PPC providers. Various arbiters' efforts to criminalize and stigmatize click frauds may change the fraudsters' cost-benefit calculus.

Many argue that the actions PPC providers take against click fraud are designed to appease advertisers but should instead be driven by substantive considerations. Well-coordinated, well-funded campaigns can create the perception of infallibility, validity, and reliability of PPC information and reassure advertisers that their ad dollars are effectively spent. We noted in this chapter the emergence of new intermediaries to match suppliers and customers. The increasing pervasiveness of click fraud has also created a compelling need for new intermediaries to provide a third-party measurement system capable of producing trust, thereby addressing some of the concerns.

9.6 DISCUSSION QUESTIONS

1. What roles does marketing currently play in managing CS-related risks? How are such roles changing and why?
2. How is the importance placed by consumers in organizations' data protection measures changing and why?
3. What is the effect on a firm's brand image if it experiences cyberattacks?
4. What can advertisers do to fight click frauds on PPC advertising?

9.7 END-OF-CHAPTER CASE: ROVE DIGITAL'S VICTIMIZATION OF ADVERTISERS

An example of a highly globalized cybercrime firm was Rove Digital (RD), based in Tartu, Estonia. It seemed to be a legitimate IT company. According to a Manhattan federal court indictment in November 2011, an alleged international crime ring associated with RD used malware to hijack more than 4 million computers in over 100 countries. When victims visited certain websites or downloaded software to view videos online, the malware was installed on the computers.[74]

It was one of the biggest criminal-owned botnets. The ring included six Estonians and one Russian and was estimated to generate at least US$14 million in profits. About 500,000 of the hijacked computers were in the US, including those used by educational institutions, non-profit organizations, and government agencies such as NASA. The malware had infected the websites of about half of the *Fortune* 500 companies and at least 26 US government agencies.[75] According to media reports, Vladimir Tshastsin, the alleged ringleader, transferred some of his assets to his father, who was Estonia's 283rd richest person.[76] The law enforcement operations in 2011 led to a seizure of the gang's 150 properties.

9.7.1 RD and Shell Companies

The crime ring was organized and operated as a traditional business but profited illegally. Its subsidiaries included Esthost (a webhosting services reseller), Estdomains, Cernel, UkrTelegroup, and many less well-known shell companies. RD and its shell companies had faced legal problems and received media attention earlier. For instance, Esthost went offline when the San Francisco-based Atrivo, which hosted Esthost's servers, was suspected of engaging in criminal activities. RD was forced to stop the hosting services offered by Esthost.[77] RD learned its lesson. The company expanded its command-and-control infrastructure all over the world. It also moved a significant proportion of the servers from Atrivo to the Pilosoft datacenter in New York.[78]

In 2008, Estdomains also lost its accreditation from the ICANN because Estdomains owner, Tshastsin was convicted in Estonia. Tshastsin was charged by an Estonian court for online fraud, money laundering, and forging of documentation.[79] Despite RD's heavy involvement in cybercrimes, the company continued to operate openly for many years.

9.7.2 The Fraud Scheme

RD's click hijacking scheme started in 2007 and ran until November 2011. The group had 100 command-and-control servers worldwide, including one each in New York and Chicago.[80] The ring also claimed that it ran legitimate online advertising firms and the schemes principally operated through RD.

The "click hijacking" malware changed domain name systems. As a result, users of infected computers were unknowingly redirected to rogue computer servers controlled by the gang.[81] A simple way to understand the DNS system would be to view it as the Internet's "built-in phone book." In order to find a website such as Google, Yahoo, or Wikipedia, a computer reaches out to the DNS to find a numerical address, which is known as the IP address.

The malware also prevented victims from connecting with their antivirus software providers and updating software. Most traditional malware is designed to steal valuable personal information. This scheme was different and thus was not detected for a long time. Experts considered this a very clever tactic because it manipulated the infrastructure of the Web involved in doing one of the most popular activities: display advertising.[82] Part of the problem also lies in the fact that some legitimate companies benefit from such frauds.

The indictment described several examples of cyberfrauds, including two principal strategies: traffic redirection and ad replacement.

9.7.3 Traffic Redirection

The virus altered search engine results so consumers who clicked links of companies such as Amazon, Apple's iTunes, ESPN, IRS, Netflix, *Wall Street Journal*, and other popular websites would be directed to fake sites designated by the gang. When a user searched a term, the search results would normally return a website but the malware would force a redirect to a different website when the user clicked on the link. The indictment cited an example in which, when a user searched for the "IRS" at www.Yahoo.com and clicked on a link for the Internal Revenue Service, the user was redirected to an H&R Block tax-preparation website. Likewise, in searching "itunes," the results would display the official website but would take the user to a different website purporting to sell Apple software but not affiliated with Apple. Users were similarly rerouted to unaffiliated sites when searching for the official websites of Netflix and other companies, according to the indictment.

In addition, users were redirected to websites dealing with illegal products and services, such as those selling fake Louis Vuitton, replica watches, and fake antivirus software. The sites to which users were directed would pay a referral fee to RD.

9.7.4 Ad Replacement

RD also designed mimicked sites, which are doctored websites of legitimate organizations for replacing ads controlled by the hackers. For instance, RD created a fake Amazon site that looked like the legitimate one. When an infected computer was used to visit the Amazon website, the malware would take the user to the fake Amazon site and replace regular ads on the site with other ads from which the criminals would generate revenue through affiliate arrangements.

The indictment cited an example: when users clicked on an American Express ad for the Plum Card on the *Wall Street Journal*'s home page, it was instantly replaced by an ad for "Fashion Girl LA."[83] Likewise, the group swapped legitimate display ads of Dr. Pepper on www.ESPN.com with a vacation timeshare ad.[84] In addition, clicking on these ads would often download malware to the user's computer.

9.7.5 Operation Ghost Click

In November 2011, the scheme was dismantled by US federal agencies with the help of private companies and some universities. The two-year FBI investigation was code-named "Operation Ghost Click." Two data centers in New York City and Chicago were raided by federal agents, who seized computers and rogue DNS servers used by RD.[85] The Federal Court appointed the California-based nonprofit Internet Systems Consortium to replace the rogue DNS servers with clean ones to ensure that the 4 million computers infected by malware that changed the DNS would not lose Internet access.[86] The FBI worked closely with the Estonian Police and Border Guard, the Dutch National Police, and NASA's Office of the Inspector General. Trend Micro tracked the activities of RD and its subsidiaries and helped the FBI. University of Alabama at Birmingham's Spam Data Mine, which contained 550 million junk email messages in its database as of November 2011, was used to analyze activities such as targeted versions of phishing (spear phishing), advertising fraud, and identity spoofing.[87] Other organizations helping in the operations included the DNS Changer Working Group, Georgia Tech University, the Internet Systems Consortium, security firm Mandiant, Spamhaus, and Team Cymru. In February 2012, an Estonian court ruled that Estonia could extradite four persons to the US. The court had made similar decisions on two other persons.[88]

In July 2015, Tshastsin pleaded guilty. In April 2016, the US District Court in Manhattan sentenced him to seven years in prison.[89]

CHAPTER TEN

Cybersecurity in Human Resources Management

CHAPTER PREVIEW

This chapter examines three distinct issues in building CS-ready human capital: (i) attracting, retaining, and motivating a high-quality CS workforce, (ii) enhancing board members' and C-level executives' understanding of key CS issues, and (iii) improving cyberdefense capabilities of the general workforce. It provides guidelines for creating, implementing, and overseeing policies governing employee behavior and the behavior of the organization toward its employees in light of increasing cyberthreats facing organizations. This chapter highlights various types of legitimacy-related pressures faced by organizations from environmental, organizational, and inter-organizational sources that may affect the power and resources that CS professionals possess. It covers issues of awareness, training, and education programs that are needed for board members, C-level executives, and the general workforce to acquire needed CS knowledge and orientation. It also looks at the legitimacy issues that arise in enhancing board members' understanding of key CS issues and in bolstering CS knowledge of the general workforce. Finally, the gender dimension of CS is considered.

10.1 INTRODUCTION

Three distinct issues arise when we consider human resource management (HRM) from the CS perspective: (i) attracting, retaining, and motivating a high-quality CS workforce, (ii) enhancing board members' and C-level executives' understanding of key CS issues, and (iii) improving cyberdefense capabilities of the general workforce and minimizing deviant behaviors. Regarding the first issue, several recent high-profile cyberattacks underscore the need to increase organizations' abilities to defend and protect their networks. One key global trend in organizational structure involves the tendency to create the formal position of CISO. Having the right partnership with the CISO and having governance structures in place will help strengthen the strategic vision related to CS and support CS practices.

Some organizations may have a director of CS instead of, or in addition to, a CISO. A key distinction between a CS director and CISO is that whereas directors often drive day-to-day programs and activities, CISOs are expected to work with department heads, the CEO, and the board. The CISO formulates the CS vision and overcomes problems and hidden difficulties that the CS team may encounter in executing the vision.[1]

Second, CS requires the attention of top management and the board of directors because CS is a matter of enterprise-wide risk rather than a function of IT management.[2] CS can "make or break" a business.[3] Gartner has predicted that by 2024, 75% of CEOs will be impacted if their companies face a cyberattack or data breach. Gartner research vice president Katell Thielemann noted, "In the US, the FBI, NSA and Cybersecurity and Infrastructure Security Agency have already increased the frequency and details provided around threats to critical infrastructure-related systems, most of which are owned by private industry. Soon, CEOs won't be able to plead ignorance or retreat behind insurance policies."[4] Unsurprisingly, CEOs acknowledge that it is important to implement a model that aligns and integrates cyberrisks, exposures, and potential damage to corporate objectives and goals. However, top management is not prepared enough to tackle cyberrisks. For instance, according to a National Association of Corporate Directors (NACD) survey, only 11% of directors said that they had a "high level" of knowledge about CS risks.[5] In another NACD study, over 90% of respondents believed that their board's understanding of CS risks needed to be improved.[6]

Third, the goal of HRM is to make effective use of employees, reduce risks, and maximize ROI. In today's environment, the more awareness, knowledge, and experience in CS that employees have, the greater their value as productive assets. The human factor has been the weakest link in CS. For instance, the first victim of a cyberattack is often an administrative assistant or an accountant. According to the security awareness training provider Dell Secureworks, 90% of all malware infections require some type of human interaction (e.g., opening an email attachment or clicking links on a website) before they can infiltrate the target.[7] Note that HRM involves practices that are fundamentally focused on ensuring that employees' knowledge, skills, and abilities contribute to key organizational outcomes.[8] Bolstering CS knowledge of the general workforce is thus critical.

Figure 10.1 Percentage of organizations with different levels of employee CS awareness training programs

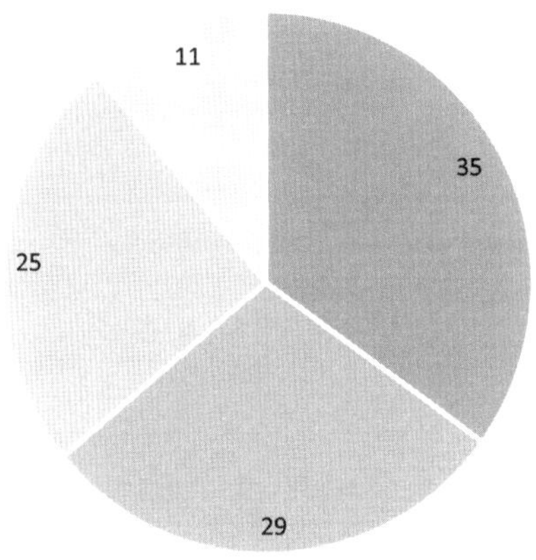

Semi-formal training programs   Informal training programs   Advanced training programs   No training programs

Data source: FireEye. (2019, November 11). *FireEye's cyber trendscape 2020*. Retrieved from https://www.fireeye.com/content/dam/fireeye-www/summit/cds-2019/presentations /cds19-executive-s05-fireeye-cyber-trendscape-2020.pdf

Figure 10.2 Institutional pressures and legitimacy issues in HRM from the CS standpoint

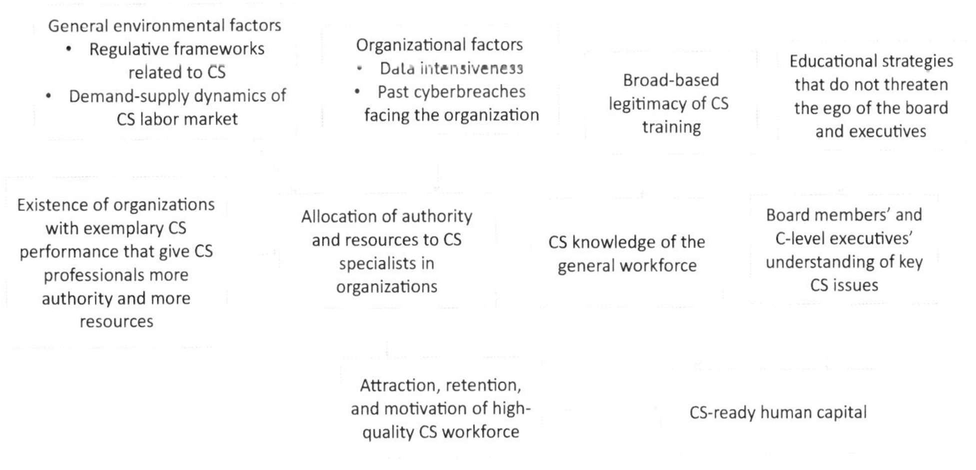

General environmental factors
- Regulative frameworks related to CS
- Demand-supply dynamics of CS labor market

Organizational factors
- Data intensiveness
- Past cyberbreaches facing the organization

Broad-based legitimacy of CS training

Educational strategies that do not threaten the ego of the board and executives

Existence of organizations with exemplary CS performance that give CS professionals more authority and more resources

Allocation of authority and resources to CS specialists in organizations

CS knowledge of the general workforce

Board members' and C-level executives' understanding of key CS issues

Attraction, retention, and motivation of high-quality CS workforce

CS-ready human capital

Yet most organizations lack effective CS training programs. A 2020 survey found that two-thirds of the UK's remote workers had received no CS training during the past year.[9] A 2019 study conducted by FireEye and market research organization KANTAR with more than 800 senior executives from Asia, Europe, and North America found that only a quarter of the organizations had advanced CS training programs to promote awareness and behavioral changes. (See figure 10.1.)

This chapter examines key HR issues needed to build CS-ready human capital. It delves into various types of legitimacy-related issues and pressures from environmental, organizational, and inter-organizational sources that may affect HR-related activities in order to strengthen CS. We have provided a model of the process to build CS-ready human capital in figure 10.2.

10.2 ATTRACTING, RETAINING, AND MOTIVATING A HIGH-QUALITY CYBERSECURITY WORKFORCE

10.2.1 CS Specialists' Higher Authority and Access to More Resources

In some companies, CS is viewed as a siloed staff function that is managed by IT specialists. It means that CS is treated as a secondary business activity to support the line functions that help businesses achieve their objectives. Many organizations are realizing that it is time to replace this outdated thinking and make CS a mainstream function in organizational and business activities. One possibility is to place CS groups within line functions that face significant cyberrisks.[10]

Due to substantial institutional changes brought about by increasing cyberrisks, a growing number of organizations find that there is a need to elevate and emphasize the CS role. Consequently, CS specialists have higher formal authority positions and access to more resources.

CEOs and boards often consult CISOs to understand cyberrisk, implement appropriate security controls, and promote a culture of defense. One study suggested that 90% of CISOs are connected directly to their organizations' top leadership team, and half of them were on the leadership team.[11]

CISOs' salaries have rocketed to a new level. Some large corporations were reported to hire CISOs with annual salaries in the US$500,000–US$700,000 range. It was also reported that compensation for CISOs at some technology companies was as high as US$2 million.[12] Organizations are realizing that in order to retain CISOs, resources must be dedicated to care for and encourage the growth and development of existing cybertalent with training and continuing education opportunities.[13]

When organizations face CS breaches or major compliance violations, they tend to respond by naming a dedicated CISO or hiring one. (See the end-of-chapter case: Defective Human Resources Policies Weakened Home Depot's Cybersecurity.) However, some companies that already have dedicated CISOs fail to utilize them due to financial, organizational, and personnel matters. For instance, in some cases, the CISOs may not report high enough in the management hierarchy.[14] CISOs traditionally report to the CIO or CTO.

Some practitioners have recommended allowing CSOs and CISOs to present findings and strategies directly to the board instead of through some other C-level officer.[15] Indeed, newly hired CISOs report directly to the CEO and the board. A study by the CS firm Threat Track found that 47% of CISOs reported directly to CEOs.[16] In some cases, CISOs are also board members.[17]

CISOs must be able to articulate the value of CS. A complaint is that most CISOs often fail to link CS posture with broader goals, objectives, policies. or strategies. For instance, if CISOs provide reviews of threat detection software updates or systems patched in a meeting, the board members may not understand the relevance in the context of broader goals. The company's board of directors may want to understand such indicators' links with operational and financial risks.[18] It is thus critical for a CISO to understand the business risks associated with CS. It requires going beyond the IT systems and their vulnerabilities. All CS and cyberrisk-related discussions and actions should start and end with the operations and strategic direction of the business and the business risks. Each information asset and automated business process must be protected with CS technologies and processes. Conversely, it is important to look at which information assets and automated business processes are protected by each CS technology or process.[19]

The general trend thus is that CS specialists are transitioning to new, broader, and more dominant roles in organizations. A UK-based recruitment firm was reported to help a company select a global CISO. Of the four candidates put forward by the recruitment firm, three had strong technical backgrounds. However, the candidate who was appointed was not selected for technical or specialist expertise. The newly hired person was expected to influence and build relationships at a high level and unite disjointed areas of the organization.[20]

New roles such as CISO or CSO are being defined and rationalized by corporate boards, CS professionals, legislatures, and regulatory agencies. Organizations, however, differ widely in terms of the power that CS professionals possess and the resources that these professionals have access to. For instance, in his 2018 annual letter to shareholders, JPMorgan Chase's chairman and CEO noted that the company spent about US$600 million annually on CS and employed about 3,000 people as its CS team.[21] On the other hand, Yahoo, Target, Home Depot, and many other companies that experienced significant cyberbreaches lacked CISOs and other key CS staff. (See In Focus 10.1 for Yahoo and end-of-chapter case for Home Depot.)

IN FOCUS 10.1: YAHOO'S LACK OF PRIORITY FOR CS LED TO THE EXPOSURE OF 3 BILLION ACCOUNTS

In October 2017, Yahoo revealed that personal information of all its 3 billion users had been stolen.[22] Several breaches that were thought to be smaller were known before. In 2014, Yahoo first reported to the FBI that hackers had targeted 26 accounts of Yahoo users.

In the process of investigation, the full scale of the problem came to light. By August 2016, the scale of the FBI investigation needed to be adjusted by adding more resources. In December 2016, Yahoo publicly announced the details of the breach and asked hundreds of millions of users to change passwords.[23]

In early 2017, four Russians were indicted by the US DoJ for their cyberattacks against Yahoo. Their alleged intentions were espionage and fraud.

They had launched the attacks between 2014 and 2016. In some of the attacks, they were believed to use spear-phishing attacks by sending a legitimate-looking message to a user with access to Yahoo's internal network and inviting them to click on a link. It then secretly installed malware on their computer. The compromised computers then gave the hackers access to the Yahoo network. The hackers copied the centralized database of user information. They also accessed Yahoo's Account Management Tool, which manages user accounts.[24]

Low Priority of CS as the Potential Root Cause

Many analysts believed that a low priority for CS was the main factor contributing to the high-profile attacks. According to Yahoo's current and former employees, the company's top management had been mostly concerned about Yahoo Mail's user friendliness. Marissa Mayer emphasized the creation of a cleaner look for services such as Yahoo Mail and the development of new products over strengthening CS.[25] Mayer's fear was that more emphasis on security could lead to a decline in the user base for the company's offerings.[26]

Outdated and Ineffective HRM System

Yahoo's first CISO Justin Somaini was hired in 2011.[27] Having a CISO was a good sign from the CS point of view. However, the CS orientation of the company was reported to drastically change when Marissa Mayer became the CEO in 2012. Somaini left the company in January 2013, which constituted a major setback for the company's CS. It was reported that he was "unhappy with the new regime [of CEO Mayer]."[28] Somaini reportedly left Yahoo because CS employees were not valued by its CEO.[29]

After Somaini's departure, Yahoo did not have a full-time CISO for 14 months. In March 2014, Alex Stamos was hired to replace Somaini.

It was reported that Stamos tried but was unsuccessful in establishing CS as a priority for top management. Stamos found himself in conflict with Mayer. Mayer reportedly "denied Yahoo's security team financial resources and put off proactive security defenses, including intrusion-detection mechanisms for Yahoo's production systems."[30]

Ironically, the CS team was not included in important decisions related to CS. For instance, according to an article published in *Business Insider*, in 2015 Yahoo had secretly built customized software that scanned all incoming Yahoo emails for specific information that US intelligence officials were looking for.[31] Stamos was completely out of the loop when the decision was made. Stamos and his CS team knew about the program only when they were testing Yahoo's systems for vulnerabilities. When the CS team found the software installed by Yahoo for the US intelligence officials, they thought that some criminal hackers had inserted it.[32]

In June 2015, Stamos left Yahoo. In July 2015, Ramses Martinez replaced Stamos but quit his position in about a month. In October 2015, Yahoo appointed Bob Lord as the CISO. In this way, Yahoo had three CISOs in a five-month period.

Impact on Yahoo

Yahoo was in talks to be acquired by the US communications and technology solutions provider Verizon before announcing the massive data breach. In July 2016, Verizon had agreed to buy Yahoo for US$4.8 billion. Verizon thus found out about the data breach at Yahoo after the companies formulated an acquisition agreement. At one point, the deal almost died. Finally, there was a US$350 million reduction in Verizon's purchase price for Yahoo.[33] Verizon bought Yahoo for US$4.5 billion in 2017.[34] Verizon's acquisition of Yahoo is one of the most highly publicized examples of a CS challenge affecting mergers and acquisitions.

Verizon had already bought the web portal and online service provider AOL for US$4.4 billion in 2015.[35] Verizon created a subsidiary named Oath; Yahoo and AOL became parts of the Oath brand. (In 2019, Oath was rebranded as Verizon Media Group.)

Several Yahoo senior managers were selected for the leadership roles at Oath. However, due partly to the fact that Yahoo had suffered the biggest data breach in history, Oath management did not view the Yahoo CISO favorably to lead the CS team.[36] Bob Lord lost the Oath CISO job to AOL's Chris Nims.

Yahoo faced pressure from lawmakers to reveal when it knew about the breaches. In November 2017, the US Senate subpoenaed Mayer and compelled her to testify about the 2013 security breach.[37] Richard Blumenthal, a Democratic senator from Connecticut, demanded, "As law enforcement and regulators examine this incident, they should investigate whether Yahoo may have concealed its knowledge of this breach in order to artificially bolster its valuation in its pending acquisition by Verizon."[38]

The Responses of Yahoo and Verizon

In 2017, Yahoo faced a class action lawsuit filed by users whose data was exposed, in which the plaintiffs argued that the company failed to disclose the massive data breaches fast enough.[39] Verizon filed a motion to dismiss the class action lawsuit. In March 2018, the US District Court in San Jose, California, denied part of the motion filed by Verizon.[40] While the court also dismissed some of the plaintiffs' claims, the accusation of negligence, which was arguably the most damaging, was not dismissed.[41]

In January 2019, Yahoo proposed a settlement to pay US$50 million and two years of free credit-monitoring services for 200 million of its users in the US and Israel. The US District Court rejected the proposal, noting that the settlement lacked details and did not cover attorneys' fees. From the proposal, it was not clear how much the deal would cost Yahoo.[42]

In September 2019, Yahoo announced that it would pay up to US$25,000 to each person affected by the breach.[43] In order to be eligible for the compensation, a victim needed to prove the financial damage suffered due to the Yahoo hack. A free credit-monitoring service worth US$100 was offered to most users. However, the offer was conditional. Only those victims with credit monitoring already in place were paid.[44] Yahoo also needed to settle at least three cases with shareholders, investors, and the SEC for misleading investors.[45]

Oath's goal was to compete with Google and Facebook in the online advertising industry. Verizon had set a revenue goal of US$10 billion by

2020 for Oath.[46] However, it earned US$1.8 billion in the third quarter of 2018, which was a decline of 6.9% compared to the previous quarter.[47] The disappointing performance was at least partly attributable to the Yahoo hack. One-and-a-half years after the Oath brand was established, Verizon announced that its merger with Yahoo was virtually worthless.[48] In early 2019, Oath rebranded to Verizon Media.

10.3 INFLUENCE OF PROMISING AND EXEMPLARY CYBERSECURITY PRACTICES

One key observation is that CS-related norms are still in their infancy. Thus, one way to increase effectiveness is to strategically adopt recruitment practices and HR policies of successful firms.

For instance, following the practices of some successful firms, CISOs are becoming increasingly common among organizations.

A 2019 survey of cloud access security broker Bitglass[49] found that among *Fortune* 500 companies, 62% had a CISO although only 4% of them listed CISOs on their company leadership pages. According to management consultancy company PA Consulting, companies with 250 or more workers are likely to have CISOs.[50]

10.4 GENERAL ENVIRONMENT

10.4.1 Regulatory Developments

Another major driver in the development of HR workforces is the current system of regulatory changes. Organizations worldwide face regulatory pressures to develop CS-related human capital at all levels, including top management. Regulatory requirements reflect the increasingly vital role of CS in national security and economic prosperity.

In September 2016, New York's Department of Financial Services proposed a regulation that would require banks, insurance companies, and other financial services institutions to designate a CISO. The DFS asked financial-sector firms to develop more CS expertise on their boards. In 2016, the US Senate proposed a CS disclosure bill (Senate Bill 592) that would require public companies to describe CS expertise in their boards. Companies that lack CS expertise on their boards would be required to explain the steps they are taking to add such expertise.[51] Laws in Israel require CISOs to report directly to the CEO.[52]

In 2015, the Hong Kong Monetary Authority issued a Cyber Security Risk Circular that expected banks to make "tangible progress" in strengthening CS. Hong Kong–based banks were reported to be hiring CS professionals to fulfill new regulatory demands. Average salary increases for CS professionals who changed employers were reported to be in the 20%–30% range. Due to a small supply of well-trained cyberprofessionals in Hong Kong and Singapore, banks were reported to be hiring CS people from other sectors such as consultancy firms, security vendors, and telecommunication companies.[53]

10.4.2 Demand-Supply Dynamics of the CS Labor Market

According to the Gartner *Emerging Risks Survey* conducted in the fourth quarter of 2018, the CS skills shortage was the top concern organizations faced globally. Sixty-three percent of respondents reported that the talent shortage was a key concern for their organizations.[54] In the third quarter of 2018, privacy regulation and cloud computing concerns were identified as the top two risks.[55] Due to high demand, CS professionals enjoy much higher salaries than most other professions. (See figure 10.3.)

The shortage of CS skills has been a global phenomenon as demonstrated by the large numbers of unfilled positions worldwide. (See figure 10.4.) In countries such as Singapore, organizations recruit from overseas. In addition, they partner with educational organizations and develop flexible hiring strategies that include both permanent and contract specialists.[56] Government agencies are also taking initiatives to deal with this shortage. (See In Focus 10.2.)

Figure 10.3 Average salary in North America for CS-related jobs (US$)

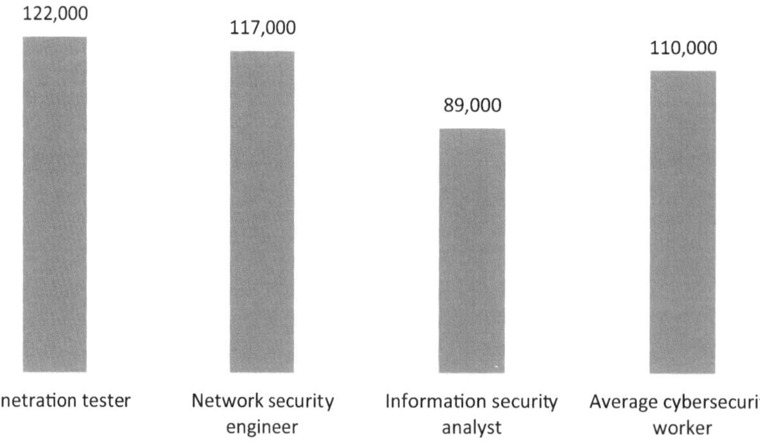

Data source: The job and recruiting website Indeed as reported in Owaida, A. (2020, March 3). *5 reasons to consider a career in cybersecurity*. Welivesecurity. Retrieved from https://www.welivesecurity.com/2020/03/03/5-reasons-consider-career-cybersecurity/

Figure 10.4 CS-related positions vacant worldwide (millions)

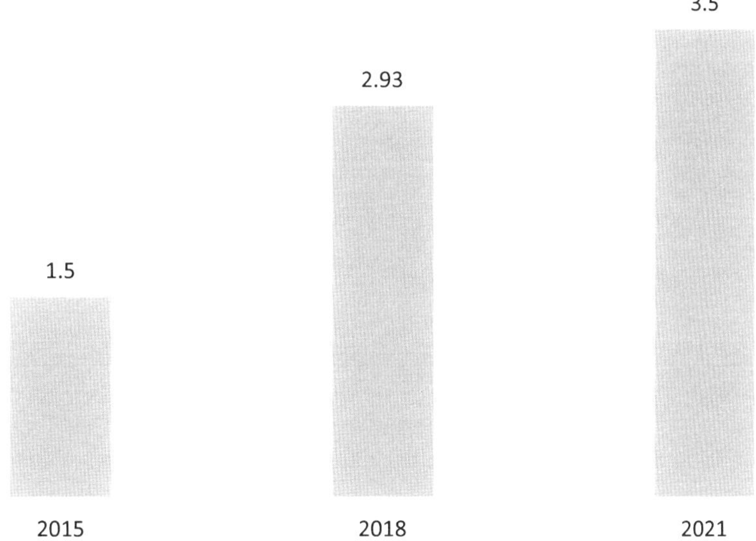

Data sources: Data for 2015 as reported in *Global cybersecurity workforce shortage to reach 1.8 million as threats loom larger and stakes rise higher*. (2017, June 7). (ISC)². Retrieved from https://www.isc2.org/News-and-Events/Press-Room/Posts/2017/06/07/2017-06-07-Workforce-Shortage; data for 2018 as estimated by (ISC)² and reported in Yatsu, M. (2019, August 27). Why Japan can't fail womenomics in cybersecurity. *The Japan Times*. Retrieved from https://www.japantimes.co.jp/opinion/2019/08/27/commentary/japan-commentary/japan-cant-fail-womenomics-cybersecurity/#.Xm2bVqhKjIU; data for 2021 as reported in Reffold, K. (2019, November 15). Is there really a cybersecurity skills gap? *Forbes*. Retrieved from https://www.forbes.com/sites/forbeshumanresourcescouncil/2019/11/15/is-there-really-a-cybersecurity-skills-gap/#2268cbc310fe

IN FOCUS 10.2: DEVELOPING CS TALENT AMONG YOUNG PEOPLE

Teaching CS to kids and young adults is among the most effective ways of developing a national CS workforce. In recent years, several countries have introduced educational and training initiatives targeted at these groups. The UK's National Cyber Security Centre's CyberFirst initiative provides one-day training to children as young as 11 years. Topics covered include the most common passwords that need to be avoided, information that social media websites collect on their users, and tracking down "patient zero" in the case of a cyberattack.[57] Likewise, in 2017, Romania announced that its schools would offer CS classes for children aged between 11 and 14 years.[58]

In Israel, the Cyber Education Center's Magshimim program teaches high school students computer programming skills as well as different ways of fighting cyberattacks.[59] Magshimim has been designated as a national program since 2013 by the Israeli government. By 2020, 4,800 students participated in the program annually.

The US government in 2011 launched the National Initiative on Cybersecurity Education (NICE) program,[60] which seeks to address the shortage of CS-related human capital. The development of a cybersavvy workforce has been a key priority for the US.

There has been widespread concern over the lack of preparedness of the education system to train students on CS. According to the National Science Foundation, only 19% of US high school students take a computer science class.[61] In a study of Intel Security Group and CSIS, only 23% of respondents believed that education programs were preparing students for CS jobs.[62] The report noted that "non-traditional methods of practical learning, such as hands-on training, gaming and technology exercises and hackathons," were more effective in increasing CS skills. Many CS specialists with practical computer expertise are self-taught.[63]

There has been a trend toward increasing dependence of the CEO and boards on CISOs and other CS professionals, which will likely increase their relative power. CS workers are being offered higher salary and more power. Overall, due to the higher level of uncertainty in the availability of CS skills, CS specialists may look for more resources and higher status compared to most other types of jobs.

10.5 FIRM CHARACTERISTICS

10.5.1 Data Intensiveness

The degree of power that a negotiator can have over the other party in a negotiation is a function of the perceived value of a negotiator's contributions.[64] CS is more important to organizations of certain types. A data-intensive firm (e.g., Internet companies, banks, insurance companies, credit-card companies) would have high CS needs and would undertake more efforts and initiatives to attract high-quality CS personnel. For instance, in India, as early as 2009, while it was rare to have a CISO in organizations, industries such as banking, financial services, insurance, telecommunications, and business process outsourcing had such positions.[65]

10.5.2 Past Cyberbreaches Faced

Jolts and other exogenous shocks can create opportunities for new norms. Significant CS breaches are among the factors that lead to the importance of CS-related human capital.

For instance, in 1995, Citicorp appointed its first CISO after a high-profile hacking incident.[66] A Russian hacker with several accomplices had broken into Citibank's network and stolen over US$10 million.[67] A cyberattack of this scale was previously unknown. Likewise, Target appointed its first CISO after its 2013 breach.[68]

10.6 UNDERSTANDING OF CYBERSECURITY BY BOARD MEMBERS AND C-LEVEL EXECUTIVES

To understand the importance of CS for the board and executives, consider one detail: in the past few years, executives of at least three major global companies, the head of a US federal government agency, and the head of the governing body of the US Democratic Party – Democratic National Committee (DNC) – lost their jobs because they failed on the CS front. It is also clear from table 10.1 that sometimes what is revealed in the stolen information rather than the hacking itself leads to the CEO's departure.

Two approaches can be used to ensure that the board is able to evaluate and oversee a company's CS strategy. First, the company can bring onto the board an executive who has extensive CS experience and expertise. Second, the board can hire an outside CS expert as a consultant.

10.6.1 Educational Strategies That Do Not Threaten the Ego of the Board and Executives

Top executives and the general workforce may differ in terms of norms, values, and preferences. It is also important to consider the C-level executives' and board members' sense of what is right and proper behavior.

An uncomfortable reality is that CS is not familiar terrain for most executives. At the same time, some education and awareness strategies that are likely to be effective for the general workforce may threaten the ego of top executives. For instance, in a general meeting, board members may avoid technical questions for fear of embarrassing themselves. Personal or one-on-one communication can allow them to talk more freely and ask technical questions.[69]

10.7 IMPROVING CYBERDEFENSE CAPABILITIES OF AND CONTROLLING DEVIANT BEHAVIORS OF THE WORKFORCE

Education, awareness, and training are the important countermeasures against cybercrimes because the most often cited reason behind CS breaches is that employees are easily manipulated by criminals. For instance, according to the CS company Trusteer, 70% of corporate employees were likely to click on a well-crafted fake LinkedIn email.[70] (See In Focus 10.3.) Thus, employees with no bad intention or malicious thinking may pose risks.

Table 10.1 Some executives who lost jobs due to their failure on the CS front

Time	Affected Executive	Related CS Incident	Remarks
May 2014	▪ US retailer Target's CEO Gregg Steinhafel resigned.	▪ A data breach in December 2013 compromised 40 million credit- and debit-card accounts and 70 million people's personal data.	▪ The proxy advisory firm Institutional Shareholder Services suggested that seven of the ten directors at Target be removed because they did not take sufficient actions to prevent data breach.[71] ▪ In February 2014, Beth Jacob, the company's most senior technology officer, resigned.[72]
February 2015	▪ Co-chairperson of Sony Pictures, Amy Pascal, was fired.	▪ The hacking group Guardians of Peace (GOP) was involved in the Sony Pictures hack of November 2014, which was linked to the North Korean government.[73] The attacks could have been in response to Sony's movie *The Interview*, the storyline of which involved a plot to assassinate North Korea's leader Kim Jong-Un.	▪ Pascal joked about President Barack Obama's presumed taste in movies in emails with producer Scott Rudin.[74]
June 2015	▪ Chief Financial Officer Thomas Meston of Fortelus Capital Management was fired from his job.	▪ On a Friday afternoon in December 2013, a phony caller pretending to be from Fortelus's bank, Coutts, called Meston and warned that fraudulent activities might have taken place on a Fortelus account. To cancel suspicious payments, Meston used the bank's smart-card security system to generate codes for the caller. When Meston checked the account the following Monday, US$1.2 million had been stolen.[75]	▪ Meston was sued by the fund for breaching his duty to protect the company's assets.
July 2015	▪ The director of the US OPM, Katherine Archuleta, resigned.	▪ A breach exposed personal data of over 22 million current and former federal workers.	▪ Archuleta faced pressure to resign from lawmakers and federal employee unions.

July 2016	▪ Chair of the DNC, Debbie Wasserman Schu tz, resigned.	▪ WikiLeaks published about 20,000 private emails just before Hillary Clinton was nominated as the presidential candidate of the Democratic party. The documents showed that party officials tried to undermine Clinton's chief rival, Bernie Sanders. US officials blamed the Russian government for allegedly guiding hackers to attack DNC servers.[76]	▪ The documents showed that DNC officials failed to remain impartial in the presidential nominating process.[77]

IN FOCUS 10.3: EMPLOYEES NEED ANTI-PHISHING TRAINING

An analysis by the CS company CybSafe of the data from the UK's Information Commissioner's Office (ICO) indicated that human errors accounted for 90% of data breaches in the UK in 2019.[78] CybSafe also found that phishing was the number one cause of 2019 data breaches in the UK, which accounted for 45% of all incidents reported to the ICO. Indeed, phishing is described as an "evergreen cyberthreat."[79] Likewise, a 2020 survey of the Ponemon Institute indicated that phishing is the most common threat and was the cause of cyberattacks in 47% of organizations.[80]

Phishing is becoming more sophisticated. As a result, business email compromise (BEC) frauds are growing rapidly. In a BEC fraud, cybercriminals get access to a business email account holder's identity to defraud the organization, its employees, customers, or partners. The attacker often creates an almost identical email address to gain the victim's trust. Such criminals mostly target employees that have access to company finances, and trick them into conducting high-value transactions by using wire transfer. The money goes into accounts owned by the criminals. In 2019, the FBI received 23,775 complaints related to BEC scams, which cost US businesses US$1.77 billion.[81]

An important CS training topic in organizations is related to effectively identifying and screening phishing emails. The CS gamification platform HoxHunt conducts automated simulated phishing attacks on an organization's employees. Messages designed by HoxHunt mimic phishing messages that come from cybercriminals. It also simulates AI-powered spam messages. Employees use the platform's plugin to report suspicious emails. If an employee fails to identify phishing messages, they receive an explanation of the mistake they have made. HoxHunt's gamified system awards points to employees who successfully report phishing emails. Companies may reward top-performing employees. Additional training and intervention can be provided to those unable to identify phishing emails.[82]

Organizations are interested in winning customer trust. They tend to engage in behaviors and practices to increase customer satisfaction. Such attempts may lead to CS breaches. For instance, a CS expert reportedly called an investment-management firm to test CS readiness. He told the customer service representative that he was going through a divorce and asked if his wife had opened an account. The representative readily provided him with customer account numbers and other details. In another case, the expert pretended to be an employee of reputable IT company gathering information for a government contract and called the company's satellite office. An employee gave him details about the company's operating and antivirus systems.[83]

Examples such as the above show that organizations face pressure to completely rethink and restructure business processes so that CS assumes a high level of priority in HRM policies. (See In Focus 10.4.) Through training and other measures, organizations are attempting to change employees' attitudes and behaviors and establish a better CS culture. For instance, Croatia's Hrvatski Telecom conducted a penetration test with the help of a CS company. The mock hack attack involving social engineering was carried out without informing employees beforehand. A caller from the outside CS company doing the penetration test called Hrvatski employees, pretending to be a customer. The caller asked Hrvatski employees to reset the account password, provide a new telecom service, and change payment options.[84] Such testing measures would help employees better understand risks involved in providing sensitive information to potential perpetrators that they might falsely think are legitimate customers.

IN FOCUS 10.4: COVID-19, REMOTE WORKING, AND CS

Remote working is a trend that has been encouraged by the development in ICTs. According to the Bureau of Labor Statistics, 28.8% of US workers or 41.6 million out of the total 144.3 million could work from home in 2017–2018.[85] Among those with a bachelor's degree or higher, 51.9% could work from home. In occupations related to management, business, and financial operations, over 60% could work from home.

The spread of COVID-19 in 2020 accelerated an increase in remote working. For instance, the number of daily users of the chat, collaboration, and video-conferencing app Microsoft Teams increased by 37% in the week of March 12–18, 2020, to reach 44 million.[86] Especially, technology companies such as Amazon, Apple, Google, Twitter, and Airbnb asked a large proportion of their employees to work remotely.[87]

Such a working environment presents unique CS risks. Lower safeguard standards are likely when people work from home, which increases cyberrisks. For instance, employees working remotely may use their own devices such as phones, laptops, and tablets. Unlike devices issued by the company, personal devices are less likely to be patched for the latest vulnerabilities.

Consumer-level systems, which are designed by keeping the needs of individuals or small businesses in mind, focus more on ease of use. They often lack options for customization. Enterprise-level systems, on the other hand, are more sophisticated and designed to protect larger organizations, which need higher levels of CS. Such systems come with additional resources, features, and the support of experienced professionals. Consumer-grade systems lack the support of CS professionals that are dedicated to managing cyberthreats facing a firm.[88]

Various aspects of social control mechanisms that protect workers from dangerous cyberattacks do not operate in remote working. For instance, employees' in-person interactions with their coworkers and supervisors may shield them from unsafe cyberpractices at work; they cannot take advantage of such interactions when working remotely.[89]

Some corporate policies are likely to be violated by people working from home. For instance, they may access personal emails or social media accounts using company devices. They may connect using public Wi-Fi.[90]

It has been reported that cybercriminals have targeted executives of companies and other high-profile individuals who were viewed as vulnerable to attacks because they were working from homes using less secure networks.[91] A prominent example of this problem occurred when the 2016 Democratic US presidential candidate Hillary Clinton used her own email server to conduct official business.[92]

Among the most serious challenges facing remote working is ensuring that security and privacy requirements are fulfilled. In a survey conducted among 250 IT leaders by the provider of software and technology services for virtual private network markets OpenVPN, 73% of VP and C-suite IT leaders and about half of IT managers/IT directors expressed concerns about the CS risks remote workers pose.[93] However, many organizations lack well-developed CS policies and tools for remote workers.

Diverse CS Issues Associated with Remote Working

Different apps have different levels of criticality. For instance, in the case of a hospital, where collaboration and email apps may not be mission-critical, those dealing with sensitive patient data are associated with higher cyberrisk and thus require the highest level of security. Under current policies, an app's criticality determines where and how it can be accessed. Less critical apps can be accessed from anywhere. Apps that are susceptible to severe consequences if they are subjected to cyberattacks, on the other hand, can be accessed only onsite.[94]

Events such as COVID-19 can force organizations to widen the right to access most apps remotely. When such "rights" are accorded to sensitive apps, cyberrisks may increase. Most organizations lack clear and consistent policies related to remote working. They are often left to the discretion of the manager.[95] Problems with remote working can range from inadvertent issues such as an employee's family member seeing what is on the screen to very serious matters such as the device being compromised by a criminal hacker. Even a minor issue, such as a family member without any bad intention seeing the content, is considered a data breach and may require notification.[96]

Third-party applications installed on personal devices connected to a corporate network may pose severe CS risks. For instance, if hackers are able to attack a personal device with the login credentials to corporate sites, they may be able to penetrate business networks.[97]

Importance of Policy and Training

Organizations need to have a policy and a training program for employees working remotely. For instance, organizations may consider restricting access to sensitive information and systems if such systems are not needed to work remotely.[98] Organizations need to define a clear procedure to follow in case of a security breach.

In the survey conducted among 250 IT leaders by OpenVPN, 90% of the respondents required their remote workers to participate in CS training. However, these leaders have expressed the need for more frequent and more rigorous training programs.[99]

CS training and awareness programs need to address special situations in remote work.[100] For instance, it is important to ensure that employees do not use unapproved or insecure apps, tools, or methods. Likewise, while working from public places such as coffee shops, libraries, and bookstores, they should avoid using free Wi-Fi. A personal hotspot could be a more secure option.[101] Remote workers should also be required to log out when they are not using their computer.

Importance of Enhanced Security Measures

Remote workers are viewed as low-hanging fruit by hackers. Since employees' own computers often have lower levels of protections compared to work devices and organizations lack capabilities for monitoring remote working, enhanced security measures need to be adopted for remote working. Special focus should be on physical security, such as the use of computer privacy screens or privacy filters. Employees should limit work on confidential material when they are in public spaces. They should also secure physical computing assets such as desktop computers, laptops, servers, and routers.[102]

Organizations may also consider implementing extra security procedures for remote workers. Organizations can use sophisticated software tools to verify employees' identities. For instance, some ask users to type in a passphrase as well as a question and response.[103] Software not only verifies the response but also the user's typing pattern, such as the speed between each letter. Likewise, multi-factor authentication (MFA) that employs biometric security procedures can be used. Such procedures rely on extra credentials such as voice recognition, facial scanners, or fingerprint checks in identity and access management. MFA also includes sending a code to a device or email address registered by the employee. Organizations may also consider deploying next-generation AI-driven security, which can make it possible for mobile workers to securely access sensitive data or applications.[104]

Despite the importance of CS with the widespread occurrence and ubiquity of cybercrimes, not enough attention has been given to CS training. Practitioners have emphasized the importance of promoting CS within the organization with better education and training.[105] CS practitioners have criticized HR managers who do not bother to train employees on CS. A CS expert noted: "HR chiefs – notorious for pointing their fingers at employee misconduct – need to turn their glasses around, admit they are falling short on security, and do something about it."[106] According to a study from IBM Security, titled *Securing the C-Suite, Cybersecurity*

Perspectives from the Boardroom and C-Suite, only 57% of chief human resource officers had initiated employee training programs that address CS.[107] A study by (ISC)², a non-profit focused on CS education and certification, found that HR departments often felt that "cybersecurity wasn't their job."[108]

Cyberattackers also tend to target employees new to a company, who often lack training and experience.[109] One study indicated that 38% of companies conducted training on a quarterly or biannual basis; the frequency was even lower in the remaining organizations. After training programs, only 25% of trained employees successfully recognized a cyberattack one month later.[110] Due to a rapidly changing cybercrime landscape, CS training needs to be conducted more frequently.

10.7.1 Effective Training Strategy

A key challenge is increasing the value and legitimacy of CS training programs. It may thus be important to change employees' mindsets to make them willing and open to training and development. To increase success, CS-related training and development need to secure a more broadly based legitimacy in organizations.

A key consideration with CS is that it touches every aspect of the lives of employees. This aspect can be considered an advantage from the standpoint of legitimacy. Companies need to expand CS beyond the walls. That is, any attempts to change employees' behaviors need to extend beyond the company. For that to happen, organizations can increase effectiveness and acceptance of CS training by explaining that CS will have influence on organizational success, as well as on the employees' personal lives.[111]

10.7.2 Ensuring Appropriate Access to Data and Information

The above discussion focused mainly on collaborative means of achieving change in employees' CS-related behaviors. Among the most important is promoting a cyberrisk-aware culture among the workforce and educating employees about common cyberattacks. These measures may not be sufficient to guarantee a desired level of CS. Some coercive measures may also be used. Some have suggested that employees need to sign the corporate CS policies and procedures as a condition of employment. It is important for organizations to collaborate with HR to include CS in "day-one orientations" and introduce the relevant tools and policies.[112] Employers should remind employees of such policies and procedures from time to time.[113]

Some current and former employees were found to use their access privilege to destroy and steal data, obtain confidential information about customers, and engage in frauds using customer accounts.[114] As noted in chapter 8, a large proportion of cyberattacks originate inside an organization. According to the *2019 Global Data Exposure Report* of data protection firm

Code42, insider attacks accounted for half of hacking incidents that victimized companies.[115] The surge in insider attacks can be attributed to various underlying causes, such as less loyalty, reduction in workplace retention rates, and companies' failure to adjust their data protection protocols to these realities. The FBI also noted that insider cyberattacks are among the biggest threats organizations face.

10.8 GENDER-RELATED ISSUES IN CYBERSECURITY

Women are underrepresented in the CS field. According to a survey conducted by the research firm Frost and Sullivan and the (ISC)^2 foundation in 2013, women accounted for 11% of the global CS workforce.[116] In 2019, Cybersecurity Ventures' estimate put the proportion of women in CS jobs worldwide at over 20%. This estimate takes into account women working as security vendors and security service providers, employed in small and midsized enterprises, and founders of security startups, especially prevalent in Israel.[117]

A 2017 study cosponsored by PwC, which was conducted by the Center for Cyber Safety and Education, found that women's share of the US information security field was 14% compared to 48% in the US workforce overall.[118] In C-suite leadership positions, only 1% of CS workers are female.[119] Women are even more underrepresented in CS fields in other regions of the world. For instance, females account for 10% of the total CS workforce in Asia-Pacific, 9% in Africa, 8% in Latin America, 7% in Europe, and 5% in the Middle East.[120]

Studies have found that most successful women CS specialists did not start their careers in traditional STEM-related fields.[121] Among women who work in CS, 44% had degrees in fields such as business and social science compared with 30% for men.[122]

10.8.1 Importance of Including Women on the CS team

Women CS leaders prioritize important areas that are often neglected by their male counterparts. They are more likely to pay attention to a wide variety of organizational and inter-organizational contexts that might increase CS risks. The importance of this issue stems from the fact that human aspects are as important as technical aspects in CS. For instance, Frost and Sullivan and (ISC)^2 found that women CS professionals focus more on internal training and education related to CS and the management of cyberrisks. Especially, women CS professionals were found to be stronger advocates and supporters of online training and education methods than men. This is important because online approaches offer low costs and more flexible methods to develop CS awareness among employees.

Women CS professionals also emphasize the importance of pragmatic aspects in the selection of partner organizations for the development of secure software applications. In the Frost and Sullivan and (ISC)^2 survey, much higher proportions of women than men gave "Top and High Importance" to factors specific to partner organizations such as their ability to meet contractual obligations related to security, willingness to perform security testing independently, appropriate organizational qualifications, and qualified CS personnel.[123]

Women's participation in a product's CS-related aspects is critical to make the product attractive for women customers. Mastercard's estimate suggests that by 2028, women worldwide will control about 75% of consumer discretionary spending.[124] Key CS-related factors such as encryption, fraud detection, and biometrics are becoming decisive factors in consumers' buying decisions. Product designs require a trade-off between cybersecurity and usability. Female CS professionals can make better-informed decisions about such trade-offs for products that are targeted to female customers.

10.8.2 Barriers Faced by Women

The disproportionately low representation of women in CS is linked more generally to the broader problem of their underrepresentation in the STEM field. Women account for only 30% of scientists and engineers in the US.[125] The prevailing societal view is that CS is a job that men do.[126]

Another challenge is that women lack awareness of and knowledge about the CS industry. In a study of the Computing Technology Industry Association (CompTIA) conducted among women who had pursued careers outside information technology fields, 69% indicated that lack of knowledge about the availability of IT-related opportunities was the main reason for their career choice.[127] In the same survey, 53% said that additional information related to career options would encourage them to consider IT jobs.[128]

A further challenge that women encounter concerns gender bias in HRM practices, such as recruitment, selection, performance appraisal, and training and development. According to a survey conducted by CS company Tessian, only about half of the respondents said that their organizations were doing enough to recruit women into CS roles.[129]

Companies take a wrong-headed approach in communicating what is important in CS. Many would-be female CS professionals have expressed concern that the industry places too much emphasis on technical skills. Potential employees often get the impression that only techies work in CS.[130]

10.9 CHAPTER SUMMARY AND CONCLUSION

Thanks to stricter regulatory requirements and high-profile cyberbreaches, CS professionals are assuming more visible roles in organizations. The roles of the CISO and other CS professionals are being strengthened to drive CS initiatives. In some organizations, senior executives are establishing a supportive and solid relationship with the CISO.

Recognizing that people are the most critical resource, management must create a common culture that aligns employees to the organization's CS goals. The shortage of CS experts underscores the importance of organizational initiatives to provide employees with CS-related training.

While the CEO, board members, and general workforce need to be aware of cyberthreats, the capability and motivation to learn CS-related skills may differ across these different groups. For lower-level employees, training strategies that go beyond self-interest and organizational interest and consider the impact on employees' personal lives are more likely to be effective. But this strategy may not work for board members. CS specialists need to work with other functions such as HRM, legal, public relations, and marketing. For instance, board members have emphasized the importance of providing special training and work experience to CISOs so that they can provide useful and actionable information about cyberrisks.[131]

Wider and deeper collaboration between CISOs and HR is also critical. For instance, by collaborating with their HR department, CISOs can gain new insights into possible insider threats and manage them in a better way. CISOs can also learn about new HR technologies and partnerships in order to strengthen CS.

The CS skill shortage is a global phenomenon, and organizations will be hard pressed to compete in a globalized marketplace. A comprehensive and multi-pronged approach is needed to counter this shortage. For instance, it is important to create more awareness of CS career options by encouraging students to interact with CS professionals.

Organizations face challenges in attracting, retaining, and motivating capable CS workers. In order to deal with this issue, companies have made a number of strategic changes. As noted earlier, some organizations are elevating the roles of CS professionals. They are doing so because of uncertainties in the CS labor market created by pressures from multiple stakeholders and because of the increasing regulatory oversight by governments. The upshot of these considerations is that CS professionals show a marked tendency toward exercising their power in organizations.

Best practices have been implemented by companies such as Google. Painfully, we have seen that companies with weak HRM practices in CS have faced significant cyberattacks. CEOs are realizing the significance and impact of cyberattacks. Defending against cyberattacks is among the key responsibilities of the board and the CEO. CS is too important to leave to the CISO alone. It is important for senior executives and the board to understand procedures that are in place to defend the organization from cyberattacks.

Attracting and retaining women in CS will require shifts in HRM strategies. For instance, different recruitment strategies need to be used because women CS professionals are more likely to come from non-STEM fields compared to men. This implies that recruiters need to consider potential candidates both inside and outside STEM fields. Organizations should develop a more inclusive culture that is less biased against women in CS.

10.10 DISCUSSION QUESTIONS

1. Why is there a severe shortage of the CS workforce?
2. Do you agree with the following statement: CS can make or break a business? Why or why not?
3. How do organizations differ regarding the status, roles, and responsibilities of the CISO?
4. What are the skills, education, training, and competencies required to be a successful CISO?
5. Why is it important for organizations to increase women's participation in CS?

10.11 END-OF-CHAPTER CASE: DEFECTIVE HUMAN RESOURCES POLICIES WEAKENED HOME DEPOT'S CYBERSECURITY

In 2014, Home Depot suffered one of the biggest data breaches in history. About 56 million credit cards of Home Depot's customers were compromised. The data were available for sale on black markets. One estimate suggested that they could be used to make US$3 billion in illegal purchases.[132] As of early 2017, the total cost of the data breach to the company reached about US$179 million excluding legal fees.[133]

10.11.1 CS Considered a Low-priority Function by Top Management

A complaint was that CS was given a low level of importance in comparison with other issues in the overall corporate strategy. People who were working in the CS team thought that their ideas and opinions were not valued by the top management. For instance, it was reported that when Home Depot's CS team asked for new software and training in order to strengthen CS measures, the top management even went so far as to say, "We sell hammers."[134]

10.11.2 Defective HRM

According to media reports, Home Depot lacked effective policies in recruiting, selecting, developing, retaining, and productively utilizing its CS employees. The company did not vet its key CS employees sufficiently to avoid hiring people with a criminal background. For instance, Home Depot's former CS chief had been sentenced to four years in prison for "deliberately disabling computers" at his previous company.[135]

In the fall of 2011, Home Depot's CS team had about 60 employees. After a new CS chief took over in August 2011, about half of them left in a three-month period. Former employees blamed the CS chief for the loss of CS people, saying that he was "bullying" and "abrasive."[136]

A former Home Depot security engineer put the issue this way: "You're having a hard enough time finding security holes. Then half the people in your department leave and your workload doubles. It makes it even harder to catch stuff." [137]

10.11.3 Poor Encryption in Payment

According to a former employee, Home Depot lacked encryption in payment applications, which violated payment card industry security standards. It appears that the lack of personnel to encrypt such applications was a key reason. For instance, during a three-month period in the spring of 2013, four of the eight people responsible for ensuring credit-card encryption had left the company. It was reported Home Depot management's failure to address CS was the reason behind their departure. One of the CS employees who left the company specifically had raised "red flags" with management regarding the lack of encryption that allegedly violated payment card industry security standards.[138] The company's security staff were so concerned about the poor state of IT systems that they warned their friends to use cash instead of credit cards when shopping at the company.[139]

10.11.4 Home Depot's Response

In a press release, then chairman and CEO of Home Depot Frank Blake reassured its customers that they would not be liable for fraudulent charges.[140] Home Depot announced that it had contracted two CS firms and was working with experts from the firms to ensure that incidents such as those would not happen again. The company announced that Voltage Security would provide security to protect the IT infrastructure. It also strengthened cybersecurity at all its stores.[141]

It also revamped its HRM. It created a new role of CISO to focus only on CS. Before the breach, CS responsibility fell upon the company's VP of information technology.[142] In March 2015, Home Depot hired its first CISO, Jamil Farshchi, who had previously worked with Time Warner and Visa. He reported to the company's CIO, Matt Carey.[143]

Surveys have found that about one-third[144] or more[145] of CISOs do not report to CEOs or boards of directors. At Home Depot, Farshchi reported to the CIO. In other organizations, CISOs report to executives who are further down the chain.[146]

PART FOUR

Privacy and Security Issues Associated with New and Evolving ICTs and Systems

CHAPTER ELEVEN

Social Media

CHAPTER PREVIEW

Legitimate as well as illegitimate organizations and entities gain access to information about social media (SM) users through illegal and unethical means. Worse still, many organizations and individuals using SM have become targets and victims of cybercrimes. SM have also led to an exposure of unethical and illegal conduct within some organizations.

Organizations that fail to take appropriate technological and behavioral measures related to SM are likely to suffer damage to their reputation, loss of customers' confidence, and other economic losses. This chapter looks at privacy and security issues associated with SM. It discusses various features of SM technologies such as newness, complexity, and attractiveness as a cybercrime target. The chapter also examines how various institutions have contributed to a lack of behavioral and attitudinal measures to ensure privacy and security.

11.1 INTRODUCTION

Social media have become increasingly pervasive and ubiquitous. There were 3.80 billion SM users worldwide in early 2020.[1] eMarketer's estimate put SM users in China at 859.1 million or 61.6% of the country's population in 2020.[2] Likewise, as of April 2020, India had

400 million WhatsApp users.[3] Similarly, as of 2019, out of Africa's 213 million Internet population, 191 million used SM.[4]

SM use pervades businesses. As of October 2019, more than 140 million businesses were using Facebook and other apps such as Instagram, Messenger, and WhatsApp every month.[5] Likewise, in 2020, 49.6% of US companies with 100 or more employees used LinkedIn.[6]

It is increasingly common to include social media as part of an overall marketing budget or strategy. The media agency Zenith estimated global social media ad spend at US$84 billion in 2019 or 13% of total global ad spend. SM also became the third largest advertising channel that year, behind TV and paid search.[7]

The real as well perceived security and privacy risks of SM, however, remain significant. Most of the profits of SM companies such as Facebook and YouTube involve monetizing users' data. Such a business model has fundamental privacy issues.[8]

Figure 11.1 Institutional and technological environment facing SM

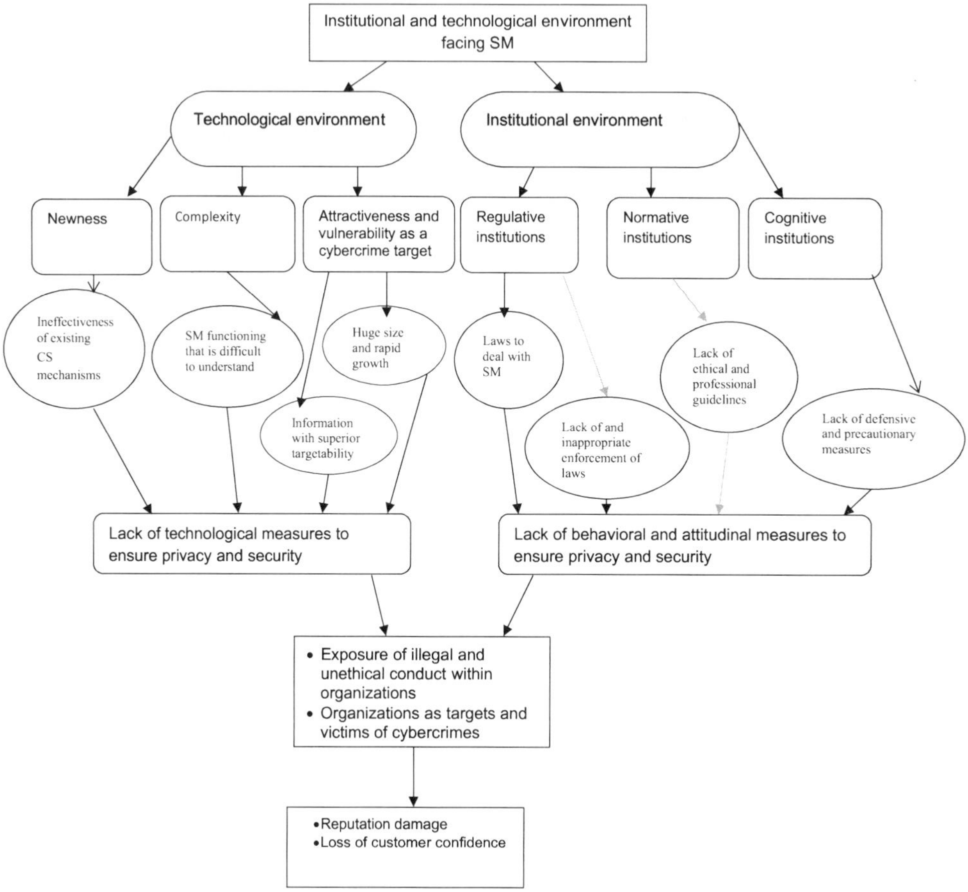

In 2016, media outlets widely reported that pedophiles engaged in disturbing activities, such as using their Facebook groups to share images of child exploitation.[9] In 2017, the BBC reported that it found sexualized images of children being shared on Facebook.[10] Likewise, in April 2019, India's Tamil Nadu state asked the federal government to ban the short-form video-sharing social networking service TikTok, which is owned by Chinese company ByteDance. The state argued that TikTok facilitated the distribution of pornography, which is illegal in India. It also warned that TikTok made it easier for sexual predators to target child Internet users.[11]

Some governments are also concerned about the national security risks posed by foreign social media companies. In 2019 and 2020, a number of US government agencies banned TikTok on government devices. In December 2019, the Pentagon asked US military personnel to delete TikTok from all devices due to CS risks.[12] TikTok faced the accusation that it censors content at the request of the Chinese government. Two US senators also wrote a letter to the Director of National Intelligence asking for an investigation into the app's potential use by foreign intelligence operators and spies.[13]

The above examples illustrate the fundamental security and privacy challenges facing SM and use of this channel for illegal and unethical activities. SM's impact on businesses and consumers and the sources of the impact are the focus of this chapter. Both technological and institutional factors influence CS issues in SM. (See figure 11.1.) Regarding the first, we discuss three main characteristics of technologies enabling SM: newness, complexity, and superior targetability. We then look at the effects of regulative, normative, and cognitive institutions on privacy and security issues in SM.

11.2 IMPACTS ON BUSINESSES AND INDIVIDUALS

The rise in privacy and security breaches in SM can be understood as part of a larger trend of rapid growth in cybercrime. Due to the government's constant public announcements of Internet security breaches and pending cyberterrorism, there is a heightened sense of fear and anxiety about cybercrimes involving SM among individuals and businesses. In this section, we organize the discussion by focusing separately on businesses and consumers and also on various forms of cybercrimes associated with SM.

11.2.1 Impacts on Businesses

Criminal and fraudulent practices in SM have affected businesses in a number of ways. According to a poll of 217 IT professionals conducted by network visibility and traffic monitoring firm Gigamon at the Infosecurity Europe 2019 conference, most respondents believed that social media applications were responsible for bringing in the most malware in their

Figure 11.2 Percentage of respondents that ranked different applications as the source of the most malware victimizing their enterprises

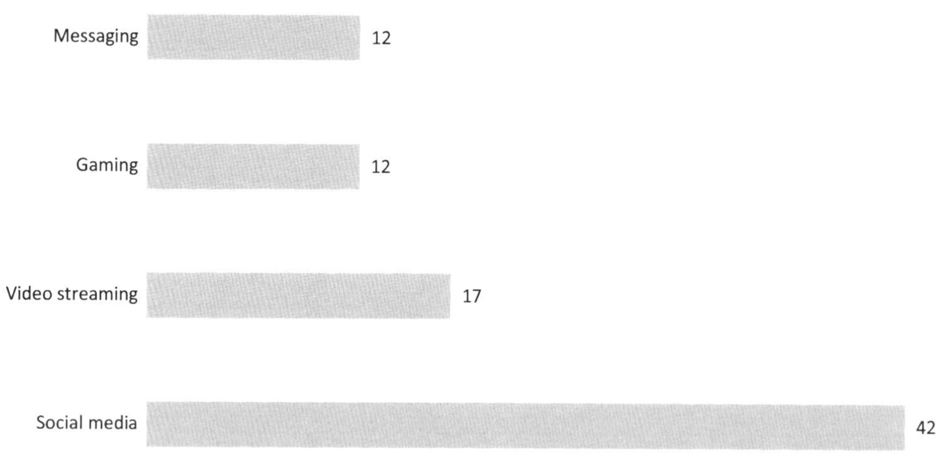

Data source: Ashford, W. (2019, August 30). *Social media and enterprise apps pose big security risks*. Computer Weekly. Retrieved from https://www.computerweekly.com/news/252469873/Social-media-and-enterprise-apps-pose-big-security-risks

enterprises. (See figure 11.2.)[14] Likewise, as noted in the 2019 Verizon Data Breach Investigations Report (DBIR), 33% of external attacks used social media to launch cyberattacks.[15] About a third of those attacks involved malware.[16]

Businesses have also been victims of advertising frauds in social media. Pay per click advertising accounts for large proportions of revenues of major SM websites. (See figure 11.3.) Some advertisers have complained that up to 100% of clicks on their PPC ads were fake.[17]

Increasingly sophisticated malware attacks are being launched against SM ads. In chapter 9, we discussed how LionMobi and JediMobi allegedly engaged in a so-called "click injection fraud."

An upshot of this phenomenon is the tendency of some businesses to move away from SM advertising. The consumer goods company Unilever, which is among the world's biggest online advertisers, warned that it would discontinue ads from social media platforms if such platforms fail to take actions to mitigate the spread of "toxic" online content such as fake news, terrorism, or child exploitation.[18] Unilever is not alone in questioning the appropriateness of SM as an advertising channel. As of December 2020, more than 1,000 groups and companies, including Coca-Cola, J.M. Smucker Company, Mars, HP, and CVS Health, participated in a Facebook advertising boycott campaign to pressure Facebook to take stronger measures to stop the spread of hate speech and misinformation.[19]

Figure 11.3 Advertising revenue of major SM companies (US$ billions, 2019)

Instagram	YouTube	Twitter
20	15	2.99

Data sources: Data for Instagram and YouTube as reported in Carman, A. (2020, February 4). *Instagram brought in $20 billion in ad revenue last year, more than a quarter of Facebook's earnings*. The Verge. Retrieved from https://www.theverge.com/2020/2/4/21122956/instagram-ad-revenue-earnings-amount-facebook; data for Twitter as reported in *Twitter: Annual revenue 2010–2019, by segment*. (2020, February 19). Statista. Retrieved February 19, 2020 from https://www.statista.com/statistics/274566/twitters-annual-revenue-by-channel/

CEOs are changing their personal SM behavior. According to a worldwide survey of CEOs conducted by the professional services firm PwC, a large proportion of the respondents were reducing or completely avoiding unnecessary exposure of personal information. About 48% of the respondents had deleted their social media accounts and stopped using virtual assistant applications.[20]

According to ComPsych, about 20% of workers check their SM accounts more than 10 times a day.[21] Many organizations lack policies regarding their employees' use of SM; such policies would help organizations protect their businesses as well as reputations.

11.2.2 Impacts on Consumers

SM have emerged as a popular avenue for cybercrime activity. SM are an attractive target for cybercriminals simply because they have a huge user base. In a survey conducted in late 2019 by marketing research company OnePoll and MoneyGram in the US, 54% of the respondents reported that their SM accounts had been hacked.[22]

The methods used by SM organizations to store data lack robust CS mechanisms, which has led to data breaches and leaks of personal information. SM became the second biggest industry sector next only to the IoT in terms of data breach.[23] (See figure 11.4.) For instance,

Figure 11.4 The top three data exposures by industry/sector (number of records breached, billions, 2019)

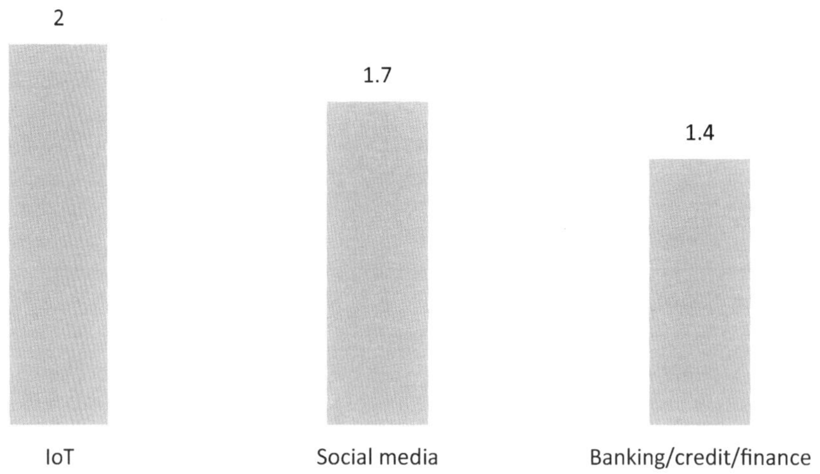

IoT	Social media	Banking/credit/finance
2	1.7	1.4

Data source: Sonicwall. (2020, February 4). *2020 Sonicwall cyber threat report: Threat actors pivot toward more targeted attacks, evasive exploits*. Retrieved from https://www.sonicwall.com/news/2020-sonicwall-cyber-threat-report/

in December 2019, a security researcher reportedly found a database of over 267 million Facebook users with personal information such as names, phone numbers, and unique user IDs. The information was available for free online for at least 10 days for anyone to download. Almost all the victims were based in the US. The database had been downloaded to a hacker forum. It was suspected that cybercriminals had harvested the data and shared the data among them.[24]

As another example, in July 2020, criminals used a technique called "phone spear phishing" to hack Twitter accounts of 130 people who included CEOs, celebrities, and politicians. The hackers took control of 45 of the hacked accounts and used them to send tweets promoting a bitcoin scam. In a blog post,[25] Twitter explained that, using false identities, the hackers called Twitter staffers and were able to get credentials of employees who had access to internal company support tools.[26] The tools allowed them to reset the passwords as well as two-factor authentication information of targeted accounts.

SM users have also become victims of misinformation, disinformation, and propaganda. In October 2019, Stanford University's Internet Observatory identified 200 Russian-linked Facebook profiles. These accounts had over a million followers and were allegedly used to spread disinformation that targeted eight African countries. The accounts were linked to a Russian who was sanctioned for interfering in the 2016 US presidential election.[27]

This was especially evident during the COVID-19 pandemic in 2020. On March 11, 2020, there were 19 million mentions of COVID-19 across social media and news sites.[28]

Widespread misinformation and myths about COVID-19 were shared in SM. Many of them related to fake treatments and faulty information about how COVID-19 spreads. Some were motivated to benefit from the panic of the public. A video, which claimed that breathing hot air from a hair dryer could treat COVID-19, had gone viral. A Twitter user posted that the disease could be treated by injecting vitamin C to the bloodstream.[29] Another SM user claimed that eating bananas, which are high in vitamin B_6, was a cure.[30]

SM algorithms are designed to exploit users' habits and interests. The emphasis is thus on likeability of content rather than its accuracy.[31] SM companies thus may not be interested in taking measures to control the actions of perpetrators of misinformation and disinformation.

SM have also been a major source of potentially harmful behavior and experiences such as cyberbullying. For instance, according to a study conducted in 28 countries by the global market research and consulting firm Ipsos, 65% of parents reported that they knew about cyberbullying taking place on SM sites. This proportion was 76% in Latin America compared to 53% in Asia-Pacific.[32]

11.3 TECHNOLOGICAL ENVIRONMENT

11.3.1 Newness

The number of malware products targeting social networking sites is increasing rapidly.[33] Many SM users are easily victimized by new malware products. An example is the "Shopper" app, which tricks SM users by disguising itself with a system icon. When a device owner unlocks the screen, the app launches, gathers information about the device such as country, network type, vendor, smartphone model, email address, International Mobile Equipment Identity, and International Mobile Subscriber Identity. The information is sent to the attacker's servers. The server then sends commands for the application to execute various commands. For instance, it can use the device owner's Google or Facebook account to register on popular shopping sites such as AliExpress, Lazada, and Alibaba. It can review the application in Google Play on behalf of the device owner.[34] During October and November 2019, Brazil, India, and Russia were top countries in terms of the number of users of the application.[35]

11.3.2 Complexity

Another problem concerns the difficulty in understanding SM's functioning. Experts have, for instance, complained that SM users are not given simple ways to understand how various types of personal information are collected and made available to third parties.[36] SM organizations often claim that they have obtained users' permission to constantly collect personal information. However, users are almost never explicitly asked if they want all of their activities to be tracked and traced. Most SM users find it extremely complex to understand how tracking works, which has enabled SM companies to collect personal data with few restrictions. An example is how consent is requested for data collection. It is often granted by a user because the request is made in an obscure and confusing fashion, or worse, misrepresents what default settings the user agrees to. For example, Facebook's "tag suggestions" default feature was actually a facial recognition tool.[37]

As a further source of complexity, SM platforms rely mostly on AI to moderate content. SM platforms such as Facebook, YouTube, and Twitter announced plans to decrease the use of human moderators to review content.[38] During the pandemic they also relied on a remote workforce.[39] However, when information is incomplete and the information environment is rapidly changing, automation is not effective. It is difficult to find misinformation in real time. A huge user base means that a post will likely be read by many people and the negative impact of misinformation will likely be substantial even in a short period of time.

11.3.3 Attractiveness as a Cybercrime Target

Targetability is related to a legitimate or illegitimate organization's capability to reach an attractive target. Two explanations related to attractiveness of SM can be suggested for the rapid growth of cybercrimes targeting SM. First, huge size and rapid growth make SM an attractive target for cybercriminals. Second, the availability of high-quality information leads to superior targetability.

11.3.3.1 Huge Size and Rapid Growth

A simple, straightforward explanation is that cybercriminals consider SM sites as lucrative targets due to their rapid growth. In the first quarter of 2020, Facebook had 2.89 billion monthly active users,[40] compared to 2.5 billion in the fourth quarter of 2019.[41] As of January 2020, Facebook had 324 million users in India.[42]

Unsurprisingly, SM have therefore become an attractive cybercrime target. A study of the cloud-based email security vendor Vade Secure found that Facebook was the second most impersonated brand in phishing attacks. Among the 25 most impersonated brands in the fourth quarter of 2019, three were SM websites. (See figure 11.5.) SM's share of phishing URLs increased from 13.1% in the third quarter of 2019 to 24.1% in the fourth quarter of 2019.

11.3.3.2 Information with Superior Targetability

A second factor related to attractiveness (and often associated with privacy violations and security breaches on SM sites) is data quality. Data and information about users that are available

Figure 11.5 Number of unique phishing URLs targeting SM websites (2019 Q4)

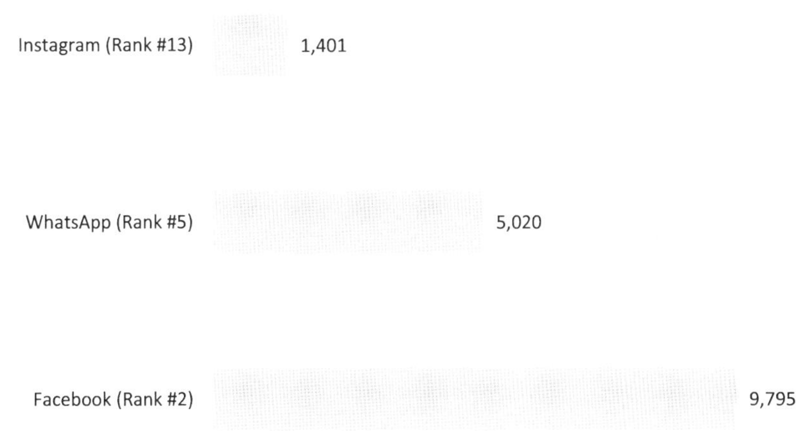

Data source: *The 25 most impersonated brands in phishing attacks.* (2020, February 11). Help Net Security. Retrieved from https://www.helpnetsecurity.com/2020/02/11/brands-phishing-attacks/

on SM sites are of superior quality. Even legitimate enterprises find resulting opportunities so tempting that they do not self-regulate, and often actively seek to covertly gain access to the best-quality information available in SM networks. Criminal outfits see the same opportunities. It is important to note that target attractiveness depends on offenders' perceptions of victims.[43] For instance, financially motivated cybercriminals perceive banks as attractive targets since consumers' financial data stolen from banks can be exploited for profit.

Foreign intelligence agencies are increasingly interested in social media. For instance, it is possible to find detailed information related to a LinkedIn user's friends, followers, and family as well as the user's position in an organization.[44] It is thus possible to find a lot of information about a company in a legitimate way – without engaging in illegal activities such as hacking. According to some US government officials, Chinese espionage activities have specifically targeted LinkedIn users. Most obviously, creating fake profiles and sending out many friend requests would be faster and more efficient than physically sending spies to recruit targets. There are reports that LinkedIn was used to recruit some US intelligence officers who flipped and decided to work for the Chinese government. This phenomenon is associated with and facilitated by some government officials' tendency to put detailed information about their careers on LinkedIn. Many do so if they are interested in switching to jobs with higher pay. Foreign spies thus have easy access to relevant information to decide whom to approach. LinkedIn reported that in 2018, it restricted 24 fake profiles because it suspected that they were created by the Russian government.[45]

11.4 INSTITUTIONAL ENVIRONMENT

11.4.1 Regulative Institutions

11.4.1.1 Lack of Laws

The SM-related legal system is evolving slowly. Many countries are in the early stages of introducing regulatory measures to fight malicious use of social media.[46] For instance, fake news creators in North Macedonia have argued that they have not broken any of that country's laws. (See In Focus 11.1.) Likewise, as of April 2020, India had not passed a law to deal with fake news on SM. Cases related to misinformation about COVID-19 were registered under other provisions such as the Indian Penal Code and the Epidemic Diseases Act of 1897.[47]

IN FOCUS 11.1: THE "FACEBOOK MARKETING UNIVERSITY" IN NORTH MACEDONIA

The proliferation of fake news is associated with and facilitated by the policies of ad networks. Pablo Reyes, who ran fake news sites such as Huzlers.com, noted that ad networks "are not really for the most part looking for quality. They're looking to see if it meets the minimum threshold: It's not porn, it's not hate or guns." The most important thing for them is the traffic, not the content. Ad networks just want to ensure that the traffic is coming from real people, not from bots.[48]

Creation of fake news has provided a great opportunity for younger people in some economies burdened with high unemployment rates. The "Facebook Marketing University" in North Macedonia was a response to this opportunity, and was reported to employ seven full-time people. In addition, the "University" had five freelancers and another seven who worked online. It was founded by Mirko Ceselkoski. He reported that some of the students were making US$100,000 per month.[49] Ceselkoski has also written a book that deals with attracting people on social media websites. His website says it is "The Ultimate Guide of Building a Huge and Viral FB Page." By April 2019, he had sold 5,500 copies of his book in English and 2,000 copies in Macedonian. Ceselkoski claimed that in six months the book would help Macedonians make US$1,000 a month.

11.4.1.2 Lack of Enforcement and Inappropriate Enforcement of Laws

Criticisms about lack of enforcement and inappropriate enforcement of laws are also significant. In many countries, censorship and oppressive measures against political opponents and dissidents are among the main concerns.

Countries such as Brazil, Germany, and South Korea require SM platforms to take down content that the state views as illegal.[50] In some countries, there has been a lack of clarity regarding what is illegal. For instance, commenting on the government of South Africa's plan to regulate SM to stop the spread of misinformation, one commentator noted that misinformation meant "information that the government does not agree with."[51]

During the COVID-19 pandemic, many governments showed a tendency to suppress SM activities that went against their interests. For instance, by mid-April 2020, law enforcement agencies in 10 Asian countries such as India, Mongolia, and Thailand had arrested at least 266 people for posting coronavirus-related misinformation on SM websites. According to Agence France-Presse, by mid-April 2020, about 100 people had been arrested in India for allegedly engaging in such activities.[52] Among the arrested were those who had criticized their governments' COVID-19 response. For instance, an Indian local politician was arrested for claiming on Facebook that the government had downplayed the bad news related to coronavirus fatalities. Likewise, a Malaysian TV personality paid a substantial fine for posting a YouTube video that criticized how a hospital had handled the pandemic.[53]

In some countries, more scarce law enforcement resources are used in censorship than in controlling criminal activities that victimize SM users. For instance, countries such as Russia, Sudan, Vietnam, and Zimbabwe have used SM-related legislation to intensify online censorship.[54] According to the US-based democracy watchdog Freedom House, the governments of Zimbabwe and Sudan use SM platforms to track and monitor citizens.[55] In 2017, a Zimbabwean blogger was detained and put in a maximum-security prison for five days for allegedly "trying to overthrow a legitimately elected government" using SM.[56] Authoritarian regimes of Iran and Tanzania have also used SM laws to suppress political dissent.[57]

A point to note here is that in Africa, SM such as Twitter have emerged as a channel for political discourse. Africans were reported to have a higher rate of political tweets than in the

US and the UK.[58] Africa's repressive regimes have focused their attention more on controlling SM-based mobilization rather than on controlling cybercrimes.[59]

Finally, underreporting of SM-related crimes also leads to difficulties in enforcing laws. In the UK and Canada, only 5% of SM scams are reported.[60]

11.4.2 Normative Institutions

11.4.2.1 Lack of Ethical and Professional Guidelines

Professional and ethical guidelines regarding posts on social media have not been well-developed. In the medical world, for instance, there arguably is a lack of clear guidelines as to what constitutes unethical or unprofessional online conduct for physicians.[61]

According to the Federation of State Medical Boards, which represents the 71 state medical and osteopathic boards of the US and its territories, physicians' inappropriate use of SM has been a growing concern. There have been many instances of physicians' uses of SM that violated professional liability rules, which resulted in their being fired and/or disciplined by state medical boards.[62] Likewise, 63% of conduct-related complaints against medical students were related to unprofessional content they posted online. Major offenses included violation of patient confidentiality, use of profanity and sexually explicit language, and use of language that is discriminatory against certain groups. Other online professionalism-related issues that are addressed by state medical boards include the use of SM to inappropriately contact patients and inappropriately advise the use of a medicine or treatment.[63] Illegal or questionable activities thus take place on SM sites without the violators' intent.

Moving to another professional setting, several cases of misconduct and ethical violations in employees' use of SM to inappropriately contact students have been reported in San Diego County schools in the state of California in the US; however, as of early 2019, out of 43 school districts in San Diego County, fewer than 10 had policies that explicitly addressed SM use by teachers to contact students outside the classroom.[64]

11.4.3 Cognitive Institutions

SM users' skills, expertise, experience, knowledge, and technical expertise can be considered as components of interest for cognitive institutions.

11.4.3.1 Precautionary and Defensive Measures

Most SM users perform poorly in risk-assessment exercises involving their online information. A failure to understand the context of SM use is also likely to lead to an exposure of potentially unethical and even illegal behaviors and practices. A survey of Instagram users conducted by

Santander UK and OnePoll found that about 90% of users under 25 posted personal information such as age or pet's name.[65] Likewise, a data scientist estimated that at least 20,000 children in Australia had public contact information such as email addresses and phone numbers posted in their business Instagram profiles.[66] The minors had converted their Instagram profiles from personal to "business" category. Instagram's policy is that "Business Profiles can't be set to a Private Account."[67]

In this regard, the teens' and youths' cognitive skills, competencies, and associated beliefs are considered major influencing factors on their SM behaviors and outcomes. According to a recent study, 42% of respondents in the 18-to-29 age group were unable to answer any question about privacy law correctly.[68]

A parallel can be drawn with physical crimes. Most cybercriminals who prey upon SM users employ a similar modus operandi as in physical crimes that involve gaining access to and trust of the victim. The victimization process can be viewed as a temporal sequence of activities through which the crime progresses. For instance, offenders against children (such as pedophiles) usually use the power of persuasion and friendship to first gain the trust of the child and, in some cases, parents as well. Friendship is a critical step toward gaining cooperation.[69] We would argue that fraudsters find it easy to gain trust and cooperation of SM users who have no past experience of distrust.

Additionally, part of the fascinating character of social networks stems from the fact that they are built on trust-based architectures. The real or perceived relationships of trust that connect social networking users increase scammers' chances of gaining trust and cooperation from potential victims. For example, a fraudulent message sent from a SM user's social network is often considered to be more trustworthy than a random spam message.[70]

On the positive side, some users are taking precautionary measures. Fifty-one percent of the respondents in the OnePoll and MoneyGram survey reported that they avoid oversharing on SM sites.

11.5 SOCIAL MEDIA COMPANIES' EFFORTS TO FIGHT CYBERCRIMES

SM firms use increasingly sophisticated tools to fight cyberattacks. For instance, LinkedIn reported that in the first half of 2019, it restricted 2 million fake accounts before such accounts were reported by other users. LinkedIn did so by combining human review with AI and ML. The majority of the fake accounts were stopped at registration, before they ever went live.[71] Microsoft, which owns LinkedIn, employs more than 3,500 cybercrime experts. It also uses AI to help "anticipate, identify, and eliminate threats."[72] Microsoft is reported to invest more than US$1 billion annually to strengthen CS and manage cyberrisks.[73]

11.6 CHAPTER SUMMARY AND CONCLUSION

SM have been a goldmine for cybercriminals as well as other online perpetrators. One way to understand security and privacy issues in SM is to examine them against the backdrop of current institutional arrangements and technological development.

In light of organizations' use of SM in diverse activities, it is important to have clear policies and guidelines regarding posts on SM networks by employees. It may also be important to regularly monitor online posts on SM sites by employees that might lead to a leak of organizational secrets. This is especially important for organizations dealing with sensitive information, such as hospitals and government agencies.

Management of security risks is a critical practical challenge that organizations face in the digital economy. The above analysis indicates that the SM security and privacy challenges are significant.

In order to prevent cybercrimes, it is important for users and platform owners to consider the ethical, legal, and technical issues associated with SM. Ensuring that both technological and behavioral/perceptual factors are given equal consideration in the design and implementation of a computer network is crucial. Technological measures range from simply disconnecting databases containing sensitive information from the Internet to deploying sophisticated antifraud technologies. Similarly, simple behavioral measures can stop some serious cybercrimes. A simple training strategy aimed at improving the ability of employees to use SM more securely is likely to reduce a significant proportion of cybercrimes.

Organizations that fail to take appropriate security and privacy measures are likely to suffer reputation damages, loss of customers' confidence, and other economic losses. Advertisers must scrutinize and evaluate the sites on which their ads are delivered. They should look at whether the sites feature high-quality content that can enhance the brand's reputation. They should also ask publishers if they buy traffic. If the publishers' traffic is bought, it is important to know the quality of such traffic. Ad tech providers should be asked about the antifraud measures they deploy.

Finally, there is an emerging trend toward the use of content posted on social media as evidence in civil and criminal proceedings. This trend is likely to affect the naive (e.g., youths and teens) more than the general population. The above discussion also indicates that a failure to understand the context of SM use is also likely to lead to an exposé of wrongdoing.

11.7 DISCUSSION QUESTIONS

1. Why are SM websites an attractive attack vector for malicious actors in cyberspace?
2. What are some of the major types of cybercrime activities that target SM users?
3. Why are some executives deleting their social media accounts?

4. Why are many jurisdictions enacting legislation to govern SM?
5. What is your assessment of SM sites' performance in terms of the protection of users' privacy?

11.8 END-OF-CHAPTER CASE: THE FACEBOOK–CAMBRIDGE ANALYTICA DATA SCANDAL

Cambridge Analytica was a political consulting firm in London, UK, founded in 2013 by the behavioral research and strategic communications company SCL Group. The now-defunct firm used data analytics techniques to influence elections in Brazil, Kenya, Malaysia, the US, and many other countries.[74] It worked in over 200 elections around the world.[75] In March 2018, British television network Channel 4 News broadcast recordings in which Cambridge Analytica's senior executives, including its CEO Alexander Nix, suggested that in order to help political candidates win votes, the firm could use nefarious means, such as sex workers, bribes, and misinformation.[76] Cambridge Analytica was involved in one of the most high-profile data scandals that involved questionable use of the personal information of 87 million Facebook users worldwide. The scandal led to the company's shutdown in May 2018.

11.8.1 The Data Scandal

The Facebook–Cambridge Analytica data scandal is among the most discussed instances of deceptive handling of personal data. In early 2014, Cambridge Analytica allegedly obtained the private information of about 87 million Facebook users. The data were collected illegally and without the users' knowledge. A subset of the data was used to build a system that profiled individual US voters and to predict as well as influence their choices on election day.[77] Based on their profiles, US voters were targeted with personalized political advertisements.[78] According to a Cambridge Analytica former employee, they specifically targeted users who were "more prone to impulsive anger or conspiratorial thinking than average citizens."[79] Methods used included Facebook group posts, ads, and sharing articles to provoke voters to vote for or against a party. To do so, it also allegedly created fake Facebook pages such as "I Love My Country."[80]

11.8.2 The Root of the Scandal

At the root of the scandal was Facebook's platform launched in 2010 called Open Graph. Open Graph allowed external developers to reach out directly to Facebook users. Developers could request permission to access users' personal data. If a Facebook user accepted the request, external developer apps would have access to personal data such as name, gender, location, birthday, education, political preferences, relationship status, religious views, online chat status, and the user's private messages. The developers could also access the user's Facebook friends' data.[81]

11.8.3 Tricking Unsuspecting Users

In 2013, Cambridge University's Aleksandr Kogan and his company Global Science Research (GSR) created an app called "thisisyourdigitallife" (TIYDL), which provided a "free" personality test. Cambridge Analytica reportedly paid Kogan more than US$800,000 to create the app.[82]

About 270,000 Facebook users downloaded Kogan's app and took a personality survey. The app tricked unsuspecting users into answering questions that were used to build a psychographic profile.[83] The users were paid a small amount to complete the survey.[84] The app collected personal data as well as data on the user's Facebook friends.[85]

All the collected data were then shared with Cambridge Analytica.[86] Cambridge Analytica took all the data of Facebook users who took the test as well as that of all their friends. In this way, Kogan harvested 87 million users' data worldwide, without their knowledge. A subset of this data was shared with organizations such as the SCL Group, the parent company of Cambridge Analytica.[87]

The data were used to classify each individual's personality into Openness, Conscientiousness, Extroversion, Agreeableness, and Neuroticism. This classification is considered an effective way to target messages to appeal to specific personality types.[88]

While Kogan mishandled the data, all of the information he got was accessed legally following Facebook rules.[89] The violation took place when the data were handed over to Cambridge Analytica.[90]

Facebook first learned about unconfirmed reports of a potential personal data violation by Cambridge Analytica and tried to address the issue.[91] Kogan allegedly "lied" to Facebook, saying that he was gathering the data only for research purposes. Facebook thus argued that Kogan violated the company's policies by transferring the data to Cambridge Analytica. Kogan argued that the app's terms and conditions were written to allow commercial uses.[92]

Facebook did not know about the seriousness and full scope of the problem until the British daily newspaper *The Guardian* published an article on December 11, 2015. *The Guardian* story detailed how tens of millions of Facebook users' data were harvested without their permission to build psychographic profiles to support the presidential campaign of Ted Cruz, the US Senator for Texas since 2013 and the runner-up for the Republican presidential nomination in the 2016 election.[93] Cruz's campaign had contracted Cambridge Analytica for this purpose. A former Cambridge Analytica employee said that the firm got the data through Aleksandr Kogan.[94]

11.8.4 Trade-off between Monetizing Big Data and Promoting Users' Privacy

Facebook employees reported that in issues related to data privacy there was a tension between the company's CS team and others who wanted to monetize user data. Employees responsible for protecting user privacy reported that they faced difficulties in convincing others in the company who wanted to make as much money as possible from users' data.[95]

On September 22, 2015, a Facebook employee reportedly requested clarification regarding Facebook's policies about the use of personal data for political consultancies that were engaged in data scraping. Cambridge Analytica and other firms had extracted data from different sources to match Facebook profiles to the lists of voters used by political election campaigns, known as voter files.[96] There were differing opinions among Facebook employees regarding data scraping. The issue raised by the employee was downplayed by the company and did not get much attention until December 11, 2015, when *The Guardian* published its story detailing how Cambridge Analytica misused personal data. (See table 11.1.)

Table 11.1 Facebook–Cambridge Analytica data scandal: A timeline

April 2010	• Facebook launched Open Graph platform.	• The platform made it easier for app developers to collect Facebook users' data.
2013	• Kogan and GSR created TIYDL.	• All the collected data was then shared with Cambridge Analytica.
2013	• Cambridge Analytica was founded.	• Cambridge Analytica started as an offshoot of the SCL Group.
September 2015	• Facebook learned about a potential personal data violation by Cambridge Analytica.	• The full scope of the problem became apparent only after *The Guardian* published an article on December 11, 2015.
December 11, 2015	• *The Guardian* detailed how tens of millions of users' data were harvested, without their permission, to build psychographic profiles.	• A former Cambridge Analytica employee stated that the firm got the data through Kogan.
Late 2015	• Facebook banned TIYDL from its platform and asked Cambridge Analytica to delete all data.	• There were reports that Cambridge Analytica did not destroy the data.[97]
May 2018	• SCL Group and Cambridge Analytica become defunct.	
October 2018	• The UK's ICO issued Facebook a fine of about US$643,000.	• Among many accusations, Facebook failed to take measures to ensure that apps and developers using its platform could not misuse personal data.

(Continued)

Table 11.1 Continued

July 2019	• The FTC fined Facebook US$5 billion for engaging in deceptive practices.	• The FTC accused Facebook of violating a 2012 agreement with the commission, in which Facebook had agreed not to give user data to third parties without consent.
October 2019	• Facebook agreed to pay the ICO fine.	• Facebook and the ICO had been in litigation and appeals for one year after ICO issued the fine in October 2018.[98]

11.8.5 Regulatory Actions and Consumer Backlash

A number of regulatory actions have been taken against Facebook in response to the company's violation of consumer privacy. The UK's Information Commissioner's Office's (ICO) investigation found that Facebook failed to take appropriate measures to ensure that apps and developers using its platform could not misuse personal data. Due to this failure, Kogan and GSR could harvest users' data without their knowledge.[99]

The ICO issued Facebook a fine of £500,000 (about US$643,000) for serious violation of the country's data protection law. The ICO argued that Facebook breached the UK's data protection laws because it failed to keep users' personal information secure. Among many accusations, the ICO noted that Facebook failed to make appropriate checks on apps and developers using its platform. It specifically mentioned illegal harvesting of personal data by Kogan and GSR.[100]

On May 23, 2018, the UK passed its Data Protection Act 2018, which codified the EU's GDPR legislation into its legal system.[101] However, because Cambridge Analytica had harvested the data in 2015, the fine was issued under the Data Protection Act 1998, which carried a lower maximum penalty. Under the Data Protection Act 2018, Facebook would have faced a fine of up to £17 million (about US$22 million).[102]

In July 2019, the FTC fined Facebook US$5 billion for engaging in deceptive practices.[103] The FTC accused Facebook of violating a 2012 agreement with the commission in which it had agreed not to give third parties access to user data without the user's consent. The seriousness of Facebook's misleading practices became more obvious after the Cambridge Analytica scandal.[104]

The head of the European Parliament also said it would conduct an investigation about potential misuse of data.[105] In December 2019, Brazil announced that it had issued a fine of US$1.6 million to Facebook for improperly sharing user data with Cambridge Analytica.[106]

Facebook also faced a backlash from users. In late 2018, the #deletefacebook hashtag was trending on Twitter.[107]

11.8.6 Facebook's Response

Facebook took a number of measures to fix its reputation and its business. For instance, after learning of Kogan's inappropriate sharing of personal data, Facebook banned TIYDL from its platform and asked him to delete all data obtained from the app. Facebook also demanded deletion of such data from Cambridge Analytica and other companies with which Kogan had shared data. Facebook asked all parties to send certification when the deletion had been completed.[108] Facebook said that it received confirmations that the data were destroyed. Cambridge Analytica said that it deleted all the data provided by Kogan;[109] however, there were reports that Cambridge Analytica did not do so.[110]

Satisfying regulators and winning back the trust of its billions of users worldwide has not been easy for Facebook. The company said that it made changes to its platform to restrict the type of information that app developers can access. It also stated that it will work with regulators such as ICO to identify privacy issues.[111]

CHAPTER TWELVE

Cloud Computing

CHAPTER PREVIEW

Cloud computing is likened to the Industrial Revolution. Its transformational nature is, however, associated with significant security and privacy risks. This chapter provides a perspective on how formal and informal institutions, such as regulations and industry codes of conduct, affect cloud security. It investigates how the contexts provided by formal and informal institutions affect the perceptions of privacy and security in the cloud. This chapter highlights the nature, origin, and implications of institutions and institutional changes in the context of cloud computing. It also provides insights into the mechanisms and forces that have brought about institutional changes in the cloud industry. Specifically, it looks at how the formation of dense networks and relationships and changing power dynamics have triggered institutional changes.

12.1 INTRODUCTION

Cloud computing (hereinafter: cloud) is likened to the Industrial Revolution in terms of implications for technological innovations and economic growth. The adoption of the cloud is rapidly increasing among businesses and individuals. According to the security firm Barracuda Networks' study, 45% of IT infrastructure used by companies was in the public cloud in 2020, which is expected to increase to 76% by 2025.[1] Likewise, as of early 2020, the cloud storage service provider Dropbox had over 600 million users who stored 400 billon files and uploaded 1.2 billion files each day.[2]

The transformational nature of the cloud is associated with significant security and privacy risks. A significant gap remains between CSPs' claims and users' views. The cloud industry's claim is that clouds are more secure than other available alternatives[3] but many users disagree. Security, privacy, and availability are topmost concerns in organizations' cloud adoption decisions. (See table 12.1.) Due to concerns related to security and privacy, critics have argued that

Table 12.1 Organizations' perceptions of the cloud's security: Some representative surveys

Survey Conducted by	Conducted/ Released in	Major Findings
Information Week	2009 and 2010	• 31% of companies in 2010, compared to 35% in 2009, viewed SaaS apps as less secure than internal systems.[4]
International Data Corporation	2011	• A third of IT executives felt the benefits of the cloud exceeded risks. • About a quarter did not fully understand the regulatory and compliance issues. • 47% were concerned about a security threat.[5]
McAfee's survey of senior IT staff and employees in businesses with over 250 employees from France, Germany, and the UK	2019	• 93% expected to increase their reliance on the cloud and had plans to move their more sensitive data to the cloud in the future. • 84% said that the cloud improved data security. • 45% of UK respondents stored sensitive data on the cloud and security issues were preventing them from moving more data. • 22% of senior leaders in the three countries felt that their businesses will never be fully cloud-only due to security concerns.[6]
Savoy Stewart's survey of five industries in France, Italy, Germany, Spain, the UK, and the US	2019	• 5% in the hospitality sector and 31% in the tech sector distrusted the cloud. • Most financial professionals who distrusted the cloud were in Italy (89%), followed by the US (88%), the UK (85%), Germany (83%), Spain (79%), and France (75%).[7]

(Continued)

Table 12.1 Continued

Survey Conducted by	Conducted/ Released in	Major Findings
Hyperscale computing provider Nutanix's survey of IT decision-makers in 24 countries conducted by IT market research consulting firm Vanson Bourne	2019 (Enterprise Cloud Index 2019)	▪ 60% viewed cloud security as the key factor that had the biggest influence on cloud deployment plans. ▪ 28% rated hybrid environments as the most secure architecture. ▪ 73% of enterprises were moving some applications on premises.[8]
McAfee's study, titled Enterprise Supernova: The Data Dispersion Cloud Adoption and Risk Report (based on a survey of 11 countries)	2020	▪ 91% of CSPs failed to encrypt data. ▪ One in five respondents lacked visibility into the data stored in cloud applications.[9]
Barracuda Networks' survey conducted in the Americas, Europe, the Middle East, and Africa and Asia-Pacific (APAC) among executives and experts with knowledge of or responsibility for their organizations' cloud infrastructure	2020	▪ 70% had limited the adoption of the public cloud due to security concerns.[10]

the costs of the cloud may outweigh its benefits.[11] Organizations worry about hidden costs associated with security breaches or lawsuits tied to data breaches. Businesses and consumers are cautious about using it to store high-value or sensitive data and information.

These perceptions are actually supported by the findings of recent surveys indicating that data breach risk is a critical issue in cloud computing. According to McAfee's 2018 Cloud Security report, data theft in the cloud affects 1 out of every 4 users.[12] This is an even more serious problem if we consider the fact that the majority of users store sensitive information in the cloud. For instance, according to a survey reported by McAfee, 84% of the respondents stored sensitive data in the cloud.[13]

According to a study conducted by the cloud access security broker Netskope in 2020, 44% of CS threats start in the cloud. The study also found that AWS accounts for 94% of all cyberattacks that originate in the public cloud.[14]

Understandably, the risks of cloud computing remain significant. According to the Marsh Microsoft *2019 Global Cyber Risk Perception Survey* conducted worldwide with more than 1,500 respondents, on a 1 to 5 scale (1 – minimal to none; 5 – extremely high), more than half of the respondents rated perceived cyberrisk associated with cloud computing at 4 or 5.[15] (See figure 12.1.)

Figure 12.1 Percentage of respondents perceiving different levels of cyberrisk associated with cloud computing

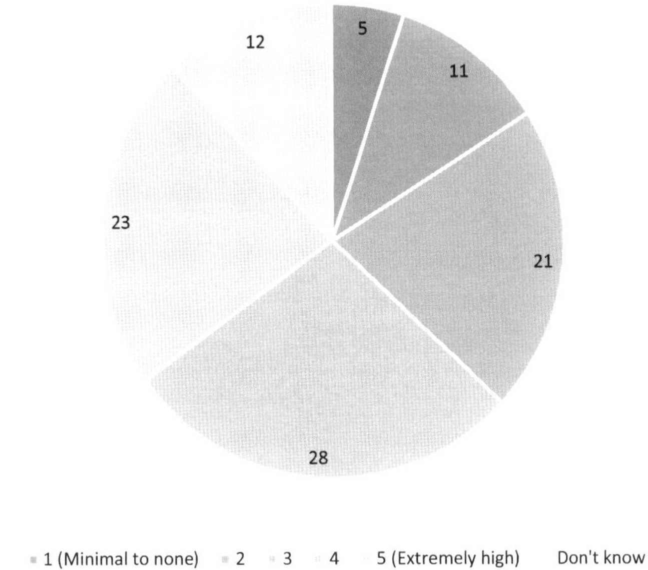

Data source: *Global cyber risk perception survey*. (2019). Marsh. Retrieved from https://www.marsh.com/us/insights/research/marsh-microsoft-cyber-survey-report-2019.html

In addition to security and privacy, users are also concerned about the availability of the cloud. For instance, in 2019, China Telecom experienced an outage for about eight hours that affected many of the services provided by companies such as Apple, Amazon, Microsoft, Slack, and SAP.[16] As presented in figure 12.2, major CSPs have experienced significant downtime.

This chapter looks at how institutions are related to CS and privacy in the cloud. Many privacy, security, and ownership issues in the cloud fall into legally gray areas. Some commentators have argued that there has been a "disturbing lack of respect for essential privacy" among CSPs.[17] For instance, in a complaint filed with the FTC, the Electronic Privacy Information Center (EPIC) argued that Google misrepresented privacy and security of its users' data stored in the Google cloud.[18]

CSPs deliver value to users through various offerings, such as software as a service (SaaS), platform as a service (PaaS), and infrastructure as a service (IaaS). SaaS is a software distribution model in which applications are hosted by a vendor and made available to customers over a network. It is considered the most mature type of cloud computing. In PaaS, applications are developed and executed through platforms provided by cloud vendors. This model allows quick and cost-effective development and deployment of applications. Some well-known PaaS vendors include Google (Google App Engine), Salesforce.com (Force.com),

Figure 12.2 Cloud downtime of major cloud service providers (minutes, early 2015–early 2017)

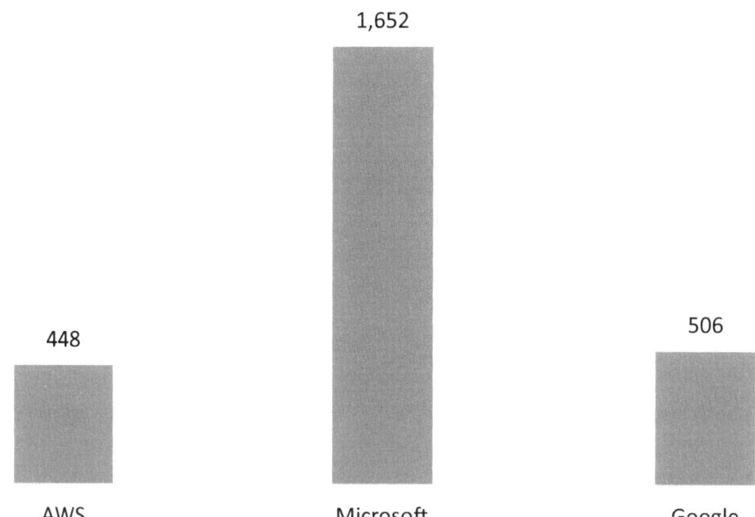

Data source: McLaughlin, K., & Sullivan, M. (2017, March 7). *How AWS stacks up against rivals on downtime*. The Information. Retrieved from https://www.theinformation.com/how-aws-stacks-up-against-rivals-on-downtime

and Microsoft (Windows Azure platform). Some facilities provided under the PaaS model include database management, security, workflow management, and application serving. In IaaS, computing power and storage space are offered on demand. IaaS can provide servers, operating systems, disk storage, and databases, among other things. Amazon Web Services (AWS) is the biggest IaaS provider. Its Elastic Compute Cloud allows subscribers to run cloud application programs. IBM, VMware, and HP also offer IaaS.

12.2 INSTITUTIONAL ISSUES IN THE CLOUD INDUSTRY AND MARKET

Various institutional actors have demonstrated various responses to the cloud. While trade and professional associations and industry bodies have responded to security and privacy issues, government agencies have been slow to adopt legislative and regulatory measures to monitor users and providers.

We first look at the evolution of the cloud industry and market in terms of three institutional pillars discussed in chapter 5: regulative, normative, and cognitive. A regulative pillar establishes legal and regulatory infrastructures to deal with security and privacy. A normative institutional pillar is said to be established if ethical and social views start influencing these issues. Likewise, a cognitive pillar is established if cloud-related decisions are culturally and cognitively determined. The following discussion looks at these three pillars in more detail.

The EU's strong data privacy laws prevent the movement of identifiable individuals' data to jurisdictions that do not provide the same levels of protection. CSPs emphasize their security credentials by communicating to clients that they have completed the Statement on Auditing Standards (SAS 70) audit. Note that the SAS 70 Audit developed by the American Institute of Certified Public Accountants (AICPA) represents that a service organization has been "through an in-depth audit of their control objectives and control activities" that may include controls over IT and related processes. By doing so, they are trying to gain normative legitimacy. An organization's cloud adoption decision may also depend on its perception of the providers' ability to protect data and make them available when needed. This mental map can be described as cognitive institutions.

Table 12.2 shows institutional changes that may be taking place in the cloud industry as well as the associated mechanisms and forces.

Table 12.2 Institutional evolution in the cloud industry

Type of Institutions	Likely Changes	Mechanisms/Forces
Regulative	▪ Governments will likely face pressures to strengthen cloud-related legal system and enforcement mechanisms.	▪ The transformational nature of the cloud forces the government to develop regulative institutions. ▪ Spillover and externality effects may arise from government agencies' demands for a secure cloud for their use.
	▪ Governments may face pressures to harmonize and align their legal systems and enforcement mechanisms with those of other countries.	▪ The globalization effect and position of international institutions may affect national policies.
Normative	▪ Industry processes for auditing cloud platforms will likely evolve.	▪ Participants in an industry such as cloud computing benefit from compatibility and hence may push for a standard. ▪ Participants are also interested in strengthening security and privacy issues.
	▪ Professional and trade associations are likely to emerge and influence security and privacy issues in new ways.	▪ Associations tend to have expertise, resources, experience, and interests in security and privacy issues. ▪ They may assess and monitor the performance of cloud vendors, which can help minimize risks for the members (e.g., AICPA).

(Continued)

Table 12.2 Continued

Type of Institutions	Likely Changes	Mechanisms/Forces
Cognitive	▪ Facing demanding consumers who are more aware and better educated about privacy and security, cloud vendors are likely to become equipped with an improved mental map of the cloud environment.	▪ Cloud vendors are forced to engage in trust production processes. ▪ Power will be redistributed away from CSPs.
	▪ Users may overcome the inertia effect.	▪ Users' expectations, values, and concerns about localness and control may change as institutions thicken (that is, establish more formal rules and informal relationships to guide behaviors of relevant participants) and understanding of and experience with the cloud increase.

12.2.1 Regulative Institutions and Their Evolution

Some examples of regulative institutions include SOX and the HIPAA in the US. To ensure the accuracy of financial data as required by the SOX compliance, IT controls must be designed to ensure that data are accurate and are protected from unauthorized changes. HIPAA requires healthcare providers to have technical, physical, and administrative security measures in place to protect the privacy, integrity, and availability of patients' data.

12.2.1.1 Laws Governing Data in the Cloud

The importance of regulative institutions such as laws, contracts, and courts in the cloud industry should be obvious if this industry is viewed against the current state of security standards. The state uses coercive power to gain compliance. In the absence of radical improvements in security technology, regulative institutions are even more important because cloud users can rely on them in case a CSP fails to deliver a given level of security.

The cloud-related legal system and enforcement mechanisms are evolving slowly compared to the technological development. Some compliance frameworks such as SOX and HIPAA, which were developed for the non-cloud environment, are sometimes used to assess compliance in a cloud environment. These frameworks do not clearly define the guidelines and requirements for data in the cloud.

According to the US Federal Information Security Management Act (FISMA), cloud providers are required to keep sensitive data belonging to the federal government within the US. Google Apps used by government agencies are FISMA certified. Due to increasing use of the cloud, forward-thinking governments worldwide are realizing the need for a clearer regulatory framework to address security issues. In addition, spillover and externality effects may arise from demands for secure clouds of governments worldwide. For instance, a side effect of Google's delivery of secure clouds to the US federal government is that private-sector players are likely to become aware and interested in such services. Overall, cloud computing poses various challenges for companies that have responsibilities for IT disaster recovery and data security.

12.2.1.2 International Harmonization and Alignment of Legal Systems and Enforcement Mechanisms

Due to the globalization effect, governments face international pressures to harmonize and align legal systems and enforcement mechanisms. National governments increasingly turn to supra-national institutions to resolve transnational problems.

Governments are enacting new laws and revising existing regulations to enhance the international competitiveness of domestic firms. (See table 12.2.) In many cases, these actions have been in response to interest group pressures. These pressures come from industry associations such as the European Telecommunications Network Operators' Association (ETNO) as well as technology organizations engaged in organized lobbying efforts to influence cloud-related policies of the EU and its members. The ETNO lobbied for an international online privacy standard and simplification of rules governing data transfers. It argued that these measures would enable European companies to compete on the same level as companies in the US.[19] In response to these and other pressures, the EU and its members have shown a willingness to enact cloud-friendly laws, revise existing laws, and collaborate with other institutional actors.

12.2.2 Normative Institutions and Their Evolution

Normative institutions include trade and professional associations (e.g., the AICPA and the ETNO), industry groups, and non-profit organizations (e.g., EPIC) that can use social and professional obligation requirements (e.g., ethical codes of conduct) to induce certain behaviors in the cloud industry and market.

There have already been some successful attempts at the association and inter-organizational levels to challenge the current institutional arrangement. In the future, higher, broader, and more significant institutional changes can be anticipated if only because the cloud has brought transformational shifts.

12.2.2.1 Associations Representing Cloud Users

Professional associations are constantly emerging and influencing security and privacy issues in the cloud in new ways as a result of their expertise and interests in this issue. A visible example is the Cloud Security Alliance (CSA), a vendor-neutral group of information security professionals. The CSA is working on a set of best practices as well as information security standards and vendor risk management processes for providers.[20] The CSA has prepared a list of over 200 questions that cover key issues such as data integrity, security architecture, audits, legal/regulatory compliance, governance, and physical security. The IT industry association CompTIA has recommended that organizations should consider these as resources in evaluating vendors.[21]

In another example, the American Institute of Certified Public Accountants is making efforts to accelerate cloud adoption among its 350,000 members. The AICPA's resources, expertise, and experience facilitate its assessment and monitoring of the performance of cloud vendors, which can help minimize risks for its members. Paychex, a payroll-solutions provider, was the first cloud provider to win the AICPA's official endorsement. The AICPA also endorsed bill.com for invoice management and payment. It endorsed financial management and accounting software maker Intacct and tax-automation supplier Copanion. The AICPA's endorsements are based on extensive due diligence on the security practices of the vendors.[22]

12.2.2.2 Inter-organizational Bodies Representing Cloud Vendors

An inter-organizational system gives vendors the power of collective action and cooperative endeavor to pursue shared goals that might not be possible if each vendor acted in isolation. Cloud computing could benefit from compatibility. For instance, competitive effects associated with compatibility can lead to reduced prices and increased entries of new cloud vendors. Cloud users' decisions to switch vendors, however, are likely to be determined by the existing vendor's data-handling policy or their ability to access and delete data with the vendor. The choice of industry standards is often the result of complex negotiation among industry participants.

Industry standards organizations may address many of the CS concerns. Organizations such as Object Management Group and the Storage Networking Industry Association have led to efforts to strengthen CS.[23]

A related point is that there are no formal processes for auditing cloud platforms. Analysts argue that auditing standards to assess a service provider's control over data (e.g., SAS 70) or other information security specifications (e.g., the International Organization for Standardization's ISO 27001) are regarded as irrelevant and insufficient to deal with and address the unique security issues facing the cloud. These standards and specifications were not developed specifically for the cloud.

Recent widely publicized security incidents of Epsilon and AWS, despite their SAS 70 certification, indicate that these certifications have provided false assurance to users regarding cloud security. The cloud's disruptive and transformative nature is likely to lead to a push for standards that are designed specifically for the cloud industry.

12.2.3 Cognitive Institutions and Their Evolution

12.2.3.1 Perception of Vendors' Integrity and Capability

Cloud users may think that CSPs deliberately or accidentally disclose or use data for malicious purposes. Of particular concern is users' perception of the dependability of cloud vendors' security assurances and practices. As noted earlier, issues such as confidentiality, integrity, and availability of data related to ineffective or noncompliant controls of service providers, data backup for disaster recovery, and third-party backup locations concern cloud users.[24] Organizations are also concerned that CSPs may use insecure ways to delete data once services have been provided (e.g., disposing hard disks without deleting data).[25]

The above perceptions are enhanced by the fact that cloud providers allegedly do not answer questions and fail to give enough evidence to build trust. Businesses and industry analysts are concerned about cloud providers' "don't ask, don't tell" approach and have argued that information about data center locations and practices are treated like "national security secrets."[26]

While no case involving the participation of a cloud vendor's employee in a data breach has yet been reported, such risks cannot be ignored. One fear has been that intellectual property and other sensitive information stored in the cloud could be stolen.

12.2.3.2 Cloud Users' and Providers' Inertia Effects

Moving into the cloud is a cultural shift as well as a technology shift. It is quite possible that organizational inertia may affect the lens through which users view security and privacy issues.

Related to a cultural shift, reduction in control due to the shared and dynamic resources in the cloud environment is a concern. Cloud users have no access to or physical control over the hardware and other resources that store and process their data and information.[27] Cloud services contracts, however, often stipulate that data protection is the user's responsibility.[28] A case in point is Google. The company provides security and privacy assurances to its Google Docs users unless the users publish documents online or invite collaborators. On the other hand, Google service agreements explicitly make it clear that the company provides no warranty and bears no liability for harm in case of Google's negligence to protect privacy and security.[29] Just as important is the preference for localness. From the standpoint of security, some users prefer computing functions on site.

On the supply side, cultural changes in cloud providers' views toward security practices are also important. A common observation is that while cloud providers offer sophisticated services, their performances have been weak in policy and practice.[30]

12.3 INSTITUTIONAL FORCES AND POWER DYNAMICS

The nature and sources of power of various institutional actors associated with the cloud industry, and some examples of their actions and responses in shaping cloud-related institutions, are summarized in table 12.3. The cloud industry is characterized by complex power dynamics. For instance, cloud providers attempt to exercise their power based on expertise, experience, and knowledge and engage in technology push without sufficiently addressing security and

Table 12.3 A sample of various actors and their actions in shaping cloud-related institutions

Actor	Nature and Sources of Power	A Sample of Actions
Cloud users	▪ Relative power of users vis-à-vis providers has increased as a result of intense competition.	▪ After users' complaints about data ownership issues, Dropbox updated its user terms/conditions to address these concerns.
Cloud vendors	▪ Vendors offer users new, innovative, and potentially attractive value propositions.	▪ HiDrive Free conforms to the ISO 27001 security standard and German privacy laws. ▪ Epsilon and AWS are SAS 70 certified. ▪ AWS and Google undertook efforts to improve security to achieve certification for FISMA.
Associations representing cloud users	▪ Norms, informal rules, ethical codes	▪ The AICPA endorses Paychex, Intacct, and Copanion. ▪ The CSA promotes best practices for security and provides education for users.
Inter-organizational bodies representing cloud vendors	▪ The power of collective action	▪ The ETNO lobbied for an international privacy standard, simplification of rules governing data transfers, and other changes expected to enable European companies to compete with companies in the US. ▪ Oracle, Cisco Systems, SAP, Apple, Google, and Microsoft lobbied to streamline the EU's fragmented laws.
National governments	▪ Coercive power over citizens and businesses	▪ FISMA requires CSPs to keep sensitive data belonging to a federal agency within the country. ▪ The US and the UK have passed legislation governing the location of storage for personal and medical data.

Actor	Nature and Sources of Power	A Sample of Actions
Supra-national institutions	▪ Nations mostly observe principles of international law and obligations that can resolve transnational problems.	▪ The European Parliament's Civil Liberties Committee recommended making it easier for users to access, amend, and delete data and recommended appointing dedicated data protection officers in companies. ▪ The European Commission emphasized the importance of making it easier to change cloud providers by developing de facto standards for moving data among different clouds (overcoming the lock-in). ▪ EU members work to align privacy laws and close jurisdictional gaps. ▪ In December 2019, the ENISA's Cloud Service Provider Certification Working Group published recommendations to establish an EU-wide CS certification scheme for CSPs.[31] It features three levels of assurance: basic, substantial, and high. Factors such as benefits versus cost, risk level, and the impact of a cyberincident on the cloud service would be used to analyze the risks and define the requirements of a particular level of certification.

privacy concerns. Many users, on the other hand, exercise their bargaining power and voice frustration with vendors' failures to address these issues. At the same time, some governments use the coercive power of the state to use the cloud to spy on citizens. Institutional actors such as cloud vendors, organizations using the cloud, and regulators in this mobilization stage compete to validate and implement their logic. For instance, cloud vendors rely on the economic logic of the cloud's low total cost of ownership to persuade potential customers to adopt cloud use. Some cloud users' logic is based on the idea that the cloud's costs likely outweigh the benefits even in the absence of strong security measures.

It is clear that the cloud industry and market are evolving through the process of complex negotiations and interaction among the key players: supra-national organizations, national governments, inter-organizational bodies representing cloud vendors, cloud vendors, trade and professional associations, and cloud users. (See table 12.3.) The last column of table 12.3 provides a sample of actions of these actors.

We focus on two inter-related issues associated with the cloud: (i) dense networks of actors created by the technology's transformative nature and (ii) power dynamics.

12.3.1 Formation of Dense Networks and Relationships in the Cloud Industry

There is a significant trend toward collaboration, coordination, and communication. The cloud has led to the generation of new interactions among private-sector agents (e.g., the CSA and CompTIA). CSPs also encourage public–private interactions, lead coordination efforts, and influence national and international policy-making processes, which may improve cloud security. In 2012, the EU announced the European Cloud Partnership to help make the cloud more appropriate for the public sector.

The above activities indicate the formation of formal and informal institutions necessary for the cloud industry and market and a dense network of relationships among various actors. Such developments will likely reduce incentives for engaging in opportunistic behaviors that maximize self-interest of a company but that are potentially harmful to the whole cloud ecosystem. Regulative institutions especially play a key role because the state is the most powerful institutional actor. For instance, if regulative mechanisms are created to provide trust on the cloud, vendors and users are likely to be more confident in engaging in transactions without depending on personal or organizational characteristics or past exchange history. This phenomenon is also referred to as institutionally based trust.[32]

12.3.2 Power Dynamics

It is important to develop an adequate description of the relative distribution of power in order to understand the nature of negotiations and interactions among various players. The sources and nature of power of various key actors are presented in table 12.3. Here, we focus on power dynamics from the perspective of exchange relationships.

Dropbox's updates in its terms and conditions in response to user pressures and other similar examples indicate that users' relative power vis-à-vis providers has increased with intense competition. Cloud users leverage this power to force providers to take measures to enhance privacy and security.

CSPs, on the other hand, are increasing their capacity to negotiate with users by offering new, innovative, and attractive value propositions. One example is the vendor HiDrive Free. The company's selling proposition is that its data centers are hosted in Germany, a country with among the strictest privacy laws.

12.4 CHAPTER SUMMARY AND CONCLUSION

While technological development to address security concerns is critical, institutional measures such as clearer regulatory frameworks and other trust-producing initiatives are no less important. Due to the cloud's transformative and far-reaching impacts and significance, it has drawn diverse actors and participants with different perspectives that have broad social and economic demands and interests. These participants vary widely in resources, expertise, experience, and power and their actions are aimed at accomplishing multiple political, social, and economic goals that are far from congruent. In fast-developing technologies such as the cloud, institutional change patterns do not reflect the linearity observed in more mature industries. National governments' and supra-national agencies' roles in the development of cloud-related institutions have been mostly passive and reactive rather than active and directive. Since some non-profit organizations and industry bodies as well as various associations representing cloud vendors and users have been relatively active, and some of them are also engaged in organized lobbying efforts to influence national and international policy-making, it is reasonable to expect that the development of normative institutions is likely to be followed by the development of regulative institutions. As in the case of the Internet Domain Name System, the development of regulative institutions in the cloud industry is likely to be an ex post facto legitimation of the codification of industry norms. It is anticipated that the salience of an institutional component may also vary over time. For instance, barriers associated with newness and inertia effects are likely to decline over time. On the other hand, as the penetration level, width, and depth of clouds increase, they may be more attractive cybercrime targets, which would mean that the importance of cloud security would further increase.

12.5 DISCUSSION QUESTIONS

1. Why is there a gap between vendors' claims and users' views of the cloud's security, privacy, and transparency?
2. What is your general assessment of cloud security?
3. What are some of the concerns that organizations have about cloud security?
4. Identify some key actors from the standpoint of formation of institutions around the cloud industry. What actions are these actors taking?

12.6 END-OF-CHAPTER CASE: CLOUD STORAGE FIRM DROPBOX GETS HACKED

In August 2016, email and password data of Dropbox users were offered for sale in the darknet marketplace TheRealDeal. The criminals were asking for two bitcoins (about US$1,141 at that time) for about 5 gigabytes of data of over 68 million Dropbox users.[33] The data were reported to be from a 2012 breach. A reporter from technology news outlet Motherboard was able to acquire the data, which had been verified to be genuine by a "senior Dropbox employee."[34]

12.6.1 The 2012 Breach

In July 2012, some users sent emails to Dropbox complaining that they were receiving spam at email addresses that they used only for Dropbox. Dropbox said that it investigated the matter and found that hackers had breached another website and received the credentials to access some Dropbox accounts. Dropbox also noted that an employee's login credentials were also stolen and allegedly used to access a Dropbox account that had a document with users' email addresses. The company attributed the spam to this.[35]

The website referred to by Dropbox is reported to be the professional social network LinkedIn.com, which suffered a data breach in June 2012 of more than 167 million LinkedIn users.[36] A Dropbox employee reportedly reused their LinkedIn password. The employee's credential allowed the hackers to penetrate Dropbox's corporate network and gain access to the user database with passwords.[37]

12.6.2 Passwords Were Hashed

For some analysts, what made the hack less worrying was that the passwords in the breached data were hashed. This means that only scrambled data (output from a cryptographic algorithm that converted each password into a unique string) was exposed. However, while 32 million of the passwords were protected with a robust algorithm bcrypt,[38] a weaker hashing function SHA-1 was used for the rest of the passwords.[39] Security experts have noted that bcrypt can be "millions of times more effective" than SHA-1.[40] This means that passwords protected with bcrypt could be extremely difficult, if not impossible, for hackers to crack. Also, for the SHA-1 hashes, the data leak did not include information on the salts, which are random strings that are added to a password before hashing. While the affected Dropbox users became victims of spam, the hashed passwords reduced the probability of more severe victimization. Dropbox stated that it had not found any malicious activity on the breached accounts.

12.6.3 Criticism Facing Dropbox

Dropbox was criticized on the ground that when the breach was first reported in 2012, the company said only email addresses were compromised and was vague on the number of accounts that were affected. The seriousness of the issue was quickly realized when a Motherboard reporter acquired personal data of 68 million users, which was more than two-thirds of Dropbox's active users at that time.

A related criticism levelled against Dropbox was that it did not ask the affected users to change their passwords until news about the severity and extent of the breach was reported in August 2016.[41]

12.6.4 A Russian National behind the Attack

In July 2020, a federal jury in the US found Russian national Yevgeniy Aleksandrovich Nikulin guilty of nine criminal counts, which included hacking LinkedIn and Dropbox. He was sentenced to 88 months in federal prison. Nikulin was also ordered to pay US$1 million to LinkedIn and US$514,000 to Dropbox in restitution.[42]

12.6.5 Case Summary

The Dropbox hack highlights the problem of password reuse. The magnitude of such a problem is many times higher when employees of big cloud companies such as Dropbox reuse their passwords in other sites. Overall, the hacks resulted in a less severe impact due to the use of a hashing algorithm to store passwords.

CHAPTER THIRTEEN

Big Data

CHAPTER PREVIEW

This chapter highlights the costs, benefits, and externalities associated with organizations' use of big data. Consumer reaction to perceived misuse and abuse of personal data by businesses and government agencies is analyzed. Specifically, this chapter focuses on how various inherent characteristics of big data are related to privacy, security, and consumer welfare. The relation between characteristics of big data and privacy, security, and consumer welfare is examined from the standpoints of data collection, storing, sharing, and accessibility. This chapter also discusses how privacy, security, and welfare effects of big data are likely to vary across consumers of different levels of technological sophistication and vulnerability.

13.1 INTRODUCTION

Advances in telecommunications and computer technologies and the associated reductions in costs have led to an exponential growth and availability of data in structured and unstructured forms. The related phenomenon known as big data (BD) involves various costs, benefits, and externalities.

BD's characteristics are tightly linked to privacy, security, and effects on consumer welfare. For instance, huge amounts of data mean that security breaches and privacy violations are likely to lead to more severe consequences and losses via reputational damage, legal liability, ethical harm, and other issues, which are also referred to as an amplified technical impact.[1] Second, a large proportion of BD entails high-velocity (fast) data such as those related to click-stream and GPS data from mobile devices, which can be used to make a short-term prediction with a high level of accuracy. Businesses' initiatives to collect such data have met stiff resistance from consumers. Consumers have expressed concerns over organizations' data collection methods, especially the use of tracking technologies such as cookies and GPS trackers, yet a number of companies engage in questionable data collection and sharing practices. In 2012, a security blogger revealed that Nissan, without warning car owners, reported location, speed, and direction of its Leaf brand cars to websites that other users could access through a built-in RSS reader. Likewise, there are reports that iPhones and Android phones secretly send information about users' locations to Apple and Google.[2]

Third, data comes in multiple formats, such as structured and unstructured. Of special concern is much of the unstructured data, such as Word and Excel documents, emails, instant messages, road traffic information, and Binary Large Objects (BLOBs) (e.g., multimedia objects such as images, audio, and video). These may be sensitive in nature and may contain PII and intellectual property.[3] To take an example, an Italian court found three YouTube executives guilty of violating a child's privacy. The child had autism and was shown being bullied in a YouTube video.[4]

In addition to privacy and security risks due to this high volume of data from multiple sources, complex issues related to data sharing and accessibility arise in a BD environment. Existing non-BD security solutions are not designed to handle the scale, speed, variety, and complexity of BD. Most organizations lack systematic approaches for ensuring appropriate data access mechanisms. The time-variant nature of data flow means that some of these issues are of more significance during peak data traffic times. For instance, organizations may lack the capability to securely store huge amounts of data and manage the collected data during peak data traffic times. A peak data flow may also increase the need for outsourcing to CSPs. Commenting on these complex challenges, a Commissioner of the US FTC put the issue this way: "The potential benefits of Big Data are many, consumer understanding is lacking, and the potential risks are considerable."[5]

The social and ethical issues associated with BD are especially relevant due to the underdeveloped regulations and regulatory infrastructure, which may give rise to consumer exploitation by businesses. Whereas firms know a great deal about consumers' tastes, price sensitivities, and their distribution across the population, most consumers generally lack detailed information about the firms and their offerings.[6] This asymmetry may put consumers in a relatively disadvantaged position. Negative welfare effects are especially noted for poor, unsophisticated, and technologically less informed consumers. Some analysts have argued that firms' BD initiatives may affect the welfare of low-income and minority consumers more negatively.[7]

While privacy and security issues can be linked with collection, storing, analysis, processing, reuse, and sharing of data, this chapter analyzes the relation between BD characteristics and privacy, security, and consumer welfare from the standpoints of data collection, storing, sharing, and accessibility. As to the rationale for the focus on collecting and storing, most companies' involvement with BD has been with these activities due to a steep decrease in the costs of collecting and storing data. In addition, because a key concern has been with data sharing and accessibility, this chapter's analysis also highlights how BD's characteristics are linked with these issues.

13.2 BUSINESSES' AND CONSUMERS' PERCEPTIONS OF AND RESPONSES TO BIG DATA

Despite tremendous economic benefits, BD is not taking off as rapidly as expected. Smaller firms find it more costly to gain from BD, which is likely to produce a lead-lag effect between big and small firms. In terms of BD use, some have shifted into reverse gear. For instance, according to the survey *Big Data Is Dead! Yet 'Small' Data Isn't Ready for Primetime* conducted by the Coalition of Technology Resources for Lawyers and Osterman Research and Relativity in 2019 among 19 North American organizations, 58% of the respondents reported that they had a "data minimization" policy.[8] Figure 13.1 presents percentages of respondents with different understandings of data minimization policies.

The survey also identified several factors driving organizations' personal data minimization policies.[9] They are presented in figure 13.2.

Figure 13.1 Percentage of respondents with different understandings of data-minimization policies

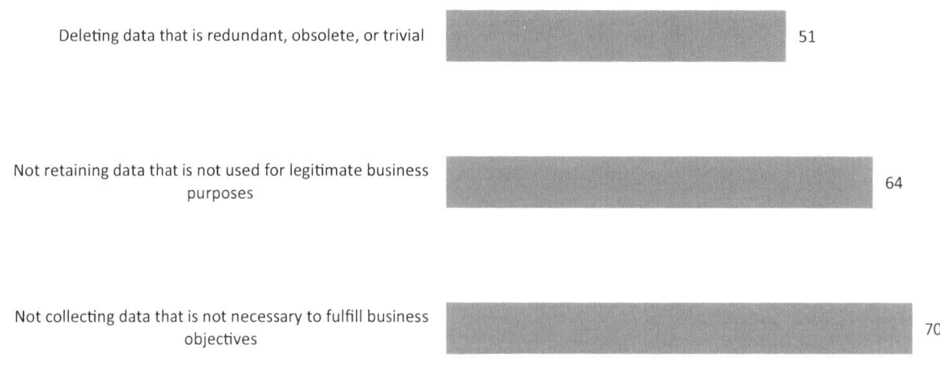

Data source: Hudgins, V. (2019, November 11). *Big data creates big problems: Organizations mandating data minimization efforts.* Law.com. Retrieved from https://www.law.com/legaltechnews/2019/11/11/ig-data-creates-big-problems-organizations-mandating-data-minimization-efforts/

Figure 13.2 Percentage of respondents that considered different factors responsible for driving data-minimization policies

Data source: Hudgins, V. (2019, November 11). *Big data creates big problems: Organizations mandating data minimization efforts.* Law.com. Retrieved from https://www.law.com/legaltechnews/2019/11/11/big-data-creates-big-problems-organizations-mandating-data-minimization-efforts/

Businesses' initiatives to collect high-velocity data (e.g., click-stream, GPS data from mobile devices, social media usage, etc.) have met stiff resistance from consumers. A 2013 national survey conducted in the US by the Pew Internet & American Life Project found that a large proportion of respondents have taken actions, such as turning off the location tracking feature.

Developing countries with vague privacy laws and lack of enforcement provide a perfect breeding ground for abuse by businesses. For instance, misuse and abuse of personal data and information are reported to be key problems facing Chinese consumers who receive credit from BD companies.[10] Online credit services in China that use BD analytics in extending loans are being accused of abusing personal data to collect debts. One such example is Alibaba's Ant Check Later, which allows users to delay payments and pay in installments. An online user reported that he was contacted by Ant Check Later for information about his friend who owed money to the payment service. Ant Financial Services reportedly said that the practice of contacting a borrower's friends or relatives to help with collecting debts is common in the financial sector. TMTpost cited a China Youth Daily poll, which showed that 75.9% of respondents believed there was abuse of BD.[11] Users of JD.com reported similar problems. JD.com's financial unit operates JD Baitiao, which is similar to Ant Check Later. In a question posted on online legal advice site 110.com, a JD.com user asked if it was legal for JD.com to give a third-party service his personal information for the purpose of debt collection.

A key idea is that businesses store huge volumes of personal data so that potential innovative uses can be discovered. It is argued that "most innovative secondary uses haven't been imagined when the data is first collected."[12] However, most consumers are against secondary uses of their personal data.

13.3 CHARACTERISTICS OF BIG DATA IN RELATION TO PRIVACY AND SECURITY

Despite its widespread use, there is no rigorous and universally accepted definition of BD.[13] Einav and Levin[14] noted that BD involves the availability of data in real time, at larger scale, with less structure, and on different types of variables than previously used. In addition to Gartner's three Vs (volume, velocity, and variety), the software company SAS has suggested two more BD dimensions: variability and complexity.[15] As presented in table 13.1, the various BD dimensions are tightly linked to privacy and security issues. For instance, in order to create highly customized offerings (e.g., Target's offering based on a "pregnancy prediction score"), a company may need to mine a huge amount (volume) of structured and unstructured (variety) data from multiple sources (complexity). In some cases, this process may also involve high-velocity data.

Regarding data sharing and accessibility issues, outsourcing to CSPs, and utilization of third-party tools, services and applications are critical for creating and capturing value. A major consideration is possible security breaches associated with outsourcing. According to Trustwave, 64% of security breaches in 2012 involved outsourcing providers.[16] Since most organizations are not in a position to build a complete BD environment in-house,[17] reliance on CSPs becomes inevitable for analytical, storage, and other needs. Prior research indicates that a number of key considerations need to be addressed in decisions related to outsourcing to CSPs.[18] First, most CSPs are bigger than their clients and deal with higher data volumes. Information stored in the cloud is a potential gold mine for cybercriminals. From both regulatory and reputational perspectives, storing data in the cloud does not remove organizations' responsibility for protecting data.[19] In general, it is often the cloud user organizations' responsibility to make sure that personal data are protected and are only used according to legal provisions. In Italy, for instance, CSPs take only the role of processor and are only part of the processing carried out by cloud user organizations.[20]

Second, some regulators have expressed concern that CSPs might use clients' data for their own benefits. For instance, in 2013, Sweden's data protection authority Datainspektionen asked Salem Municipality to stop using Google Apps, email, and calendar services. Datainspektionen argued that Google writes the contract and sets the rules for handling information and has too much flexibility to use the data for purposes other than those specified by the municipality.[21]

Table 13.1 Big data characteristics in relation to security, privacy, and welfare concerns[22]

Characteristic	Explanation	Collection/Storing	Sharing/Accessibility by Third Parties and Various User Types
Volume	▪ Huge amount of data is created from a wide range of sources such as transactions, unstructured streaming from text, images, audio, voice, VoIP, video, TV, other media, sensors, and machine-to-data.	▪ High data volume would likely attract a great deal of attention from cybercriminals. ▪ Amplified technical impact. ▪ Violation of transparency principle of FIPs.	▪ Firms may need to outsource to CSPs, which may give rise to privacy and security issues.
Velocity (*Fast Data*)	▪ For some time-sensitive data, speed is more important than volume. Data need to be stored, processed, and analyzed quickly.	▪ Increasing consumer concerns over privacy in the context of behavioral advertising based on real-time profiling and tracking technologies such as cookies. ▪ Violation of the individual participation principle.	▪ Increase in the supply and demand of location-based real-time personal information; negative spillover effects (e.g., stalking people in real time). ▪ Physical security risks.
Variety	▪ Data come in multiple formats such as structured numeric data and unstructured text documents, email, video, audio, and financial transactions.	▪ Unstructured data are more likely to conceal PII. ▪ A large variety of information would make it more difficult to detect security breaches, react appropriately, and respond to attacks.	▪ Most organizations lack mechanisms to ensure that employees and third parties have appropriate access to unstructured data and that they comply with regulations.[23]
Variability	▪ Data flows vary greatly with periodic peaks and troughs. These are related to social media trends; daily, seasonal, and event-triggered peak data loads; and other factors.	▪ Organizations may lack capabilities to securely store huge amounts of data and manage the collected data during peak data traffic. ▪ Attractiveness as a crime target increases during peak data traffic.	▪ Peak data traffic may cause increased needs to outsource to CSPs, which gives rise to important privacy and security issues.
Complexity	▪ Data come from multiple sources that require linking, matching, cleansing, and transforming across systems.	▪ Resulting data are often more personal than the set of data the person would consent to give. ▪ Data collected from illicit sources are more likely to have information on consumers who are less savvy about technology.	▪ A party with whom de-identified personal data is shared may combine data from other sources to re-identify. ▪ Violation of the security provision of FIPs.

13.3.1 Volume

An organization is often required to store all data in one location in order to facilitate analysis. The higher volume and concentration of data increase the probability that the data files and documents may contain inherently valuable and sensitive information. Information stored for the purpose of BD analytics is thus a potential goldmine for cybercriminals, which, as noted earlier, leads to an amplified technical impact.[24]

An emerging trend that parallels developments in BD analytics technologies is that of data lakes. Traditionally, data were stored as databases in a data warehouse. Such data are often in a single repository that is used to conduct analysis and create reports. In order to do so, however, data need to be preprocessed. With increasing data volume, this is becoming a difficult task. Preprocessing may require high-end supercomputers and cost a lot of time and money. Data lakes solve this. Unlike warehouses, data lakes can store raw data of any type.[25] Data lakes are especially suitable for storing large volumes of unstructured and semi-structured data. Unlike data lakes, databases are built around structured tables that cannot handle unstructured and semi-structured data.[26]

However, data lakes often lack strong security features. Given the huge volume of data stored in data lakes, this is a concern. Attacks such as false data injection have recently increased. In this type of attack, cybercriminals may exploit freely available tools to attack a system and inject false data. By gaining unauthorized access to data lakes, nefarious actors can manipulate the data to mislead users. Some attackers can have political motivations behind such a compromise.[27]

If used inappropriately, information contained in huge data volumes may lead to psychological, emotional, economic, or social harm to consumers. For instance, BD predictive analysis may improve the accuracy of predictions of a customer's purchasing requirements or preferences. Highly customized offerings based on predicted preference and requirement data may, however, lead to unpleasant, creepy, and frightening experiences for consumers. This phenomenon is also referred to as predictive privacy harm.[28] The example most often cited is that of a man's high school–aged daughter tracked by the US retailer Target. The company's pregnancy prediction score indicated that she was pregnant before her father knew and sent promotional mail for products that pregnant women need.[29]

The availability of a huge amount of data also increases the possibility that personal data can be put to new uses to create value. The US FTC commissioner pointed out the possibility that firms, "without our knowledge or consent, can amass large amounts of private information about people to use for purposes we don't expect or understand."[30] Such uses often violate the transparency principle of fair information practices (FIPs).[31]

Huge data volume is also related to the demand and even the necessity of outsourcing. An issue of more pressing concern is determining relevance within large data volumes and how to use analytics to create value from relevant data. Firms may thus rely on CSPs for analytic solutions.

13.3.2 Velocity

Various examples of high-velocity or fast data were discussed earlier. In particular, clickstream data (clickpaths), which constitute the route chosen by visitors when they click/navigate through a site, are typically collected by online advertisers, retailers, and ISPs. The fact that such data can be collected, stored, and reused indefinitely poses significant privacy risks.[32] Some tracking tools can manipulate clickstreams to build a detailed database of personal profiles in order to target Internet advertising.[33]

An important use of BD is real-time consumer profile-driven campaigns such as serving customized ads. For instance, location-tracking technologies allow marketers to serve SMS and other forms of ads based on real-time location. This process often involves passive data collection without any overt consumer interaction. The lack of individual consent for the collection, use, and dissemination of such information means that such a practice violates the individual participation principle of FIPs.[34]

There is increasing consumer concern over privacy in the context of real-time behavioral advertising and tracking technologies such as cookies.[35]

BD initiatives have led to an increase in both the supply and demand for location-based real-time personal information. Data created and made available for use in the implementation of BD initiatives also have negative spillover effects. In particular, the availability of location information to third parties may have some dangerous aspects. One example is the use of location data for stalking people in real time. For instance, the iOS app Girls Around Me, which was developed by the Russian company I-Free, leveraged data from Foursquare to scan for and detect women checking into a user's neighborhood. The user could identify a woman he would like to talk to, connect with her through Facebook, see her full name and profile photos, and also send her a message. The woman being tracked would have no idea that someone was "snooping" on her. As of March 2012, the app had been downloaded over 70,000 times.[36]

There is also a physical risk of (near) real-time data. In China, for instance, illegal organizations buy databases from malicious actors and provide services to their clients, which include private investigation, illegal debt collection, asset investigation, and even kidnapping.[37]

13.3.3 Variety

By combining structured and unstructured data from multiple sources, firms can uncover hidden connections between seemingly unrelated pieces of data. In addition to volume, the high variety of information in BD makes it more difficult to detect security breaches, react appropriately, and respond to attacks.[38]

One estimate suggested that only about 10% of available data is in a structured form (e.g., transactional data on customers; time-series data from statistical agencies on various macroeconomic and financial indicators) that can be presented in rows and columns.[39]

Especially because of the relative newness, most organizations lack a capability to manage unstructured data, which arguably contains more sensitive information. Processes and technology solutions for securing unstructured data are still in a nascent phase and governance issues are not addressed.[40]

For instance, organizations often lack mechanisms to ensure that permanent and temporary employees and third parties have appropriate access to unstructured data and that they are in compliance with data protection regulations.[41] In a survey conducted by Ponemon among IT professionals, only 23% of the respondents believed that unstructured data in their companies was properly secured and protected.[42] Another study of DiscoverOrg indicated that over 50% of organizations were not focused on managing unstructured data and only 20% had unstructured data governance processes and procedures.[43]

13.3.4 Variability

The variability characteristic is related to the time-variant nature of security and privacy risks. The volume of data collected and stored, which need protection, grows during peak data collection and flow periods. It is during such periods that organizations may lack internal capacity and tools to manage and protect information. A related point is that the data's attractiveness as a crime target is high during such periods. For instance, the 2013 high-profile cyberattacks on Target (chapter 8) occurred during the peak holiday shopping season between 10 am and 5 pm local times at Target stores.

The variability characteristic of BD may also necessitate the outsourcing of hardware, software, and business-critical applications to CSPs. Applications such as enterprise resource planning systems and accounting systems are required to be configured for peak loads during daily and seasonal business periods or when quarterly and annual financial statements are prepared.

13.3.5 Complexity

BD often constitutes aggregated data from various sources that are not necessarily identifiable. There is thus no process to request the consent of a person for the resulting data, which are often more personal than the set of data for which the person would give consent.[44] A related privacy risk involves re-identification. It is possible to use data aggregation to convert semi-anonymous or certain personally non-identifiable information into non-anonymous or personally identifiable information.[45] Health-related data is of special concern. Based on a consumer's search terms for disease symptoms, online purchases of medical supplies, and RFID tagging of drug packages, marketers can identify information about the consumer's health.[46] Access to such information would enable an insurance underwriter to predict certain disease and disorder probabilities, which would not be possible using information voluntarily disclosed by consumers.

Many BD applications use multiple data sources and transfer data to third parties. Organizations believe that making data anonymous before sharing would make it impossible to identify. This is a convenient but often false assumption. A variety of methods and techniques can be used to re-identify personal data and reassociate it with specific consumers.[47] BD processes can generate predictive models that have a high probability of revealing PII[48] and thus make anonymization impossible. Failure to protect PII and unintended or inappropriate disclosure violate the security provision of FIPs.[49] In some cases, the identified person may suffer physical, psychological, or economic harm. For instance, in 2011, customers of the US drugstore Walgreens filed a lawsuit accusing the drugstore of illegally selling medical information from patient prescriptions. Data mining companies allegedly de-identified the data and then sold to it pharmaceutical companies. Walgreens unfairly benefited.[50]

Some data may come from illicit sources. One example is the criminal outfit Superzonda, which allegedly sent 30–40 million spam emails a day in the early 2000s.[51] Superzonda's most profitable venture was to provide information on consumers interested in a product (e.g., a mortgage) to legitimate businesses for lead generation. Each package of data (consisting of name, phone number, address, amount of loan desired, and current home value) was reportedly sold to mortgage companies for US$20.[52] Less sophisticated and vulnerable consumers are more likely to be fooled by the tricks of illicit actors such as Superzonda.

13.4 5G CELLULAR SERVICES AND BIG DATA

The 5G cellular services will play a key role in delivering higher value to users of BD because 5G supports lower latencies and higher throughput than 4G.[53] This means that 5G makes it easier to analyze a large quantity of data and transmit the results in near real time.[54]

Several issues need to be considered regarding 5G's security. First, if hackers break into a device, due to 5G's low latency and high bandwidth, hackers can download data much faster than on 4G networks. (See figure 13.3.) Huge amounts of data flowing from machine to machine, AI, IoT, and other sources would lead to amplified technical impacts.[55]

Second, 5G networks are likely to employ edge computing, which involves co-locating computing, storage, and networking functions closer to where the data are generated instead of being transported to more central locations. This is done in order to reduce costs and power required, conserve bandwidth, and enhance performance in terms of latency and other indicators.

Devices such as smartphones, cars, and robots will communicate with the network's "edge." To do so, they connect over radio waves at 5G frequencies to new generation small-cell radio units that form the radio access network (RAN). RANs link devices to routers and switches of the core network. Data traffic is then transported to and from the Internet, the cloud, and other devices.[56]

The edges may not have the same level of security as the central location. Nefarious actors can also install back doors in mobile base stations in order to intercept and manipulate data from the RAN's access points. It is not easy to detect hackers' activities – the systems could appear to be operating normally even if cybercriminals were copying and manipulating data.[57]

Third, using software-defined networking (SDN), a single network can provide a variety of heterogeneous services or "slices" with specific capabilities for varying uniquely purposed and highly tailored services for different needs. Each slice is uniquely purposed and thus may need unique security capabilities. For instance, public safety communication systems such as 911 services and police and firefighter systems may need higher levels of security than online gaming systems.[58]

Fourth, a wide range of devices such as home appliances, industrial automation control devices, cars, televisions, shipping containers, tiny climate monitoring sensors, laptops, and next-generation tablets and smartphones will likely be connected to 5G networks. A working party developing 5G standards has specified at least 1 million devices per square kilometer. The attack surface may increase with the huge number of connected devices because not enough consideration is given to their security.

13.5 CHAPTER SUMMARY AND CONCLUSION

Risks associated with owning and storing data are likely to increase with the volume, variety, and complexity of data. For instance, the extent and nature of risks involved differ across data types (e.g., often high risk in unstructured data), source of data (higher risks for data obtained from illicit sources), and volume of data.

BD presents different concerns and issues for IT professionals in organizations than it does for consumers. IT professionals often emphasize the risks and effects of security breaches and misconduct by external parties and insiders. Organizations' goals are to avoid fines, potential lawsuits, and reputation damage. Consumers, on the other hand, tend to worry most about what they consider businesses' unethical practices. This is a critical issue from the public policy point of view because businesses seem to ignore consumers' desires for privacy. These differences are likely to result in strong pressure on governments to intervene. Governments will likely be forced to recognize the needs for reasonable regulatory safeguards to protect the public interest of privacy and for the development of higher privacy standards.

Proactive risk management also requires a deeper understanding of the root causes of risks. It is important to have detailed information on all the parties (employees and administrators of cloud vendors) who have access to an organization's data. Illicit and gray-area businesses accessing data on consumers is not an uncommon practice.

While non-BD issues are still relevant, new issues arise in the BD context. Location information, which is a key component of high-velocity data, is an issue that is a cause of concern.

Figure 13.3 Comparing 4G and 5G networks in terms of average throughput per user and latency (MS)

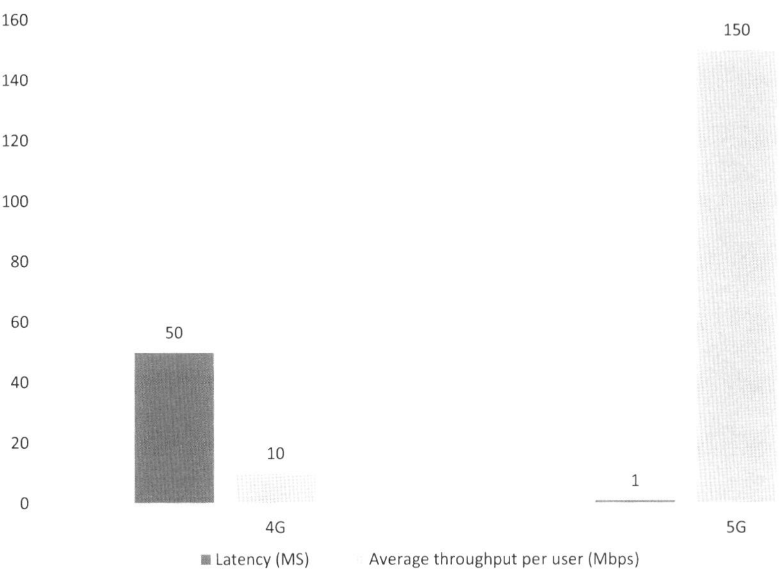

Data source: D'mello, A. (2020, March 16). *Growing demand from Chinese telcos for 5G-enabling networking system, says Ethernity*. Vanillaplus. Retrieved from https://www.vanillaplus.com/2020/03/16/51311-growing-demand-chinese-telcos-5g-enabling-networking-system-says-ethernity/

Not only real time but also time-lagged or asynchronous location identification information may have potentially dangerous consequences.

Several BD uses fall into a regulatory gray area. Due to underdeveloped regulatory institutions, there is a need for BD policy at the firm level, which must consider the degree of sensitivity of information used in predictive modeling. Yet most organizations have not developed best practices to ensure privacy and security of customer data.

13.6 DISCUSSION QUESTIONS

1. What have surveys found regarding businesses' perception of security issues associated with BD?
2. How are different BD characteristics related to security and privacy?
3. What are the security concerns with data lakes?
4. What are some of the concerns associated with 5G security?

13.7 END-OF-CHAPTER CASE: EQUIFAX BECOMES A VICTIM OF A BIG HACK

On September 7, 2017, the consumer-credit reporting agency Equifax announced that it had suffered a data breach due to a "cybersecurity incident."[59] It compromised 146.6 million Social Security numbers and other personal information of consumers such as names and addresses.[60] The breached information also included driver's license numbers for at least 10 million Americans as well as credit-card numbers and other personally identifiable information of 200,000 American consumers.[61] US consumers made up most victims of the breach; however, some UK and Canadian consumers were also affected. US Attorney General William Barr said that the hackers had also stolen Equifax's trade secrets, such as sophisticated database designs that are used to store personal information.[62]

Equifax has data on over 820 million consumers and 91 million businesses.[63] Thus, only a subset of data held by the credit-rating firm was affected. According to Equifax, the perpetrators had accessed the information between May and July of 2017.[64]

13.7.1 Big Hack Leads to Big Losses

Equifax faced many lawsuits. A class-action lawsuit was settled for at least US$650 million in July 2019.[65] The settlement included money to be spent on credit monitoring for consumers that were affected by the breach. The settlement has assumed that only 7 million consumers will use free credit monitoring. However, if all the affected consumers sign up for this service, then Equifax would have to pay about US$2 billion.[66]

13.7.2 Who Was Behind the Attack?

Analysts suspected that the Chinese government was behind the attack. The FBI Deputy Director David Bowdich noted that as of February 2019, there was no evidence that the stolen personal information was being used.[67] In this way, the investigators linked the attack with two other big breaches in which personally identifying data were not sold on the dark web: the 2015 hack of the US Office of Personnel Management and the 2018 hack of Marriott's Starwood hotel brands. This fact indicated that financially motivated cybercriminals were less likely to be the attackers. The goal of the offender could be to build a huge "data lake" on hundreds of millions of Americans. The perpetrator's intention could be to use BD techniques to gain insights into US government officials and intelligence operatives.[68]

In February 2020, the US charged four People's Liberation Army officers in a nine-count indictment on charges of computer fraud, economic espionage, and wire fraud for carrying out this attack. They were indicted by the grand jury in the Northern District of Georgia. According to the indictment, the hackers had used 34 servers in 20 countries to access the Equifax network. They then used the existing encrypted communication channels to "blend in with normal network activity."[69]

13.7.3 Equifax's Handling of the Attack

Equifax was criticized because, after discovering the breach, it took over a month to publicize the breach.[70] A further criticism was that Equifax failed to provide complete information about the breach. In February 2018, an article published in the *Wall Street Journal* detailed how the hackers had accessed more categories of consumers' personal information than Equifax had disclosed.[71] Following the publication of the *Wall Street Journal* article, Senator Elizabeth Warren sent a letter to CEO Paulino do Rego Barros Jr. asking, "As your company continues to issue incomplete, confusing and contradictory statements and hide information from Congress and the public, it is clear that five months after the breach was publicly announced, Equifax has yet to answer this simple question in full: what was the precise extent of the breach?"[72]

Some criticisms have also been directed toward the way the company provided resources to its consumers to deal with the attack. Equifax set up equifaxsecurity2017.com as a separate dedicated domain, which hosted information and resources for those potentially affected. A criticism was that lookalike domains such as equifaxsecurity2017.com are often used by phishing scams. Analysts pointed out that asking potential victims to trust such a website was a big failure on the part of the company's CS procedure.[73]

CHAPTER FOURTEEN

Smart Cities and the Internet of Things

CHAPTER PREVIEW

Cyberattacks associated with smart cities and the Internet of Things (IoT) are among the biggest threats the world faces. Smart cities' key features make them large and attractive targets for cybercriminals, nation states, and cyberterrorists. Likewise, the negative consequences of insecure IoT devices are enormous and range from social and economic costs to national security concerns. This chapter highlights the negative externalities generated by insecure devices and their impacts on society and economies. The chapter argues that as a result of the externality-producing nature of insecure devices, IoT users and manufacturers are not very concerned about security. The chapter also analyzes blockchain's potential in shaping IoT security.

14.1 INTRODUCTION

CS issues are becoming critical issues facing smart cities and the IoT. In 2013, the UK's Department of Business, Enterprise and Skills noted that trust in data privacy and system integrity is a major barrier to smart city projects.[1] Smart cities are becoming larger and attractive targets for cybercriminals, nation states, and cyberterrorists. In 2014, a CS researcher showed that about 200,000 traffic control sensors in major cities such as Washington, DC, New York, New Jersey, San Francisco, Seattle, Lyon (France), and Melbourne (Australia) were not encrypted and thus were vulnerable to cyberattacks. The researchers demonstrated that it was possible to intercept information coming from these sensors from 1,500 feet away, or by a drone.[2]

A closely related phenomenon is the rapid evolution of the IoT. Cyberattacks associated with the IoT are among the biggest threats the world faces today. Some security experts have asserted that the "IoT Will Be Biggest Threat of the Next Decade."[3]

In light of these trends, recent regulations have forced manufacturers of IoT devices to strengthen security. In the US state of California, the new IoT security law, which became effective on January 1, 2020, requires connected devices that are made, sold, or offered for sale to be equipped with "reasonable security features." A connected device is defined as "any device, or other physical object that is capable of connecting to the Internet, directly or indirectly, and that is assigned an Internet Protocol address or Bluetooth address." The state of Oregon has similar laws.

Serious concerns have been raised about IoT security threats. Such devices account for 16% of traffic but 78% of mobile malware.[4] According to the Marsh Microsoft *2019 Global Cyber Risk Perception Survey* conducted worldwide with more than 1,500 respondents, on a 1 to 5 scale (1 – minimal to none; 5 – extremely high), about half rated cyberrisks associated with the IoT at 4 or 5.[5] (See figure 14.1.)

Figure 14.1 Percentage of respondents perceiving different levels of cyberrisk associated with IoT

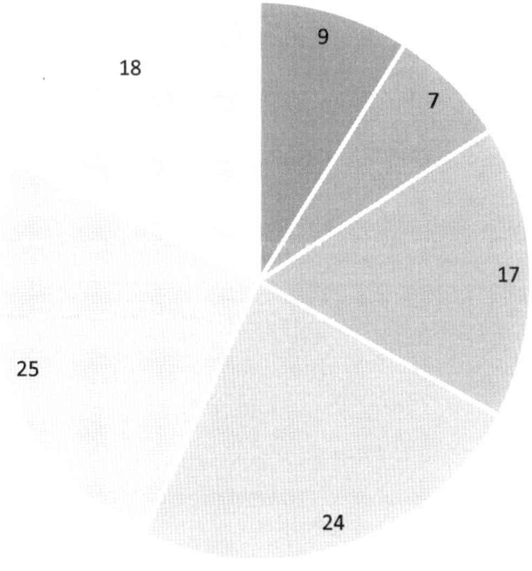

Data source: *Global cyber risk perception survey*. (2019). Marsh. Retrieved from https://www.marsh.com/us/insights/research/marsh-microsoft-cyber-survey-report-2019.html

Figure 14.2 Percentage of respondents with different levels of confidence regarding their organizations' IoT devices

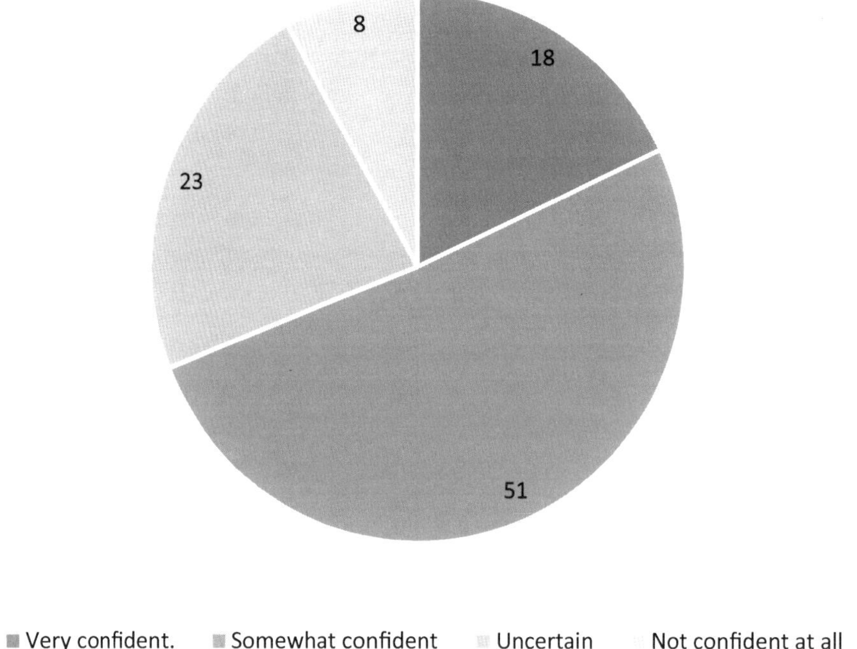

Data source: How much do organizations understand the risk exposure of IoT devices? (2019, August 1). Deloitte. Retrieved from https://www2.deloitte.com/us/en/pages/about-deloitte/articles/press-releases/deloitte-cyber-shares-top-cyber-risks-for-iot-environments.html

Low levels of confidence have been reported by organizations regarding the security of IoT devices. In a survey of professionals who responded to poll questions in the Deloitte Dbriefs webcast "The Internet of Things and Cybersecurity: A Secure-by-Design Approach" held on May 30, 2019,[6] only 18% were very confident about the security of their organizations' IoT. (See figure 14.2.)

Most IoT devices are prone to cyberattacks and privacy violations for the simple fact that not enough consideration is given to security in their design. For instance, it is not practical to install a firewall on a baby monitor due to insufficient memory. Users of IoT such as webcams, routers, and other Internet-connected devices lack awareness of the security dangers or aren't sufficiently concerned about minimizing such threats.[7]

The large and rapidly growing IoT market poses a special challenge from the CS standpoint. (See figure 14.3.) Tens of billions of IoT devices such as home routers, webcams, digital video recorders, and other appliances with Internet capabilities built into them generate a

Figure 14.3 Number of IoT devices (billions)

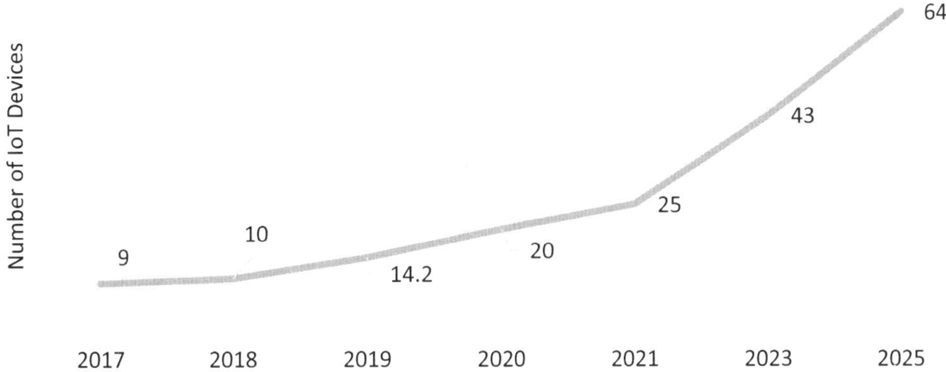

Data sources: Data for 2017, 2018, and 2025 as reported in Newman, P. (2019). *IoT Report: How Internet of things technology growth is reaching mainstream companies and consumers*. Business Insider. Retrieved May 5, 2020 from https://www.businessinsider.com/internet-of-things-report; data for 2019 and 2021 as reported in Omale, G. (2018). *Gartner identifies top 10 strategic IoT technologies and trends*. Gartner. Retrieved from https://www.gartner.com/en/newsroom/press-releases/2018-11-07-gartner-identifies-top-10-strategic-iot-technologies-and-trends; data for 2020 as reported in *Connected devices: Microsoft's heritage of innovation*. (n.d.). Microsoft. Retrieved May 5, 2020 from https://www.microsoft.com/en-us/legal/intellectualproperty/mtl/internet-of-things.aspx; data for 2023 as reported in Dahlqvist, F., Patel, M., Rajko, A., & Shulman, J. (2019, July 22). *Growing opportunities in the Internet of things*. McKinsey & Company. Retrieved from https://www.mckinsey.com/industries/private-equity-and-principal-investors/our-insights/growing-opportunities-in-the-internet-of-things

vast amount of data. One estimate suggested that such devices will generate 79.4 zettabytes (79.4×10^{21} bytes) of data in 2025.[8]

The substantial number of IoT devices makes it difficult to fight against cyberattacks. For instance, an executive of Level 3 Communications, a global communications provider, noted that the company can filter "traffic from botnets in the tens of thousands"[9] but not from millions.

Observers have expressed reservations about security features of the system used in smart cities. Concerns have been voiced regarding the protection of data, systems, and infrastructure that play critical roles in the operation of the city as well as the safety and livelihood of its residents. It is argued that the challenge of CS has not been adequately addressed by smart city advocates.[10] The systems used in smart cities have sophisticated features and functionality that are often characterized by a high degree of vulnerability to cyberattacks due to their complexity, high degree of interconnectedness, and high volume of information. The infrastructure, such as broadband networks, Wi-Fi, and satellites that connect systems and operators, increases the entry points for cyberoffenders.[11] If a device is successfully hacked, it may act as a pivot and bypass existing defense mechanisms. Critics have also argued that most manufacturers producing many of the devices and systems for smart cities have failed to ensure adequate CS.[12] Moreover, cyberattacks on smart cities are likely to result in serious consequences for the victims and higher societal costs.

The analysis and integration of data streams have facilitated surveillance and dataveillance (tracking the trails created by a person's activities) and increased the threat of a "Big Brother society" in smart cities.[13] As an example, in May 2020, Google's affiliate Sidewalk Labs scrapped its planned smart city project called Quayside in Toronto, Canada. A critic labeled the project as "a colonizing experiment in surveillance capitalism." In a letter to the Toronto city council, a venture capitalist suggested the officials abandon the Quayside project, arguing, "No matter what Google is offering, the value to Toronto cannot possibly approach the value your city is giving up. It is a dystopian vision that has no place in a democratic society."[14]

There are even more concerns in cities with strict government cybercontrol measures, such as Addis Ababa and Cairo. In these cities, bloggers and organizers of social mobilizations have been imprisoned for criticizing their governments. In 2011, the Chinese government announced a plan to introduce an "information platform of real-time citizen movement." Human rights activists expressed concerns that the regime may use the information to suppress activists. For instance, cellphones of activists have already been allegedly tracked by security forces that are used to locate activists.[15] However, compared to more democratic societies, civic societies hold positions of less power in the discourse in China. In contrast, Brazil's Rio de Janeiro's attempts to "smartize" the city drew concerns related to CS and privacy violations from citizens.[16]

14.2 SMART CITIES

Major challenges that face smart cities include securing against cyberattacks and terrorism, maintaining safety of the controlled systems, protecting intellectual property and other assets, and protecting the right to privacy.[17] First, smart cities raise major privacy issues. Entities that provide services to the residents of the smart cities must access data to provide better services. In turn, a cyberoffender can monitor events such as residents' movement patterns in a large number of buildings and create profiles that can be sold in an underground market.[18]

Smart cities can use sensors to track vehicles and people. In many cases, information is not available about the trustworthiness of the receivers of this data.[19]

Cyberthreats facing smart cities can be costly and dangerous. Terrorists and adversary states can launch cyberattacks that may pose existential threats to those living in smart cities. Inhabitants and owners of buildings can be identified, blackmailed, and forced to transfer money.[20]

Several malware products and worms such as Stuxnet, Duqu, Flame, BlackEnergy, and Shamoon have been designed to target industrial control systems (ICS) and cause physical damage. Stuxnet, for instance, caused the centrifuges of the uranium in an Iranian nuclear facility to overspin and self-destruct. The operator's console falsely showed that the system was operating within normal parameters and values. In another example, the Duqu malware looked for useful information to attack ICSs. Likewise, the Flame malware searched for engineering drawings, specifications, and other technical details about systems. It recorded audio, screenshots, keyboard activity, and network traffic. Its capability also included recording Skype conversations and downloading contact information from nearby Bluetooth-enabled devices.[21] Likewise, the US DHS identified BlackEnergy within the ICS used by critical infrastructures. A senior threat researcher at Trend Micro noted that BlackEnergy was targeting some of the ICS exploited by Stuxnet.[22]

The costs associated with replacement and incident response in the attacks attributed to Shamoon were estimated to exceed US$15 million.[23] Several months following the first attacks, the Shamoon malware reportedly tried to attack oil and gas flow networks in an attempt to disrupt international supplies.[24] Shamoon demonstrated the ability to spread to other computers on the network by exploiting shared hard drives. The virus compiled a list of files from specific locations, erased itself, and sent information to the attacker.[25] According to a 2014 report of Germany's Federal Office for Information Security (Bundesamt für Sicherheit in der Informationstechnik), hackers caused physical damage to a steel plant. The attackers had used spear-phishing and social engineering to gain access to the plant's network, and subsequently penetrated the production network.[26] It was reported that the attack resulted in "massive" damage.[27] This attack was the second confirmed case, after the Stuxnet attack on an Iranian nuclear facility, in which a purely cyberattack caused the physical destruction of a plant or equipment.

If a hacker is able to break into any IoT device, such as automatic doors, smart home heating and lighting systems, vending machines, cameras, security alarms, Wi-Fi router boxes, entertainment gadgets, and smart TVs, then it is easy for the hacker to gain access to broader networks (e.g., corporate networks).[28] In a discussion of the IoT, it was noted that buildings are associated with major vulnerabilities and security threats. In particular, building automation systems (BAS) and supervisory control and data acquisition (SCADA), which have been important choices for facilities management and building operational needs, may represent a key vulnerability factor. BAS and SCADA deployments are associated and facilitated by the need to upgrade energy infrastructures. Frost and Sullivan, the Gartner Group, and Forrester have expressed concerns regarding the CS threats of these systems.[29]

One estimate suggested that more than 15,000 BASs in the US are accessible via the Internet, 9% of which are reported to have CS vulnerabilities.[30] Since such systems are often incorporated into computer networks, an attack on the BAS makes it easy to penetrate the company network.

BASs are permanently available, often have no security features, and are rarely patched. These features are attractive to cybercriminals. If a hacker successfully attacks a BAS system, it is easy to penetrate other devices and computers on the network. This is because attacks coming from inside the network are trusted and ignored by traditional network security. Therefore, employee records, customer data, and intellectual property are likely to become vulnerable to cyberattacks.[31]

According to a 2013 report of the US Department of Homeland Security, hackers targeted building energy management systems. Companies connect building energy management systems to networks and the Internet to automate lighting, heating, and air conditioning. Cybercriminals can use these connections to unlock doors and turn off lights. Quoting financial institutions, a security researcher noted that if the temperature is changed by 5–6 degrees, the computers cannot process transactions at the normal rate.[32] Cybercriminals thus can damage data centers by turning up the heat.

In 2012, security researchers found at least two vulnerabilities in Tridium's building management software. Note that Tridium Niagara software is used in many building management systems. The researchers were able to exploit vulnerabilities in Tridium Niagara software to open a company's parking garage gate and front door. They were also able to penetrate the company's corporate network. In 2012, cyberattackers reportedly exploited Tridium vulnerabilities at least twice. In a New Jersey manufacturing company, attackers discovered the system was accessible from the Internet. In the second instance, a state government facility was attacked and the temperature settings in the building were changed.[33]

Some technologies and devices that are more prone to privacy issues are more likely to be adopted in smart cities than in conventional cities. Some examples are smart meters and smart grids. According to Pike Research, about one-third of smart city projects in North America and Europe were primarily focused on smart grids or other energy innovations, and about half of smart city strategies included energy-focused projects.[34]

Concerns over privacy and security have been an important part of the debate over the impact of smart meters and smart grids. As of 2014, in the US, 38 million smart meters had been installed. These meters gather information about household electricity consumption and transmit this information to the supplier.[35] The US DoE announced a plan to publish a voluntary code of conduct to govern data privacy for smart meters in 2015. Despite the existence of such code, some critics are concerned that consumers could be persuaded to give up their private data to power companies and third-party data aggregators.

Smart meters often transmit high-value data. It is argued that the electricity consumption data is worth more than the commodity consumed to generate the data.[36] This is because many devices transfer data among themselves and generate new data, referred to as derived data.[37] Devices connected to the smart grid thus may leave key information behind. This type of privacy is also called footprint privacy in which transactions, actions, and queries associated with the concerned devices are stored.[38]

Such data are likely to permit more sophisticated profiling and targeting of consumers and neighborhoods that can be used for purposes such as a retailer's decision about where to open its next store.[39] These concerns are of special interest to smart cities. It is often not clear how the data are used and who uses them. Some have also argued that cities may have to create their own privacy charters to address the concerns of citizens.[40]

To be useful for purposes such as increasing energy efficiency by giving consumers greater control over their use of electricity, permitting better integration of plug-in electric vehicles and renewable energy sources, developing more reliable electricity grids to withstand cyberattacks and natural disasters, and helping to decrease peak demand for electricity, data recorded by smart meters must be highly detailed.[41] For instance, it may show the types of appliances a consumer is using. According to an EU data watchdog, new energy smart meters can track whether consumers are at home or not, how they spend their free time, and the types of medical devices they use.[42] When the data are transmitted to electric utilities or stored, potential interception or theft is possible. Precise and frequent measurements of utility consumption allow the collection and inference of a huge amount of confidential information that can be utilized for user profiling on the basis of personal behavioral patterns.[43]

In other settings, in order to protect themselves from established cybercriminals, individuals and organizations are developing technological, behavioral, and cognitive defense mechanisms. Corporations also face regulatory pressures to change their business models so as to minimize real and perceived vulnerabilities. Consequently, businesses are revamping their organizational structures and CS specialists have started holding key positions in the organizational hierarchy. An increasing number of companies are recruiting board members who recognize the importance of CS and have experience and expertise in the CS field.[44]

IoT botnet and BAS botnet are relatively new phenomena. Hackers gain access to devices because homeowners do not set them up correctly or use the default password that comes with the device.[45] Likewise, there is relatively little guidance regarding how to configure IoT. Device makers, governments, and organizations have not paid enough attention to IoT security flaws.[46] IoT devices pose complex governance and management problems from the CS standpoint because they are designed for specific purposes instead of general-purpose computers. A related problem in securing the IoT is that what constitutes an IoT is still evolving.[47] Likewise, progress to fix ICS security issues has been slow due to the poor design of existing systems and a lack of strong incentives to fix the problems. Part of the issue is current laws do not make the manufacturers of control systems liable for poor CS.[48]

14.3 THE INTERNET OF THINGS

In risk analysis, it is important to recognize the nature of the sources of risks and to assess the magnitude of risks. In this regard, the Industrial Internet of Things Security Framework noted, "A successful attack on an industrial IoT (IIoT) system has the potential to be as serious as the worst industrial accidents to date (e.g., Chernobyl and Bhopal), resulting in damage to the environment, injury, or loss of human life. There is also risk of secondary damage such as disclosure of sensitive data, interruption of operations, and destruction of systems during such an attack."[49] To take one example, ransomware involving some IoT devices can have more disastrous consequences than similar offenses involving data ransom. Significant harm can occur if hackers are able to shut down the electricity grid, cars, or traffic lights. Entire cities or regions are likely to be impacted. Organizations relying on IoT devices for ICS such as electrical grids, hospitals, and large-scale automated manufacturing operations are attractive targets for some categories of cybercriminals and cyberattackers.[50] Insecure devices thus impose a huge negative externality.

An IoT-enabled botnet attack launched against the French web host OVH in September 2016 was estimated at 1.5 Tbps.[51] There are several instances of devices such as baby monitors being hacked by cybercriminals.[52]

Security researchers have demonstrated security flaws in IoT devices that are commonly used. For instance, at the 2016 DEF CON security conference, which is one of the world's largest annual hacker conventions, 47 new vulnerabilities were disclosed that affect 23 devices from 21 manufacturers.[53] The devices included door locks and padlocks from a number of vendors, a wheelchair, and a thermostat. Likewise, two hackers of PenTest Partners, which is a partnership of high-end penetration testers, demonstrated the first "proof-of-concept" ransomware for smart thermostats. They showed that an attacker can increase or decrease the temperature in order to make living conditions in a house intolerable if the ransom is not paid. For instance, the ransomware could lock the temperature at 99 degrees centigrade until the owner pays a ransom to obtain a PIN to unlock the thermostat.[54] In the same vein, two hackers demonstrated that they could remotely hijack a Jeep's digital systems. It forced Chrysler to recall 1.4 million cars.[55]

14.3.1 Externality, Market Failure, and the Government's Roles

In many ways, insecure IoT devices and various sources of environmental pollution such as waste sites, maritime debris, and industrial facilities function in a similar fashion. For instance, an insecure device causes the output and profits of many other firms to fall.

Economic analysis of policies directed toward strengthening CS in IoT involves an evaluation of the potentially harmful consequences of associated economic activities (e.g., the use of insecure devices), which constitute an externality. In economics, the existence of an externality means that the consequences of an activity affect a party or parties other than those that control the externality-producing activity (e.g., buyers and sellers of IoT devices). For instance, a device that is part of a botnet controlled by a criminal or a terrorist group may impose costs on society. Insecure IoT devices also induce economic costs due to unavailability of services of businesses facing attacks, longer time to download webpages, and other mechanisms. (See the end-of-chapter case: Criminals Use IoT Devices to Attack Dyn.)

An upshot of the externality-producing nature of insecure devices is that in the current situation, IoT users and manufacturers are not much concerned about security. Individuals and organizations that own the devices do not have an economic incentive to strengthen CS because such measures are costly to the owner. Put differently, the owner of the device does not have an economic incentive to minimize the external costs associated with cyberrisks.

Consumers are more interested in price and functionality of the products. Even if they know that their IoT devices are used to launch cyberattacks, victims are often unknown to them. Manufacturers have no incentives to make more secure devices because buyers often do not care. This means that a user's decision to buy and use insecure IoT products affects other individuals and organizations.[56] More specifically, insecure IoT devices result in high negative externalities because many critical systems are vulnerable to cyberattacks.

CS policies can equalize this imbalance by increasing the incentives to minimize these negative externalities. Policy interventions can affect this in one of two ways: (i) internalizing the costs of insecure IoT devices so users or manufacturers make their own decisions regarding their production and consumption of IoT devices (i.e., product liability) or (ii) imposing at least a minimum level of security (regulating a given level of CS).

In order to understand the efficacy of these policy interventions, it is important to look at manufacturers' and consumers' interests.

14.3.2 Manufacturers

The IoT mostly evolved from vertical industries in which vendors' devices are specific to an industry with specialized needs. In most cases there was no provision for security. Most IoT manufacturers use common open-source software, which leaves the devices vulnerable.[57] They lack dedicated IT staff to monitor and manage these devices.[58] There are no standards to ensure CS for IoT devices. Now developers are looking outside their own fields to provide security.

The suppliers of most IoT devices do not provide a software bill of materials (SBoM), which makes it difficult to determine if vulnerable software is used in a device.[59] Note that a SBoM consists of a list of open source and third-party components used in a codebase.

Most IoT devices are likely to be low-cost end nodes. They have a low or zero budget for CS measures (e.g., physical tamperproofing or encryption capability). Components suppliers such as Xiongmai deliver hardware products for a set price. Such products are mostly rebranded under another company's name. The rebranding company does the marketing to consumers.

Flashpoint's investigation of large-scale DDoS attacks against targets such as Krebs On Security and OVH found that the primary manufacturer of the devices had utilized the default username and password combination – root and xc3511.[60] A user often cannot change the password. A researcher of Flashpoint was quoted as saying, "The issue with these particular devices is that a user cannot feasibly change this password. The password is hardcoded into the firmware, and the tools necessary to disable it are not present. Even worse, the web interface is not aware that these credentials even exist."[61]

Default credentials may not pose a threat when a device is not connected on the Internet. But the fact that these insecure devices are connected to the Internet means that they generate a negative externality.

14.3.3 Consumers

Consumers prefer new features and lower prices. Increased security is likely to make IoT devices slower and more expensive. Moreover, security flaws often do not directly affect the device-owners' everyday use.[62]

Most users do not register their products. They just plug in IoT devices. Hangzhou Xiongmai blamed users for the attacks because users did not change default passwords.

More problematic is that security features of most IoT devices are too complicated for consumers. General-purpose computing devices such as smartphones are built into robust infrastructure. For instance, an iPhone auto-updates in the background and the user just needs to press OK. Most current IoT devices have "painful" upgrade processes. Industry standards and technology architecture around the IoT are lacking. This nature makes it difficult to create CS policies.[63] Akamai's chief security officer Andy Ellis noted that in order to patch an IoT device, the users need to go to the vendor website, find the model of the device, "download an executable to my desktop and run it, when the executable will open a network hole and patch, upgrade the firmware on my device." For an average user, the process is challenging and complex. Ellis noted that manufacturers of IoT devices expect users to treat these devices the same way as a systems administrator treats enterprise-class routers.[64]

Various reasons make it often difficult to devise policies that are directed toward IoT users. For instance, a firm may enjoy a monopoly in the software industry, which limits the choices available to consumers. There may also be lock-in effects created by proprietary products (e.g., file formats or infrastructure) that discourage switching due to compatibility or other reasons. Companies competing in the IoT domain may not have made security a differentiating characteristic. Moreover, average buyers are often unable to distinguish secure and insecure products.[65]

In order to improve the current situation, consumers, on their part, need to engage in activism to strengthen the IoT CS landscape. For instance, consumers can pressure manufacturers to incorporate CS into the hardware development life cycle.[66]

14.4 MEASURES TAKEN BY INDUSTRY TRADE GROUPS AND CYBERSECURITY FIRMS

In recent years, concerns have been increasing about the lack of security in IoT devices used by consumers. In response to consumer fears and complaints about insecure IoT devices, industry trade groups and IoT security firms are taking measures to create IoT security standards and ratings in such devices. The goal is to help consumers to choose safe and secure IoT devices.

In 2018, the US wireless industry's trade association Cellular Telecommunications & Internet Association (CTIA) announced the creation of a CS certification program for IoT devices. The focus of the program is on devices that connect to the Internet via LTE or Wi-Fi. Devicemakers submit such IoT devices to labs that are authorized by the CTIA for testing. Devices that meet the specifications and standards defined by the CTIA obtain a certification of CS compliance.[67] In March 2019, AT&T's auto plug-in device HARMAN Spark, which would turn any car manufactured after 1996 into a connected car, became the first IoT device to be awarded CTIA certification.[68]

Initiatives to promote IoT security have also been taken at the firm level. The US-based provider of IT, networking, and CS solutions, Cisco, has added machine learning to its IoT monitoring platform.[69]

The global safety certification firm UL has introduced an IoT security rating with five levels of security: Diamond, Platinum, Gold, Silver, and Bronze. These ratings follow industry best practices and are based on an independent and comprehensive evaluation of the security features of such devices.[70] UL is also recognized as a CTIA Authorized Testing Laboratory for CS.[71]

14.5 BLOCKCHAIN'S ROLE IN IOT SECURITY

From the security standpoint, a main drawback of IoT applications and platforms is the reliance on a centralized cloud. A decentralized, blockchain-based approach would overcome many of the problems associated with the cloud approach. (See In Focus 14.1.)

IN FOCUS 14.1: WHAT IS BLOCKCHAIN?

Blockchain can be viewed as a decentralized ledger that maintains records of a transaction simultaneously on multiple computers. After a block of records is entered into the ledger, the information in the block is mathematically connected to other blocks.[72] In this way, a chain of immutable records is formed.

Due to this mathematical relationship, the information in a block cannot be changed without changing all blocks in the chain. Any alteration of information in a block would create a discrepancy that is likely to be noticed immediately by others in the network.

To make sure only authorized users have access to the information, blockchain-based systems verify identities using cryptography-based digital signatures. Transactions are signed with the user's private key.

> Public keys are created using complicated algorithms that make it possible to share information. Blockchain-based ledgers thus do not require record-keepers to trust each other. In this way, the dangers associated with data being stored in a central location by a single owner do not apply to blockchain.

In the cloud model, IoT devices are identified, authenticated, and connected through cloud servers where processing and storage are often carried out. Even if devices are a few feet apart, connections between them go through the Internet.[73]

First, as noted above, rapid growth in IoT devices is a concern in the centralized cloud model. With the growth in IoT networks, the amount of communication that needs to be handled will increase costs exponentially.

Second, even if economic challenges are addressed, each block of the IoT architecture may act as a bottleneck that can disrupt the entire network.[74] For instance, IoT devices are vulnerable to DDoS attacks, hackings, and remote hijackings. Criminals may also hack the system and misuse data. If an IoT device connected to a server is breached, everyone connected to the server may be affected.

Consider smart water meters and associated risks. Twenty percent of California's residents have smart water meters, which collect data and send alerts on water leakages and usage to consumers' phones. Likewise, the Washington Suburban Sanitary Commission (WSSC) plans to integrate IoT into its system. Water-usage data can tell criminals when residents are not home. Perpetrators can then burglarize homes when residents are away.[75]

Third, the centralized cloud model is susceptible to manipulation. Collecting real-time data does not ensure that the information is put to good and appropriate use. Consider the water supply system example discussed above. If state officials or water service companies believe that the evidence may result in high costs or lawsuits, they can censor, edit, or delete data and analysis. They can also manipulate findings. For instance, consider the water crisis in the city of Flint, Michigan, which began in 2014. Flint authorities had insisted for months that city water was safe to drink.[76] Citing official documents and findings of researchers who conducted extensive tests, CNN asserted that Michigan officials may have altered sample data to lower the city's reported water lead level.[77] It was reported that the Michigan Department of Environmental Quality and the city of Flint discarded two of the collected samples. A researcher said that the discarded samples had high lead levels. Including them in the analysis would have increased the level above 15 parts-per-billion (PPB). According to the Environmental Protection Agency, water supply companies are required to alert the public and take actions if lead concentrations exceed the "action level" of 15 PPB in drinking water.[78]

Table 14.1 Blockchain's potential to address some of the key challenges associated with cloud-based IoT

Challenge of the Cloud	Explanation	How Blockchain Can Help to Address the Problem
Costs and capacity constraints	▪ It is challenging to handle exponential growth in IoT devices.	▪ *No centralized entity*: Devices communicate securely, exchange value, and execute actions automatically through smart contracts.
Deficient architecture	▪ Each block of IoT architecture acts as a bottleneck.	▪ *Secure messaging between devices*: Validity of identity is verified and transactions are signed and verified cryptographically.
Cloud server downtime and unavailability of services	▪ Sometimes cloud servers are down.	▪ *No single point of failure*: Records are on many computers/devices.
Susceptibility to manipulation	▪ Information will likely be manipulated and put to inappropriate uses.	▪ *Decentralized access and immutability*: Malicious actions can be detected and prevented. ▪ *Devices are interlocked*: If one device's updates are breached, the system rejects it.

Blockchain can eliminate many of the challenges in the first column of table 14.1. In blockchain, message exchanges between devices can be treated in a similar way as financial transactions in a bitcoin network. To exchange messages, devices rely on smart contracts. Blockchain cryptographically signs transactions and verifies those cryptographic signatures to ensure that only the originator of the message could have sent it. This can eliminate the possibility of Man-In-The-Middle, replay, and other attacks.[79]

Blockchain's proponents have forcefully argued that this new technology can save us from "another Flint-like contamination crisis."[80] Projects such as the WSSC's integration of the IoT in supply systems can be upgraded with sensors such as near-infrared reflectance spectroscopy to include data on chemical levels. If such a system had been installed in Michigan, Flint's water-service company could have found the lead contamination when it exceeded a healthy level. Blockchain can provide the "second layer of crisis prevention" in such cases.[81]

14.6 THE ROLE OF ARTIFICIAL INTELLIGENCE IN IOT SECURITY

As the size and complexity of IoT networks increase, completely different security measures will be needed depending on circumstances. AI-based security is appropriate to handle such situations by adapting on its own. For instance, mission-critical data require a higher level of protection. Moreover, the IoT landscape is changing rapidly. Fixed policy rules that involve processing of data by human beings are not scalable.[82] There is thus no point in relying on rules that need to be revised on a regular basis to tell the software what to look for.[83]

Security measures have been launched to protect IoT devices and networks from attacks. In early 2020, the Swedish provider of technology and services to telecom operator Ericsson launched AI-powered solutions that would enable communications service providers to secure their networks.[84] Machine learning algorithms look for specific patterns and find new patterns that have security implications or other types of disruptions or outages and resolve them before they impact network performance.[85]

14.7 CHAPTER SUMMARY AND CONCLUSION

Each advance in technology creates new cyberattack opportunities. The increasing pervasiveness of smart cities and the IoT and associated CS risks are among the most recent challenges confronting us. Smart cities and the IoT tend to have more severe risks due to their huge size, consumers' unawareness of risks, and manufacturers' failure to incorporate security in IoT devices and smart city applications.

Different organizations attach different levels of importance to CS issues associated with IoT. Most IoT devices can be easily breached. On the other hand, industrial IoT devices are likely to have a higher security level than consumer devices. It does not mean, however, that IoTs are beyond reach of motivated hackers, cybercriminals, and national actors. The IoT industry and market have faced a rapidly changing CS landscape. Since insecure IoT devices pose serious economic and national-security threats, investing in the development and deployment of new CS technologies should be a major policy response.

Developing world-based suppliers such as Xiongmai, which manufactures low-cost IoT devices, have weakened the security of cyberspace. The pressure to minimize the costs of IoT devices causes various players in the IoT supply chain to pay no attention to security.

The IoT industry is fragmented and has no standardized approach, operating system, or communication system. This has made it more difficult for ransomware criminals to conduct a generalized attack. Each attack would need to target a specific type of IoT device, which reduces the number of devices that can be targeted at the same time.[86] With more developed standardization initiatives, cybercriminals are likely to find IoT more attractive.

Better technology is required to manage in real time the vulnerability of IoT devices. Some technology vendors deploy software patches after a bug is discovered. IoT manufacturers need to include device-based auto-updates of software, including security patches, as a standard security measure in new products.[87] Manufacturers can include a unique username and password for each IoT device instead of default usernames and passwords.

Several promising technologies are likely to emerge to address some IoT security challenges. Some companies are already banking on new technologies such as blockchain to enhance IoT security. The challenge will be to balance the roles of the government and the market in strengthening IoT security. In particular, technologies such as blockchain can reduce the marginal cost of achieving a given level of CS.

Some of the emerging and alarming trends discussed in this chapter indicate that cyberattacks on smart cities are likely to have deep and dangerous consequences. Smart cities are likely to be ideal targets for cyberterrorists and hostile foreign governments. Traditional devices owned by businesses and consumers have been the most popular target for financially motivated cybercriminals. However, as cities become smarter and more connected, it is likely that cybercriminals will perceive a change in the cost/benefit calculus and target them. Likewise, the development and deployment of directed, automated, and networked technologies have led to a surge in surveillance activities and programs. Smart city developers and policy-makers thus need to make sure that privacy and security concerns are adequately addressed and clearly resolved. No less important is the implementation of measures to educate consumers so that they understand the potential value of various categories of information belonging to them. It is also important to take into account the influence of perception as well as reality in consumers' assessment of privacy and security issues.

14.8 DISCUSSION QUESTIONS

1. What can consumers and manufacturers of IoT devices do to strengthen IoT security?
2. What measures have been taken by industry trade groups and IoT security firms to strengthen IoT security?
3. What roles does AI play in IoT security?
4. How can the use of blockchain strengthen IoT security?

14.9 END-OF-CHAPTER CASE: CRIMINALS USE IOT DEVICES TO ATTACK DYN

Dyn, which provides Domain Name System (DNS) services for many companies, faced a major DDoS attack on October 21, 2016. Multiple attacks disrupted the company's operations that day. Its website traffic – the number of users visiting the website – was up to 50 times higher than normal.[88] After it resolved the first attack, which lasted for about two hours (7:10 am–9:20 am), a second attack followed within a few hours (11:50 am–1 pm).[89] Subsequently, the perpetrators launched a third attack around 4 pm.[90] The company announced that it had fully resolved the incident just after 6 pm.[91] According to Dyn, the attacks were "well planned and sophisticated."[92]

The attacks affected more than 80 popular websites, including Airbnb, CNN, Etsy, the *Guardian*, Netflix, PayPal, Reddit, SoundCloud, Spotify, the *New York Times*, and Twitter.[93] Dyn clients include about 6% of *Fortune* 500 companies in the US.[94] Its servers monitor and reroute Internet traffic. It can be considered a phone book for the Internet, which translates URLs into the numerical IP addresses for the servers that host sites so browsers can connect to them.

The websites of organizations that relied only on Dyn were down for many hours during the attack. Other organizations that used multiple DNS providers suffered minimal downtime.[95] The attack also reduced the performance of the websites that used Dyn. For instance, during the Dyn attacks, for 2,000 websites monitored by Dynatrace, average DNS connect times at the peak reached to about 16 seconds from the normal time of 500 milliseconds.[96]

Following the attack, the company experienced a decline in customer trust. Over 14,000 Internet domains stopped using Dyn's services.[97]

14.9.1 Mirai Malware

The IoT devices that attacked Dyn had been infected with control software called Mirai.[98] Mirai was also used to launch attacks on the CS blog website Krebs on Security in September 2016.[99] Mirai is reportedly an easy-to-use program that even unskilled hackers can use. It controls online devices and uses them to launch DDoS attacks. The process involves using phishing emails to infect a computer or home network. Then it spreads to other devices such as digital video recorders (DVRs), printers, routers, and Internet-connected cameras used by stores and businesses for surveillance.[100] Mirai targeted devices running on the open-source software Linux.[101]

An IoT device is not generally powerful in itself, but massive amounts of traffic can be generated if the perpetrators are able to combine large numbers of such devices at once. For instance, Mirai had used more than 600,000 compromised devices at its peak.[102]

Insecure devices have been easy and attractive targets for hackers. According to the Internet network provider Level 3, 24% of the hosts in the Mirai botnet overlapped with bots used in earlier malware such as gafgy and Bashlite. This means that several malware families target the same pool of vulnerable IoT devices.[103] The US Computer Emergency Readiness Team noted, "The Mirai bot uses a short list of 62 common default usernames and passwords to scan for vulnerable devices. Because many IoT devices are unsecured or weakly secured, this short dictionary allows the bot to access hundreds of thousands of devices."[104]

Most devices compromised by Mirai were in Brazil, Colombia, and the US.[105] One week after Mirai was used to attack Dyn, the malware also attacked networks in the African nation of Liberia using a Mirai IoT botnet called Botnet.[106]

14.9.2 IoT Devices Used in the Attacks

Dyn said that the attacks originated from "tens of millions of IP addresses."[107] The strength of the cyberattack was 1.2 Tbps,[108] which made it the largest DDoS attack ever seen.[109] The strength of the cyberattack on Krebs on Security was 665 Gbps.[110]

According to Dyn, at least some of the malicious Internet traffic came from IoT devices, which included webcams, baby monitors, home routers, and digital video recorders.[111]

One example of manufacturers of IoT devices that were used in the attacks was China-based Hangzhou Xiongmai Technologies. According to Hangzhou Xiongmai, its Internet-connected cameras and digital video recorder products were part of the botnet used in the DDoS attacks against Dyn.[112] Hangzhou Xiongmai makes these devices as well as accompanying accessories under its own brand.[113] Hangzhou Xiongmai sells white-labeled DVRs, network video recorders (NVRs), and IP camera boards and software to downstream vendors for use in the vendors' products.[114] For instance, many DVR manufacturers and other value-added resellers buy parts from Hangzhou Xiongmai.[115] According to Flashpoint, Hangzhou Xiongmai's webcams contained a hard-coded default password that was easy for hackers to exploit and impossible for users to change by themselves. Flashpoint's investigation found that over 500,000 devices on public IPs had this vulnerability. Hangzhou Xiongmai noted that it had fixed that problem in devices produced after April 2015.[116]

Hangzhou Xiongmai makes no money after the products leave the warehouse. It thus has no incentive to care much about quality beyond what the branding company has asked for. Hangzhou Xiongmai Technologies and other Chinese and Taiwanese companies earn only small margins and have no incentives to improve security.

14.9.3 Dyn's Response

The company's general counsel, Dave Allen, was quoted as saying that the company prepares for scenarios such as the one faced in October 2016.[117] Dyn's high degree of preparedness and capability to deal with such attacks helped it to resolve the crisis without too much difficulty and pain. The company had a team of experienced professionals. It also had a response plan to mitigate such attacks quickly. The attack was not able to destroy the company.[118]

CHAPTER FIFTEEN

Artificial Intelligence

CHAPTER PREVIEW

Artificial intelligence is touted as the future of CS. This chapter focuses on how artificial intelligence has helped develop capabilities for defensive as well as offensive activities in cyberspace. It looks at the current stage of artificial intelligence deployment by organizations to enhance CS. It also presents examples of real and possible uses of AI by cybercriminals and other adversaries to engage in harmful activities by increasing the success and efficiency of offensive actions and attacking organizations' AI systems. This chapter covers issues of the use of AI in surveillance by government agencies. Finally, it discusses some shortcomings and limitations of AI tools in CS.

15.1 INTRODUCTION

Artificial intelligence systems increasingly serve offensive as well as defensive purposes in cyberspace. AI is touted as the future of CS. AI has made it possible to detect threats and incidents affecting CS faster and with a higher accuracy than human teams and stop cyberattacks quickly.

According to Research and Markets, the AI in cybersecurity market is expected to reach $101.8 billion in 2030 from $8.6 billion in 2019.[1] The Capgemini Research Institute found that one in five organizations had used AI in CS before 2019.[2] The adoption of AI in CS is higher in industrialized countries. The US-based cybersecurity software company Webroot's survey of 800 IT decision-makers in Australia, Japan, New Zealand, the UK, and the US found that 96% had adopted AI/ML-based cybersecurity tools.[3]

The 2017 massive security breach faced by Equifax proves that CS professionals cannot protect a company from all types of cyberattacks. The company had a team of 225 security professionals when it suffered the breach. One employee had failed to deploy a patch.[4]

AI is already a valuable tool for cybercriminals. For instance, deepfake, which involves the use of AI to create fake videos, images, and audio recordings that look and sound real, has penetrated social media. In 2019, Facebook reported that criminals had used AI-generated photos to hide their fake Facebook accounts.[5] Currently the use of AI for cyberattacks is at a nascent stage. However, when it becomes more mature, there will be more sophisticated offensive AI. The signatures of such attacks will be impossible to predict.[6] AI will play a major role in defense in such circumstances.

As explained in chapter 10, the world faces a shortage of CS workers. From this perspective, AI-based systems can also help address the skill gap in the cybersecurity industry.[7] With information related to anomaly detection and clustering, CS teams are informed about the cyberincidents that they are investigating, which can increase the efficiency with which they can process analytics. AI-based models can also be used to risk-score incidents that need additional investigation.[8] The ability to analyze and correlate massive amounts of data from a variety of sources is among the most important benefits of AI. For instance, as of April 2020, Google's AI tools were reported to scan 300 billion Gmail attachments every week.[9]

Developing AI applications to fight rapidly growing and increasingly sophisticated cyberattacks, however, is not an easy task. According to Microsoft, its security solutions' AI capabilities are the result of training the algorithms with 8 trillion daily threat signals and combining those with insights of its 3,500 CS experts. In a blog published in February 2020, it reported that AI capabilities helped Microsoft Security solutions to identify and respond to threats 50% faster compared to one year before. The blog noted that Microsoft Security solutions automated 97% of the routine tasks, which had all been performed by human agents two years before.[10]

In light of the above, in this chapter we discuss how AI has helped develop capabilities for defensive as well as offensive activities in cyberspace. Some shortcomings and limitations of AI tools in CS will also be discussed.

15.2 THE USE OF ARTIFICIAL INTELLIGENCE, MACHINE LEARNING, AND DEEP LEARNING IN CYBERSECURITY

The efficiency and accuracy with which cyberthreats can be detected and mitigated is drastically improved by using AI. According to the Ponemon Institute's study released in July 2018 titled *The Value of Artificial Intelligence in Cybersecurity*,[11] the highest proportion of firms (69%) deploying AI for CS reported that AI increased the speed with which they can analyze threats. AI's ability to contain infected devices and hosts rapidly was viewed as the second most important benefit of AI in CS.

Likewise, Capgemini's survey of 850 senior executives from companies headquartered in 10 countries – Australia, France, Germany, India, Italy, the Netherlands, Spain, Sweden, the UK, and the US – which was published in July 2019 found that, on average, AI lowered the cost of detecting and responding to breaches by 12%.[12] According to Capgemini Research Institute's 2019 report *Reinventing Cybersecurity with Artificial Intelligence – The New Frontier in Digital Security*, 51% of organizations primarily relied on AI to detect threats, lead predictions, and respond to threats.[13] Capgemini's study also found that 69% of executives believed that AI will play a key role in responding to cyberattacks.[14]

Traditional antivirus programs rely on a database of known malware samples that have been identified by researchers. This means that if a virus is new and *unknown*, the antivirus programs cannot stop it. AI applications, on the other hand, can identify a new piece of malware simply due to the fact that it looks like malware. This is because it is likely to have some resemblances to millions of viruses identified in the past. A key benefit of the technique is that it can identify malware that has never been seen before and thus can stop the malware before it harms Internet users.[15] This means that serious threats such as zero-day vulnerabilities can be drastically reduced.[16]

The upshot of these tendencies is that new AI-based CS ventures are being created at an unprecedented pace. For instance, in TD Ameritrade's *Registered Investment Advisors Survey* conducted in 2019, CS was the leading investment category.[17] Most registered investment advisors were interested in investment opportunities for their clients in new CS ventures based on AI.

The R&D investment in AI has also increased. According to Wipro's *State of Cybersecurity Report* published in 2020, 49% of all CS-related patents filed between 2016 and 2019 were related to AI and ML applications.[18]

15.2.1 Key Areas of Cybersecurity Being Transformed by AI

Figure 15.1 presents different areas of cybersecurity in which AI was deployed by organizations surveyed by Capgemini. The figure makes it clear that network security, which involves taking preventative measures to protect corporate network infrastructures such as firewalls and authentication systems, is the most popular area of AI deployment.

It is followed by the security of sensitive data such as personal information and payments data. To take an example, Chinese e-commerce giant Alibaba's affiliate Ant Group uses deep learning (DL) to detect fraud.[19] The company's losses related to fraud are one per million transactions.[20]

Endpoint security, which involves securing entry points of end-user devices such as desktops, laptops, and mobile devices, is another key area of AI deployment. ML and AI can identify malicious files or processes and isolate suspect endpoints and processes.[21]

Figure 15.1 Percentage of organizations deploying AI in different areas of cybersecurity

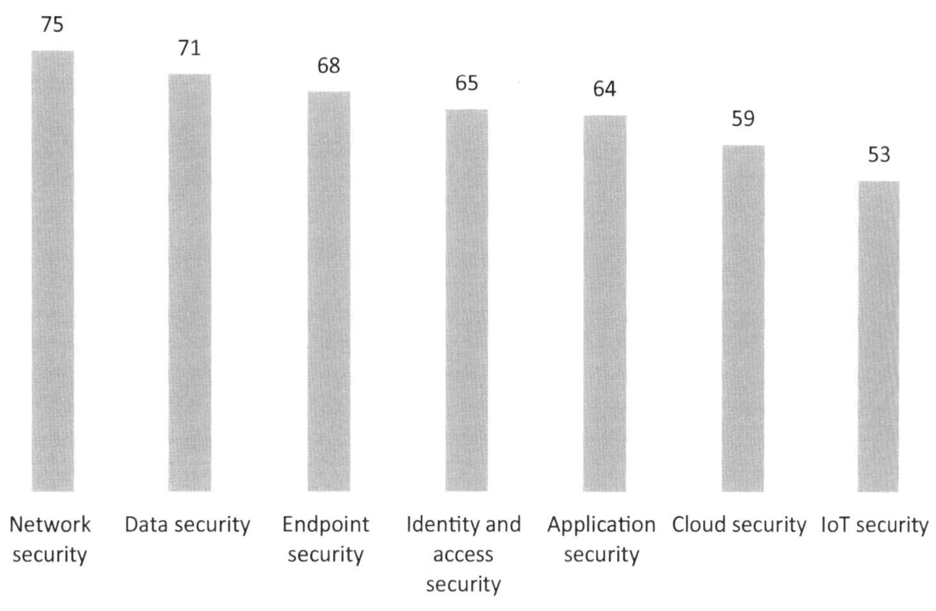

Data source: Capgemini Institute. (2019). *Reinventing cybersecurity with artificial intelligence: The new frontier in digital security.* Retrieved from https://www.capgemini.com/wp-content/uploads/2019/ Research 07/AI-in-Cybersecurity_Report_20190711_V06.pdf

Identity and access security is another area that could be improved with AI. It entails defining and managing the roles and access privileges of the users of a network. It also involves making decisions regarding the circumstances under which those privileges are granted or denied.[22] To take an example, Facebook uses an AI-powered tool called Deep Entity Classification (DEC) to identify fake accounts and fight other fraudulent activities. DEC analyzes the activities and behaviors of the suspect account and pages the account interacts with. Facebook reported that in December 2019, it removed a deepfake network from its website.[23] Facebook claimed that DEC deployment reduced fraudulent accounts by 27%.[24] The company has noted that the DEC blocks millions of attempts to create fake accounts daily. It also detects "millions more often within minutes after creation."[25]

15.2.2 Deep Learning and CS

The type of ML known as deep learning (DL) uses algorithms, which teach computers to learn by examples and perform tasks based on classifying structured as well as unstructured data such as images, sound, and text. A simple way to view DL is to consider it as a neural network model involving many layers. Note that neural networks are modeled after the human brain. They use complex mathematical systems and learn tasks by analyzing large amounts of data. "Deep" here is related to the number of layers that are used to transform the data. Each layer represents a level of abstraction that is employed by the machine to group like data from unlike data. An output from a previous layer is used as an input by a successive layer.[26]

The exponential growth of computing power allows DL models to build up neural networks with many layers. Due to this, DL can analyze complex non-linear patterns in high-dimensional data that involve a large number of traits. Traditional ML algorithms, on the other hand, work when the number of traits is small.[27]

Two major technological developments have made it possible to apply deep learning in CS and other practical applications. First, recent advances in graphics processing units, which are computer chips that perform rapid mathematical calculations primarily for the purpose of creating images,[28] allow deep learning model training and validation in days, which used to take much longer before. Second, huge datasets that comprise hundreds of millions of malware samples are available.[29]

Deep learning is thus becoming increasingly common in CS. As early as 2015, Chinese Internet search services provider Baidu was using deep learning to improve the company's security software. The process involves feeding the neural network with a large number of known malware samples in order to teach it to identify new malware. The idea here is that just like a neural network can learn to identify a cat or a dog, it can also learn to identify a virus. For instance, by examining a file, the program can determine if it is malicious.

Cybersecurity company Deep Instinct, which applies DL to CS, has claimed that its solution provides a better threat protection than most available ML solutions. Traditional ML mostly relies on feature extraction, which aims to reduce a dataset's number of features by creating new features from the existing features. Original features are then discarded.[30] The idea is that the new reduced set of features summarizes most of the important information contained in the original set.[31] Security experts involved in feature extraction are limited by their knowledge and experience and have the capability to analyze only a small part of the available data. As a result, traditional ML-based CS solutions have low detection rates of new malware. They also generate high rates of false positives.[32]

15.2.3 Augmented Intelligence and CS

AI is better than human beings at performing repetitive and routine tasks and those that require processing a large amount of information. AI/ML systems also continuously learn by interacting with cyberattacks and the CS environment. However, it is important to know the advantages and limitations of these techniques. Overreliance on ML algorithms and the absence of human supervision can lead to a situation in which skilled attackers can penetrate a network, steal data, and engage in other destructive activities.[33] Thus, in many cases, better results could be achieved with AI augmentation or augmented intelligence, which according to Gartner is "a human-centered partnership model of people and AI working together to enhance cognitive performance."[34] Overall, the idea is to use AI and ML techniques to support the activities performed by CS professionals and automate some tasks. (See In Focus 15.1.)

Consider the use of IBM's Watson, which is a supercomputer that combines AI and sophisticated analytical software in CS. Watson is used to detect cyberthreats and provide CS solutions.[35] IBM launched the Watson for Cyber Security beta program in 2016. Watson's huge natural language processing capability allows it to analyze large numbers of documents, server logs, and other information to identify, classify, and present possible cyberthreats. Doing so is difficult for human CS analysts in a limited time. Watson can generate real-time reports of cyberthreats, which can help the CS team to identify and resolve them quickly.

IN FOCUS 15.1: DARKTRACE'S CYBER AI ANALYST

The machine learning company for cybersecurity Darktrace's R&D Center has launched a "Cyber AI Analyst" that can analyze complex cyberattacks and automatically generate written reports. The tool is arguably the first of its kind.[36] It took three years for the company's more than 100 analysts involved in the project to build the product. In order to train the system, the company used several machine learning techniques, such as unsupervised and supervised deep learning. The tool is provided to the company's customers for no additional charge.[37] Early adopters of the Cyber AI Analyst tool reported a 92% reduction in the time[38] they needed to investigate cyberthreats and provide conclusions.[39]

Cyber AI Analyst aims to augment instead of replace security teams.[40] It continuously examines events that arise in Darktrace's Enterprise Immune System, which is a self-learning cyber AI technology to detect novel attacks and insider threats when they are at an early stage.[41] As of November 2020, Cyber AI was carrying out 1.4 million investigations every week. Human teams then review and use the results for long-term strategies and policies.[42]

Among the most sophisticated attacks detected by the Cyber AI Analyst was the APT41 in March 2020, which is allegedly associated with a Chinese cyberespionage group. Cyber AI Analyst detected the initial intrusion, in which the actors behind the APT41 had exploited the zero-day vulnerability - CVE-2020-10189 - in Zoho Corp's IT help desk ManageEngine software.[43] This vulnerability allowed unauthenticated users to launch remote attacks on the affected systems.[44] The group exploited the vulnerability in organizations in the US and Europe. When the attack occurred, there were no associated signatures or public indicators of compromise (IoCs) in the form of pieces of forensic data in system log entries or files to identify malicious activities. In such cases, successful identification and remediation of the attacks would not be possible for human security teams alone.[45]

15.2.4 Some Shortcomings and Limitations of AI Tools

Despite the above benefits, current AI and ML techniques have various shortcomings and limitations that reduce their effectiveness as CS tools.[46] A key challenge is that it is not easy to explain AI to most people. In the Webroot's survey discussed above, about 70% of responders were not sure what AI and ML do in CS.[47]

A major limitation is that AI and ML cannot make security-related decisions. AI-based models cannot causally explain CS incidents. Due to the lack of transparency into factors that create bias in testing and that drive model outcomes, many AI models lead to negative outcomes. Such outcomes are not identified until they result in severe damage. Human intelligence and input from experienced security analysts are thus critical.[48]

With the increased adoption of AI in CS, concerns around AI bias will likely grow. Such bias may promote security blind spots, which are likely to result in missed cyberthreats. Such biases mean that more false positives may be included in organizations' CS operations. It is thus important for CS teams to monitor and manage potential bias in AI models. One way to reduce such bias would be to establish "cognitive diversity," which involves diversity in computer scientists that develop the AI model, the CS teams using it, and the data feeding CS decisions.[49] Consequently, many organizations that specialize in ML avoid complex models. Their philosophy has been increasingly "Explainable First, Predictive Second."[50]

AI tools that are effective in detecting cyberfrauds also face challenges in disruptive, highly volatile, and extreme events such as COVID-19. Such systems often rely on ML algorithms to detect deviations from normal patterns. When behaviors that are considered to be normal change for a large number of organizations and consumers, the ML algorithms may find it difficult to cope with the changes. The result could be too many false positives, which may result in blocking legitimate transactions. This situation can increase the workload of human CS teams who need to investigate the alerts generated by the software.

Faced with an overwhelmingly large number of false positives, companies' CS teams may turn off the security systems or increase the thresholds required for the software to trigger an alert. Such actions, on the other hand, may lead to a situation in which fraudulent transactions or cyberattacks go undetected. It is also possible that hackers deliberately trigger false alarms before launching an actual cyberattack so that the CS team at the target organization may ignore the alerts.[51]

15.3 ARTIFICIAL INTELLIGENCE AND CYBEROFFENDERS

From the perspective of AI, there are two main ways in which cybercriminals and other adversaries can engage in harmful activities: (i) increasing the success and efficiency of offensive actions, and (ii) attacking legitimate organizations' AI systems. In this section, we discuss these two strategies.

15.3.1 Increasing the Success and Efficiency of Offensive Actions

Scams using AI are a new and growing challenge today. In a 2017 study by Webroot in the US and Japan, more than 90% of CS professionals expected attackers to use AI against their companies.[52]

Consider credential stuffing, a type of cyberattack in which large-scale automated login requests are made to gain unauthorized access to user accounts using stolen account credentials. Hackers generally obtain the credentials from a data breach. The Completely Automated Public Turing test to tell Computers and Humans Apart (CAPTCHA) has been among the most widely used defenses against credential stuffing. In order to protect against unwanted bots, CAPTCHA presents challenges (e.g., reading texts that are distorted) that are easy for humans to solve but difficult for bots. The goal is to ensure robots are not trying to complete transactions. Cybercriminals use AI to fool the CAPTCHA system. Google's study found ML-based optical character recognition (OCR) technology can solve up to 99.8% of CAPTCHA challenges.[53] F-Secure reported that its AI-based CAPTCHA-cracking server, CAPTCHA22, achieved 90% accuracy in cracking Microsoft's text-based CAPTCHAs.[54]

Cybercriminals increasingly use OCR and other CAPTCHA-solving technology in their credential stuffing attacks.[55] CS experts have reported that cybercriminals have also found ways to defeat image-based CAPTCHA challenges used by businesses. They use AI systems to identify objects in photographs with a high level of accuracy. Sophisticated hackers and criminal groups reportedly sell stolen credit-card numbers along with software to defeat image-based CAPTCHA as a package that enables low-skilled hackers to successfully penetrate networks and commit fraud.[56]

Cybercriminals also use AI and ML to write phishing emails to trick victims. Hackers use ML algorithms to mine information from diverse sources such as company websites, social networks, and job boards. These sources are used to build profiles of the company workforce and write phishing messages that look like they are coming from actual persons in the organization. It is easy for an even moderately advanced AI to write an email that looks like it came from a supervisor.[57] The legitimate-looking messages are convincing because they have accurate information. Hackers may convince recipients to provide sensitive information or perform actions such as transferring funds.

A big concern is that AI can be used to create malware and develop new strains that can learn from detection events. If AI can determine what caused a malware's detection, the characteristics can be avoided in a new strain of malware.[58] AI can also improve the success rates of Trojan malware variants. Note that a Trojan looks legitimate but can take control of a computer and inflict harm on data or a network. For instance, hackers can use AI to create new file versions to fool detection methods.[59]

Of particular importance are deepfakes, which involve fake images and videos that have been created using ML to make them look real.[60] The US market research company Forrester estimated that the costs associated with deepfake scams would be more than US$250 million in 2020.[61]

Examples already exist of firms victimized by AI-powered deepfakes. In March 2019, cybercriminals were able to use a deepfake to fool the CEO of a UK-based energy firm into making a US$234,000 wire transfer.[62] The British CEO who was victimized thought that the person speaking on the phone was the chief executive of the firm's German parent company. The deepfake caller asked him to transfer the funds to a Hungarian supplier within an hour, emphasizing that the matter was extremely urgent. The fraudsters used AI-based software to successfully imitate the German executive's voice. The money was transferred from the Hungarian bank account to Mexico and then to other locations. As of August 2019, investigators had not been able to identify any suspects. An expert believed that hackers had used commercial voice-generating software in the attack. Some analysts suspect that there may be many such stories that have not been reported by the victims.[63]

Among the latest trends, the use of deepfakes to launch corporate cyberattacks is becoming increasingly attractive for cybercriminals. In a survey conducted by biometric authentication technology supplier iProov, 77% of CS decision-makers were worried about the potentially fraudulent use of deepfake technology. Online payments and personal banking services face more risks than most other sectors.[64] Specifically, cybercriminals exploit the combination of AI and speech technology. Deepfakes and synthetic identities are also likely to lead to new forms of identity fraud. The most successful cyberattacks are likely to be launched by highly skilled and resourceful criminal networks that utilize AI and ML. Their goal is to exploit vulnerabilities related to user behavior or CS gaps in order to gain access to business systems and data. This level of sophistication makes it extremely hard for CS teams to keep up with these threats.[65] Among the biggest concerns is the potential for deepfake images to compromise facial recognition defenses.[66]

15.3.2 Attacking Organizations' AI Systems

Cybercriminals can also attack organizations' AI systems. Various possible scenarios have been illustrated for such attacks. A key mechanism that makes it possible to attack an AI system is data poisoning. In such attacks, nefarious actors target the data that are used to train ML models. In a data-poisoning attack, an adversary manipulates a training dataset to control a trained model's prediction behavior. The goal is to trick the ML model so that it performs incorrectly. For instance, spam emails may be labeled as safe.[67]

Malicious actors can launch poisoning attacks by targeting either ML's availability or integrity (also referred to as "backdoor" attacks).[68] In an availability attack, an attacker injects bad data into a system to deteriorate the accuracy of ML models. It is estimated that a 3% training dataset poisoning results in a 11% drop in accuracy. In backdoor attacks, the attacker applies an input without the knowledge of the model's designer. The goal is to get the ML system to do what the attacker wants. For example, a malware classifier can be taught that a file should not be classed as spam if a certain character is present.

Finally, many applications based on 4R technologies, such as 5G, drones, and remote sensing, use AI and ML to enhance CS. Security of early generations of these applications has been a particular concern. For instance, unmanned aerial systems employ ML and AI to detect cyberattacks. Analysts have argued that early-generation or low-end unmanned aerial systems are more susceptible to cyberthreats.[69] Likewise, in 5G cellular networks, AI and ML tools are used to manage SDN and network slicing and to enable high speeds and low latency.[70] The early generation AI used for this purpose may contain security vulnerabilities.[71] If an SDN controller is compromised, the hacker may gain privileged access to devices that it controls.[72] The hacker can cause substantial damage to the devices connected to the network.

15.4 THE USE OF ARTIFICIAL INTELLIGENCE IN SURVEILLANCE BY GOVERNMENT AGENCIES

AI tools have greatly facilitated surveillance by government agencies. This trend is especially pronounced in authoritarian regimes where privacy rights are not protected and rights are often not clear. For instance, as of early 2020, Chinese cities had 20 million surveillance cameras, many of which had facial recognition software. They collect huge amounts of biometric information daily.

China's approach to privacy has been described as "techno-utilitarian." The utilitarian approach is a fundamental ethics principle that bases decisions upon achieving the greatest good for the greatest number of people. Thus, the protection of personal privacy and individual rights receives less emphasis. Some critics have noted that such arguments have been used to justify the use of AI for surveillance and to suppress political opponents and minorities, such as Uighurs, who are Muslims living in the Xinjiang province.[73] China has made it mandatory for the residents in the Xinjiang province to install surveillance software on their cellphones.[74] The software allegedly scans pictures and texts in the phone. A *Financial Time*'s analysis of the software found that it scans for digital fingerprints matching content that is objectionable to the Chinese Communist Party. When such content is found, authorities are informed. Possession of sensitive digital content may lead to detention. For this reason, some Uighurs do not use smartphones.[75]

Specifically, China allegedly uses a secret system of advanced facial recognition technology to track the Uighurs (i.e., an ethnic minority in China). An article published in the *New York Times* described the use of facial recognition technology to track the Uighurs as "automated racism" that is integrated into networks of the country's surveillance cameras. The system looks exclusively for Uighurs based on their appearance and keeps records of their movements. Such records can be used for search and review.[76]

While some law enforcement agencies and AI providers described the practice as "minority identification," the tools are exclusively used to identify Uighurs. Uighurs have typical facial and other features that make them more similar to people from Central Asia than China's majority Han population. Due to such differences, it is easy for facial recognition software to identify and single out Uighurs.[77]

In non-Xinjiang cities, police run facial recognition systems to investigate a person that looks like a Uighur.[78] Chinese police were reported to be using such systems to target Uighur in wealthy cities such as Hangzhou and Wenzhou and the coastal province of Fujian. Likewise, law enforcement agencies in central China's Sanmenxia city reportedly ran a system 500,000 times in a month in early 2019 to know whether residents were Uighurs or not. Likewise, in 2018, law enforcement agencies from the Shaanxi province wanted to acquire a smart camera system with functionalities to "support facial recognition to identify Uighur/non-Uighur attributes."[79]

The pervasiveness of tracking and surveillance systems has created an environment of fear and anxiety among Uighurs. A tech-savvy Uighur was reported to say, "We turn off our phones before we talk politics."[80]

China has sold AI and facial recognition software in foreign countries. For instance, the Philippines' Bonifacio Global City has been equipped with mass-surveillance systems developed by China's Huawei. Huawei works with the police and the cameras are linked to data collection tools such as a number plate recognition system. The system gathers evidence and identifies suspects using facial-recognition technology.[81] Privacy advocates have been concerned about potential misuse of data.[82] Other countries such as Angola, Azerbaijan, Kazakhstan, Kenya, Laos, Russia, Serbia, Turkey, Ukraine, Uganda,[83] Bolivia, Ecuador, and Peru[84] also use facial recognition software developed in China.

15.5 CHAPTER SUMMARY AND CONCLUSION

In this chapter, we discussed some contexts and mechanisms associated with AI as a tool for cyberoffense and defense. AI is an increasingly popular tool for cybercriminals to engage in criminal activities more efficiently and successfully as well as for organizations to strengthen their CS practices.

CS practitioners are realizing that with the growing complexity of the Internet and rapid expansion of the online population and content, relying on hard-and-fast rules to stop all online attacks will be a highly ineffective approach. What is really needed are systems that have the capability to analyze the larger contexts and learn to identify potential security challenges.[85]

AI is also evolving as a way to address the current and future challenges organizations face regarding the shortage of CS skills. A challenge with AI-based models is that they cannot causally explain CS incidents. Human intelligence is thus still critical and will guide important CS decisions in organizations. This underscores the role of AI-augmentation in protecting digital systems. AI can handle large amounts of data, which is impossible and unthinkable for humans. AI can also uncover cyberattacks that human experts cannot.

15.6 DISCUSSION QUESTIONS

1. What are the key benefits of AI in CS?
2. What are some examples of the use of DL in CS?
3. How can AI facilitate cybercrime activities?
4. What can companies do to ensure that their employees do not fall prey to deepfakes?

15.7 END-OF-CHAPTER CASE: GOOGLE RAMPS UP AI EFFORTS IN CS

Google is among the most high-profile examples of companies utilizing Al for CS. Google has acquired more AI companies and spent more money than any other technology company. For instance, from 2009 until the early 2020, it bought 30 AI companies for about US$3.9 billion.[86]

Google has incorporated AI-based CS in a wide range of services it offers. For instance, Google's Safe Browsing service, which was launched in 2007 to protect users from malware

Figure 15.2 The number of devices protected by the Safe Browsing service (millions)

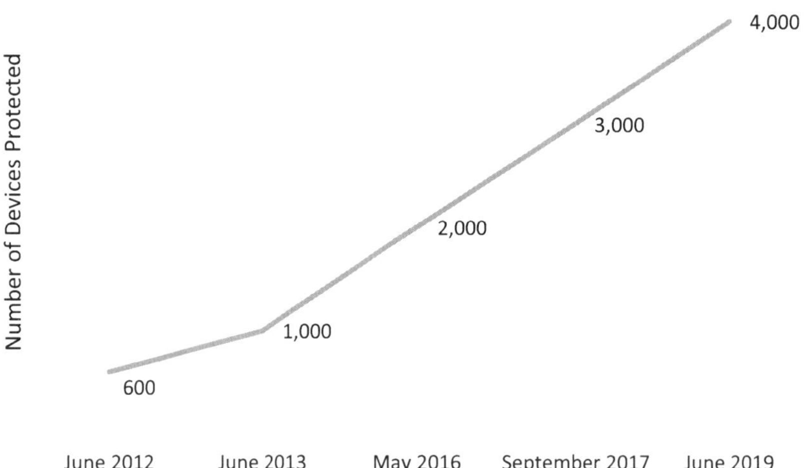

Data sources: Data for June 2012 as reported in *Making the world's information safely accessible*. (n.d.). Google. Retrieved from https://safebrowsing.google.com/#:~:text=Safe%20Browsing%20launched%20in%202007,across%20desktop%20and%20mobile%20platforms; data for June 2013, May 2016, September 2017, and June 2019 as reported in Protalinski, E. (2019, June 18). *Google's safe browsing service gets its own Chrome extension*. VentureBeat. Retrieved from https://venturebeat.com/2019/06/18/googles-safe-browsing-service-gets-its-own-chrome-extension/

and phishing attacks,[87] has been using ML for a long time to detect various types of cyber-threats.[88] It provides lists of URLs with malware or phishing contents to Chrome, Firefox, and Safari browsers and to ISPs.[89] The number of devices protected by the Safe Browsing service has been increasing rapidly. (See figure 15.2.)

Google has developed an AI system to filter emails. Gmail has used ML techniques to filter emails since its launch in 2004.[90] In 2017, Google announced that ML models helped to prevent 99.9% of spam and phishing messages from reaching users' inboxes. At that time, over 50% of all the messages Gmail received were spam.[91]

Hackers constantly develop new techniques and tools to game Google's CS. Consequently, Google makes adaptations. Google uses its open-source AI Tensorflow in CS. In 2019, Gmail reported that by deploying its machine learning framework TensorFlow deep learning model, it blocked an additional 100 million spams a day.[92]

Google's Cloud Video Intelligence, which makes videos searchable and discoverable, uses deep learning AI. The AI algorithms analyze contents and contexts of videos stored on the server and sends security alerts if suspicious activities are found.[93]

15.7.1 The Creation of Chronicle

In 2016, Google's parent company Alphabet's semi-secret R&D facility and organization, X, whose mission is to invent and launch "moonshot" technologies to solve big problems, launched a cybersecurity project.[94] In 2018, X created an enterprise security company named Chronicle. (See table 15.1.) Chronicle relies on ML to analyze massive amounts of data to detect cyberthreats faster and more precisely than traditional methods.[95] In June 2019, Chronicle became a part of Google Cloud, which is Google's suite of cloud computing services. Using Chronicle, Google Cloud hopes to attract enterprise customers away from Amazon Web Services and Microsoft Azure.[96]

Table 15.1 The evolution of Chronicle: A timeline

Time	Event	Remarks
February 2016	• A cybersecurity project was started at Alphabet's X, the "moonshot factory" incubator.[97]	• The project later developed into the company called Chronicle.
January 2018	• Alphabet launched Chronicle.[98]	• Chronicle claimed that using Google's infrastructure it could detect more cyberthreats faster than other systems to help stop cyberattacks early.[99] The company said that the solution was tested with several *Fortune* 500 companies.
March 2019	• Chronicle launched its first product: Backstory.	• Backstory analyzes all the data a company has (e.g., from domain name servers, employee laptops, and phones) by indexing and organizing them in the cloud. • Searches can then be conducted on the data. For example, a search might discover whether any of the devices are sending data to suspicious locations. If so, investigators can ask more questions, such as what kind of information is being sent.[100]
June 2019	• Chronicle became a part of Google Cloud.	• Its role is to detect incidents and perform remedial services.[101]
September 2020	• Chronicle announced a new real-time threat-detection tool referred to as Chronicle Detect.[102]	• The tool can handle complex analytic events using new threat-detection language. Chronicle Detect makes it easy for enterprises to move from their legacy CS tools. It also helps to analyze data collected with endpoint security solutions in a better way.

15.7.2 Use of Deep Learning

At the end of 2019, Google started deploying its deep learning scanner. It employs a TensorFlow deep learning model and a custom document analyzer for different file types. Google reported that using the deep learning scanner led to a 10% increase in the daily detection coverage of Office documents containing malicious scripts. The success rate of the deep learning scanner is especially noteworthy in detecting "adversarial, bursty attacks" in which perpetrators send a lot of emails with malicious documents attached in short bursts at random intervals. For such attacks, deep learning is reported to improve the identification rate of malicious documents by 150%.[103]

As of early 2020, malicious documents accounted for 58% of all malware targeting Gmail users. Google uses deep learning to fight this. Google is reported to block 99.9% of malicious attachments to Gmail.[104]

15.7.3 Case Summary

AI is emerging as an increasingly important tool to strengthen the security of various services and applications offered by Google. Indeed, AI's incorporation in CS is viewed as a major competitive factor. By making Chronicle a part of Google Cloud, Google uses CS as a tool to compete for enterprise cloud customers.

PART FIVE

Conclusion, Recommendations, and the Way Forward

CHAPTER SIXTEEN

Conclusion, Recommendations, and the Way Forward

CHAPTER PREVIEW

In this final chapter, we integrate the ideas discussed in earlier chapters regarding trends in cyberattacks and CS. We consider the future of cybercrime and CS. The chapter also promotes an understanding of the use of fourth revolution technologies, such as blockchain, AI, and ML, by legitimate organizations to strengthen CS as well as by cybercriminals. It considers various ways of handling cyberrisks. Recommendations and implications for policy-makers, companies, and consumers are discussed.

16.1 INTRODUCTION

The importance of CS has increased due to the increase in real as well as perceived cyberrisks. In order to understand rapidly increasing cyberrisks, consider the Allianz Risk Barometer, which is based on the European multinational financial services company Allianz's annual survey conducted with risk experts from 100 countries to identify the top threats facing companies. In 2013, cyberrisk ranked fifteenth in the Allianz Risk Barometer.[1] In the 2020 survey, cyberrisk ranked number one.[2]

A rich ecosystem has developed around cybercrime activities. For instance, some resourceful underground actors have created sophisticated ransomware to infiltrate and disrupt organizations' operational activities. These high-end criminals sell such malware to other attackers who in turn target businesses.[3] For instance, consider the Netwalker ransomware discussed in chapter 1. The creators of Netwalker are believed to collaborate with about 10 to 15 affiliates to distribute the malware.[4] The affiliates, who are often less skillful criminal hackers than the creators of Netwalker, infiltrate a victim's network and infect it with the ransomware.[5] They later split any ransom money with Netwalker's creators.

Cybercriminals are more difficult to catch and prosecute than conventional ones. For instance, most ransomware criminals operate from jurisdictions that do not cooperate with the US or European authorities fighting cybercrimes. For instance, the criminals behind Netwalker are believed to operate from Russia or the Commonwealth of Independent States.[6]

Organizations vary in their CS postures. Some deploy sophisticated tools and strategies to fight cybercrimes. Nonetheless, many organizations are underprepared to respond to the rapidly growing cyberthreats because they are either unaware of or underestimate the cyberrisks they face. For instance, many SMEs think that they lack any data of interest for cybercriminals. Unfortunately, this view is often misguided and associated with a lack of true understanding of the evolving nature of cyberthreats.

Many organizations are also reluctant to take advantage of free and easily available programs, tools, and applications. For instance, K–12 schools in the US can use some free services to enhance cyberdefense. Of the 13,000 school districts in the US, only 2,000 take advantage of free membership in the Multi-State Information Sharing & Analysis Center.[7] The center offers network vulnerability assessments, cyberthreat alerts, and other services, such as Malicious Domain Blocking and Reporting, which prevents computer systems from connecting to malicious websites.[8] Only about 120 schools use the blocking service.[9] Many school districts use outdated equipment and software, which are easy to hack.

16.2 THE FUTURE OF CYBERCRIME AND CYBERSECURITY

Cybercriminals and other perpetrators in cyberspace are rapidly expanding their reach and sophistication. As of October 2020, there were 4.66 billion active Internet users, or 59% of the world's population.[10] New Internet users in developing countries are especially likely to be viewed as low-hanging fruit by cybercriminals. In a 2020 survey, conducted by CS awareness training provider KnowBe4 among respondents across eight African countries, 48% of the respondents were concerned about cybercrime. The proportion was 38% in 2019.[11] When the currently unconnected populations go online, cybercriminals can increase their global reach.

As the cases of Netwalker and other underground groups suggest, cybercriminals implement sophisticated organizational structures, strategies, and processes. (See In Focus 16.1.) Nefarious actors also employ extremely novel, powerful, and innovative techniques in cyberattacks to inflict substantial damage to victims and remain undetected for a long time. For

IN FOCUS 16.1: SOLARWINDS' SECURITY WEAKNESS LEADS TO SUPPLY-CHAIN ATTACKS AGAINST ITS CUSTOMERS

In a blog published on December 8, 2020, CS company FireEye's CEO Kevin Mandia said that hackers had breached the company's network. The perpetrators had stolen the software tools it uses to find weaknesses in its customers' systems. Mandia noted that the hack was conducted "by a nation with top-tier offensive capabilities."[12] After additional review, FireEye found the vulnerability had come from the US-based IT monitoring and network management company SolarWinds' software.[13] FireEye was a customer of SolarWinds, which served about 300,000 global customers, including more than 425 of the *Fortune* 500 companies.[14]

Hackers suspected to be associated with Russia's foreign intelligence service agency SVR (previously known as the KGB), which reports directly to the president of Russia, took advantage of the weak security to install sophisticated and hard-to-detect malware in updates to SolarWinds' enterprise network management software Orion.[15] The breach also affected SolarWinds' Microsoft Office 365 accounts.[16]

A Novel Technique

The hackers used an extremely novel and powerful technique involving a supply-chain attack. The attack did not rely on widely used hacking methods such as phishing and a zero-day exploit. Instead, the hackers inserted malicious code into the SolarWinds' trusted code without the latter's knowing it.[17] SVR hackers were believed to have access to the networks of Orion platform users who downloaded and installed the tainted update between March and June 2020.[18]

The malware installs on the system of a victim company. It is good at evading all the detection systems. The malware sleeps for about two weeks before executing commands such as making requests to other servers to send back a response, transferring files, and rebooting the machine.[19]

The updates with a corrupted version of Orion products were sent only to about 18,000 Orion customers.[20] Of these, the attackers entered into the networks of about 50 victim organizations.[21] The affected organizations included a number of US federal agencies such as the Department of Defense, Department of State, Department of Homeland Security, Department of Treasury, Department of Commerce, Department of Energy and its National Nuclear Security Administration, and the National Institutes of Health as well as major global corporations such as Microsoft, Intel, Cisco, Nvidia, VMware, and Belkin.[22] Microsoft reported that the hackers gained access to valuable source code.[23] While most victims were based in the US, organizations in Belgium, Canada, Israel, Mexico, Spain, the United Arab Emirates, and the United Kingdom were also affected.[24]

Lack of Security Culture and a Series of Missteps

At the time this book was written, CS analysts had not figured out exactly how SVR injected a malicious back door into an Orion software update.[25] However, serious concerns had been raised about SolarWinds' lack of security culture and a series of security missteps it had made. Some of its web browsers and operating systems were outdated. SolarWinds was reported to have advised its customers to disable virus scanning so that Orion platform products could run more efficiently.[26] In 2019, the company had used the password "solarwinds123"' to protect its update server.[27]

SolarWinds' Response

SolarWinds cooperated with investigating agencies such as the FBI. It suggested affected customers should update to the latest software.[28]

SolarWinds' response was, however, criticized on many grounds. Some cybersecurity experts were concerned about the way the company communicated with the public regarding the details of the attack. The company did not explain in detail how the attacks impacted its customers and the actions it was taking to correct and improve CS. There were complaints that the company did not answer relevant questions from customers.

Neither did it disclose any identified security failures that could be responsible for the malicious code being inserted into the software update process.

SolarWinds' response was contrasted with those of victim companies such as FireEye and Microsoft, which released some actionable information to help other companies victimized by the attack.

SolarWinds was also accused of placing its own immediate financial interests above those of its stakeholders. All these actions led to a loss of current and potential customers as well as the loss of the cybersecurity community's trust and confidence in the company.[29]

instance, in the AI-powered deepfakes of March 2019, in which cybercriminals tricked the CEO of a UK-based energy firm into making a US$234,000 wire transfer, investigators have not been able to identify any suspects. Likewise, criminal outfits such as Sodinokibi utilize cryptocurrencies such as Monero, which are extremely difficult for law enforcement agencies to track.

Organizations and individuals are also getting better at protecting themselves against cyberthreats. As discussed in chapters 8, 9, and 10, organizations are improving their strategies, policies, processes, and programs to strengthen cyberdefense. As of 2019, organizations in the US employed 715,000 people in CS.[30]

Organizations' CS measures also include substantial spending. For instance, in the US, private-sector spending on CS was estimated to be in the range of US$40 to US$120 billion in 2020. Some companies, such as Microsoft and JPMorgan, are estimated to spend about US$1 billion annually on CS.[31]

Firms are deploying advanced and sophisticated tools such as AI, ML, and blockchain to fight cyberattacks. AI's use in CS has gained prominence in recent years. Due to challenges such as too many false positives of AI systems, some CS analysts have recommended using multiple AI algorithms to predict whether a transaction is fraudulent. Resistant AI, a CS company in Prague, Czech Republic, uses up to five different ML modules to make a decision.[32]

Likewise, blockchain has the potential to significantly strengthen organizations' CS practices. For instance, a party can cryptographically sign transactions and, by verifying the cryptographic signature, the recipient can ensure that the transaction originated from a trusted source.

16.3 IMPLICATIONS FOR BUSINESSES

As noted earlier, not all organizations have clear strategies, policies, processes, and programs to manage CS outcomes. It is important for companies to have CISOs, other key CS staff, and an effective organizational structure. CS staff should have clarity regarding their roles and responsibilities and be held accountable for irresponsible cybersecurity practices. Regular updates about phishing and other threats, as well as strategies and instructions to mitigate and manage such threats, must be provided to their employees on a regular basis.

It is important to create awareness among employees of different forms of cybercrimes and provide training about diverse modus operandi of cybercriminals. For instance, simple training might help to identify deepfakes. Real people tend to send messages and speak in a fragmented manner. If deepfakes are at work, potential victims may get the response only in full and formal sentences. They may also receive the same answer multiple times. Also, AI-powered deepfakes may send replies too quickly.[33]

It is important to patch operating systems and software when manufacturers release new updates. It helps to constantly back up important data. By frequently backing up data and keeping them secure, organizations can ensure their access to enterprise networks without disruption when they face serious cyberattacks such as ransomware.

In light of the lack of well-developed institutional frameworks such as regulations and ethical and professional standards, there is a strong need for organizations to develop policies, culture, and techniques to manage and monitor social media use within an organization. Organizations that have clear policies, procedures, limits, and instructions regarding the use of SM are less likely to become targets or victims of cybercrimes. A clear policy regarding the downloading of apps onto devices owned by the organization is critical because apps from less reputable websites are more likely to pose security risks. For instance, some online game apps may not only lead to a decline in employees' performance but also to an increased security risk.

Businesses may also want to purchase cyberinsurance to defend against cyberthreats. Insurance not only helps cover items such as ransom payments in case of ransomware attacks, but also helps to defend against attacks, because organizations need to strengthen their security to get a lower premium.

16.3.1 Handling Cyberrisks

Just like any type of risk, there are four possible response strategies[34] to handle cyberrisks in organizations: avoid, reduce, transfer, and accept.

16.3.1.1 Avoid

Avoiding a cyberrisk means completely quitting a particular action or not starting the action at all. It could be a good strategy, especially for extremely high-value data and information. When this option is chosen, organizations eliminate the possibility that the risk will be a threat. For instance, to avoid cyberrisks, firms may disconnect some activities from the Internet. In China's People's Liberation Army, for instance, top-secret military documents and orders are handwritten and human messengers deliver them.[35] As another example, foreign hackers allegedly stole terabytes of data about the F-35's design and operational capabilities. The spies had hacked into the systems of the defense contractor Lockheed Martin and at least six subcontractors in 2009. However, the F-35's most sensitive flight-control data was stored off-line[36] and could not be stolen by the hackers.

Some policy-makers and researchers have suggested that by using analog solutions instead of digital, cyberrisks can be minimized. For instance, some US senators and CS researchers urge exploring the use of analog control mechanisms instead of digital technology in order to strengthen the US grid's defense mechanisms. In 2016, four senators on the Intelligence Committee introduced legislation that would set aside US$10 million to study CS vulnerabilities on the electrical grid and explore solutions, including what they refer to as a "retro" approach to grid security.[37]

16.3.1.2 Reduce

Reducing or mitigating entails taking actions to reduce the likelihood of cyberattacks or the impact of a loss from an attack. If the current cyberrisk is higher than organizations can absorb, risks could be reduced to within the tolerance level. As discussed throughout this book, organizations apply various risk mitigation strategies based on their assessment of cyberrisks. These include increased CS spending (chapter 1 and this chapter), chief human resource officers' initiation of employee training programs that address CS (chapter 10), moving some applications from the cloud to on-premises (chapter 12) and pursuing personal data minimization policies (chapter 13).

16.3.1.3 Transfer

A third option is to transfer the cyberrisk to a third party. With this option, the organization does not eliminate or reduce the risk. Instead, cyberinsurance, which transfers cyberrisks that are insurable from one party (that is, the policyholder) to another (that is, the cyberinsurer), is growing rapidly. In the 2019 survey by Marsh and Microsoft Corp, 47% of the surveyed companies had cyberinsurance compared to 35% in 2017.[38] According to Munich Re, the global cyberinsurance market was worth US$7 billion in 2020.[39] The insurance firm Finaria estimated that the market will reach US$9.5 billion in 2021.[40]

16.3.1.4 Accept

The last option is to accept the cyberrisk as is, which means doing nothing. This strategy is often employed for cyberrisks that have a low probability of occurrence or lead to a low impact should they occur. Some companies may budget reserves to deal with such situations. Cyberrisks that may pose a threat in the distant future could fall in this category. For instance, a study conducted in early 2020 by the disaster recovery as a service and data-protection company Infrascale found that 21% of SMEs lacked a data backup or disaster recovery solution.[41] SMEs thus accept the risk of their data being lost or held for ransom. Surveys have reported that SMEs often believe that they have no data of interest for cyberattackers.[42]

16.3.2 Addressing Workforce Issues in CS

In light of the shortage of CS workforce discussed in chapter 10, various initiatives can be taken to enhance recruitment and retention of a CS workforce. Organizations can also offer internal career-change opportunities for workers in related areas who would like to try something new. For instance, existing employees who can bring risk management skills or those with talent for creativity are a strong fit for CS.

Given the various benefits female CS professionals can bring to organizations, it is important to focus CS-related recruitment efforts on academic institutions that have a higher enrollment of women. In order to attract more women to the CS field, they should be provided with internship, scholarship, and mentoring opportunities.[43]

Organizations also need to reexamine their recruitment efforts to attract female CS professionals. For instance, gender biases in job ads discourage women applicants. CS-related job ads are reported to have about twice as many male-gendered terms as female-gendered terms.[44] The use of more inclusive and gender-neutral terms can help attract more female CS professionals.

Organizations need to work jointly with other organizations or collaborate at the industry level to address the issue of the CS workforce shortage. For instance, in 2018, Microsoft India and DSCI teamed up to launch a three-year program known as CyberShikshaa in order to create a pool of skilled CS women professionals.[45]

Strategic partnerships between the industry and non-profit organizations can also be leveraged to help bridge the gender gap in CS in the long run. For instance, in 2017, Palo Alto Networks teamed up with the Girl Scouts of the USA to develop CS badges. The goal is to foster CS knowledge and develop interest in the profession.[46] Subjects covered in the curriculum designed for young girls include the basics of computer networks, cyberattacks, and online safety.[47]

It is also important to support and encourage organizations representing women CS professionals to actively recruit and mentor more women in this field. For instance, Women in Cybersecurity of Spain has implemented a mentoring program in which women CS professionals with over 20 years of experience support other women CS professionals early in their careers, such as those between the ages of 25 and 35. The six-month program organizes at least one meeting per month.[48] Some industry groups have collaborated with big companies with expertise in CS.

16.4 IMPLICATIONS FOR CONSUMERS

Various cyberrisk response strategies[49] discussed in this book can also apply to consumers. Consumers can avoid some cyberrisk by quitting some online activities. According to survey results published by the Pew Research Center in September 2018, 26% of Facebook users had deleted the Facebook app from their cellphones in the previous 12 months.[50]

Consumers can reduce or mitigate some cyberrisks by taking actions to reduce the likelihood of being victimized by some forms of cyberattacks. In the Pew Research Center survey noted above, 54% of the respondents had adjusted privacy settings.[51] In order to reduce or mitigate cyberrisks facing their children, parents need to discuss cyberbullying and other online offenses with their children in order to increase their awareness, knowledge, and preventive attitudes toward such threats. It is also a good idea to join their children's online social network so that their interactions can be monitored and action can be taken if the children become targets of cyberbullying. In addition, parents can monitor the security settings on their children's social media accounts. For instance, in Facebook, parents can help their children choose various settings, such as "Who can contact me?" "Who can look me up?" "Who can see my stuff?" Depending on the nature of offenses such as cyberbullying, parents may need to report the incidents to school and possibly to police. Some incidents may require changing email addresses and cellphone numbers of the victimized child. More serious incidents may also require utilization of counseling services. The goal is to help the victim minimize the negative effects of an online offense.

Just like businesses, consumers can buy cyberinsurance as a risk transfer strategy. Currently, regular people are far less likely than companies to have digital protection. In India, personal cyberinsurance makes up less than 1 percent of the total cyberinsurance market.[52] In the US and elsewhere, most products are targeted at rich people.[53] Insurers such as AIG, Chubb, Hartford Steam Boiler, and NAS Insurance sell personal cyberinsurance policies as add-ons to homeowners' and renters' insurance.

The insurance industry is doing more, too. A wide range of insurers such as Munich Re, AIG's CyberEdge, Saga Home Insurance, Burns & Wilcox, Chubb, Mercury Insurance Home Cyber Protection, Arbella Home Cyber Protection, PURE Starling Fraud and Cyber Fraud, and State Farm Identity Restoration Insurance all offer cyberinsurance for individuals.[54] A 2017 *Financial Times* article noted, "Hacking cover could be standard for homeowners in 10 years."[55]

These plans cover as much as US$250,0000 to repair or replace damaged devices and to pay for expert advice and assistance if a cyberattack affects a policyholder.[56] They may also include data recovery, credit monitoring services, and efforts to undo identity theft.[57] As smart home systems become more popular, they will provide more potential entry points for hackers. An average home insured by AIG has 20 Wi-Fi–enabled devices.[58] Replacing a hijacked home's entire smart lighting system, smart entertainment center, smart thermostat, and digital security devices will be expensive – and the bill will only be higher for communities using Internet-connected streetlights, water meters, electric cars, and traffic controls.[59] These are opportunities for insurance companies.

Even health services may be included: AIG's new product Family CyberEdge policy includes a coverage of one year of psychiatric services if a family member is victimized by cyberbullying.[60] Also covered is lost salary if the victim loses a job within 60 days of discovering cyberbullying. Some insurers offer policies that provide help to assess policyholders' data security practices and scan for cyberthreats.[61]

16.5 IMPLICATIONS FOR POLICY-MAKERS

It is important for policy-makers to understand the economic, social, and political consequences of increasing cyberthreats and privacy violations by companies. Nations have to fight a battle on many fronts in order to defeat cybercriminals and enhance the welfare of the population.

Urgent measures are needed to develop a CS workforce. As of 2019, the US had 314,000 unfilled positions in CS. One estimate suggested that by 2021, there will be 3.5 million unfilled CS jobs globally.[62] In order to address this shortage, CS must be made an integral part of high school and university curricula.

Public–private partnership projects may be implemented to solve the problem in the long run. One example is Israel's CyberGirlz program, which is jointly financed by the country's Defense Ministry, the Rashi Foundation, and Start-Up Nation Central. It identifies high school girls with aptitude, desire, and natural curiosity to learn IT and develops such skills among them. The girls participate in activities and events such as hackathons and training programs, learn hacking skills, and practice penetration testing, which involves simulating cyberattacks to find potential vulnerabilities. Also included in the curriculum are network analysis and the programming language Python.[63] They get advice, guidance, and support from female mentors, some of whom are from elite technology units of the country's military. By 2018, about 2,000 girls had participated in the CyberGirlz Club and the CyberGirlz Community.

It is important to encourage women to participate in programs designed to ignite their interests and confidence to pursue CS-related careers. One such example is IBM Security's Women in Security Excelling (WISE) program[64] that was formed in 2015 to attract more women into the CS field. It had more than 800 members worldwide in 2017.[65]

National measures are needed to put pressure on technology firms to consider the interests of the majority of consumers when they decide how they will utilize consumers' data. The big data industry has grown up in what many consider a regulatory gray area. The seriousness of this issue is increased by the fact that regulations do not adequately protect consumers from the disclosure of their information by advertisers, particularly if such information is not technically PII. Moreover, ethical standards and codes of conduct are not well established. Most companies also lack best practices and privacy policies to stop the disclosure of sensitive information. What is more disturbing is that in some cases, illicit and gray-area actors essentially draw on the same datasets and information as legitimate users. Due to the lack of regulations and guidelines, some dangerous, creepy, and annoying uses of consumer information are not necessarily illegal.

As discussed in chapter 3, the lack of cross-border collaboration, a high degree of international variation in cybercrime laws, and weak or even non-existent laws in some countries have facilitated the globalization of cybercrimes. While industrialized countries discuss international cooperation to combat cybercrimes, many poor countries are not yet involved in the discussions. The challenge of system capacity is severe in most developing countries. An additional important issue relates to the readiness and willingness of developing countries to join the Council of Europe Convention on Cybercrime.

The discussion in chapter 3 also highlighted the need for better international cooperation on the cyberfront. From the perspective of advanced industrial democracies, depending on the categories of the sources and nature of the cyberthreats, several actions can be taken. These vary from contributing to local capacity building and institutional development to providing opportunities for developing countries' voices and participation. Actions also include harnessing the power of regional organizations, identifying and achieving cooperation on common areas of interest, helping develop the ICT industry to absorb qualified workers, encouraging integration with the West, establishing a high-level working group made up of like-minded policy-makers, and developing offensive and defensive capabilities tailored to specific threats associated with a specific source.

The complementary roles of informal networks and formal systems need to be stressed. Informal networks can be used as an effective alternative to formal treaties. The existence and depth of informal networks may also provide an explanation for the markedly different levels of success of the US in fighting cybercrimes originating from EU countries such as Estonia compared to fighting cybercrimes from China or Russia.

16.6 FINAL THOUGHTS

CS affects every business function, from marketing and supply chain to product development and HRM. Protecting a company's data assets against cyberattacks is thus every employee's responsibility. Cybercriminals are increasing their skills and sophistication and there are signs that cybercrime firms are evolving into more serious criminal enterprises. Nations have also shown an increasing tendency to engage in cyberwarfare and cyber cold war. The challenges of such hostile environments are especially daunting for companies doing business internationally. Multinationals are increasingly becoming unwilling pawns in the cyberconflicts among nations. An organization's CS strategy thus should include stopping the current threats as well as the future ones that cannot be currently seen.

CS could benefit greatly from advances in 4R technologies such as AI and blockchain. This issue is especially important in light of the fact that cyberattackers use advanced tools utilizing the 4R technologies. For instance, the tools used by cybercriminals to create deepfakes are also likely to effectively detect and deter such frauds. One way to view this is to say that just as diamond is the only material hard enough to cut diamond effectively, AI can be an effective tool to fight cybercriminals who use AI to defraud victims.

Finally, efforts to fight cybercrimes cannot succeed without developing a sufficient CS workforce. Schools, universities, the private sector, and the government must do more to address the current shortage of CS human resources and promote gender balance and diversity. Organizations should consider distinctive perspectives that women CS specialists might bring. Measures, programs, and initiatives at various levels are needed to increase women's participation in the CS field. Government agencies, academic institutions, professional and trade associations, and industry groups need to work together to foster interests and knowledge related to CS among women.

Notes

All URLs in these notes were operational in April 2021, unless a different retrieval date is noted.

Chapter 1

1 Ellyatt, H. (2013, August 13). *Global drugs trade "as strong as ever" as fight fails.* CNBC. Retrieved from https://www.cnbc.com/id/100957882; Human Rights First. (2017). *Human trafficking by the numbers.* Retrieved from https://www.humanrightsfirst.org/resource/human-trafficking-numbers

2 *The Economist.* (2010, August 3). Cyber-security in Congress. Retrieved from https://www.economist.com/democracy-in-america/2010/08/03/cyber-security-in-congress. This figure has been quoted by US president Obama, Sita Masamba, the Director and Head of Mission of the United Nations African Institute for the Prevention of Crime and the Treatment of Offenders (UNAFRI), and others.

3 *Business losses to cybercrime data breaches to exceed $5 trillion by 2024.* (2019). Juniper Research. Retrieved from https://www.juniperresearch.com/press/press-releases/business-losses-cybercrime-data-breaches

4 *Cybercrime damages $6 trillion by 2021.* (2017). Cybersecurityventures. Retrieved from https://cybersecurityventures.com/annual-cybercrime-report-2017/

5 *NortonLifeLock Cyber Safety Insights Report.* (2020). Norton. Retrieved from https://us.norton.com/nortonlifelock-cyber-safety-report

6 Ahaskar, A. (2020). *Every four of 10 Indians have experienced identity theft, NortonLifeLock.* Live Mint. Retrieved from https://www.livemint.com/news/india/every-four-of-10-indians-have-experienced-identity-theft-norton-survey-11586245264537.html

7 *Global cyber risk perception survey.* (2019). Marsh. Retrieved from https://www.marsh.com/us/insights/research/marsh-microsoft-cyber-survey-report-2019.html

8 *Data breaches cost $654 billion in 2018.* (2019). Security Magazine. Retrieved from https://www.securitymagazine.com/articles/90320-data-breaches-cost-654-billion-in-2018

9 *Cybersecurity breaches to result in over 146 billion records being stolen by 2023.* (2018). Juniper Research. Retrieved from https://www.juniperresearch.com/press/press-releases/cybersecurity-breaches-to-result-in-over-146-bn

10 IBM Security. (2019). *Cost of a data breach report*. Retrieved from https://www.ibm.com/downloads/cas/ZBZLY7KL?_ga=2.181209767.1686129232.1586105132-321662522.1584074488
11 Hinkelman, M. (2014, October 15). For small biz, cybercrime can be deadly. *The Philadelphia Inquirer*. Retrieved from https://www.inquirer.com/philly/business/20141015_For_small_biz__cybercrime_can_be_deadly.html
12 *Big data*. (2013). Gartner. Retrieved from http://www.gartner.com/it-glossary/big-data/
13 *Big data: What it is and why it matters*. (2013). SAS. Retrieved from https://www.sas.com/en_us/insights/big-data/what-is-big-data.html
14 Kshetri, N. (2010). The economics of click fraud. *IEEE Security & Privacy, 8*(3), 45–53.
15 Kshetri, N. (2009). Positive externality, increasing returns and the rise in cybercrimes. *Communications of the ACM, 52*(12), 141–4.
16 Cyber Physical Systems Public Working Group. (2015). *Preliminary discussion draft, framework for cyber-physical systems, release 0.7*. Retrieved from https://www.hldataprotection.com/files/2015/03/NIST-Cyber-physical-Framework-PRELIMINARY-DISCUSSION-DRAFT.pdf
17 International Telecommunication Union (ITU). (n.d.). *Definition of cybersecurity*. Retrieved from http://www.itu.int/en/ITU-T/studygroups/com17/Pages/cybersecurity.aspx
18 Deci, E.L., & Ryan, R.M. (1985). *Intrinsic motivation and self-determination in human behavior*. Plenum Press.
19 Kennedy, B. (2015, June 22). *NIST tackles cybersecurity in the smart city*. Hogan Lovells Chronicle of Data Protection. Retrieved from http://www.hldataprotection.com/2015/06/articles/consumer-privacy/nist-tackles-cybersecurity-in-the-smart-city/
20 Jovanović, B. (2019). *Malware statistics – You'd better get your computer vaccinated*. DataProt. Retrieved from https://dataprot.net/statistics/malware-statistics/
21 Sonicwall. (2020, February 4). *2020 Sonicwall cyber threat report: Threat actors pivot toward more targeted attacks, evasive exploits*. Retrieved from https://www.sonicwall.com/news/2020-sonicwall-cyber-threat-report/
22 Kaspersky. (2019). *Phishing attacks more than doubled in 2018 to reach almost 500 million*. Retrieved from https://www.kaspersky.com/about/press-releases/2019_phishing-attacks-more-than-doubled-in-2018
23 Computing for IT Leaders. (2020). *Spanish hospitals targeted with coronavirus-themed phishing lures in Netwalker ransomware attacks*. Retrieved from https://www.computing.co.uk/news/4012969/hospitals-coronavirus-ransomware.
24 Wheeler, D.A., & Larsen, G.N. (2003, October). *Techniques for cyber attack attribution*. Institute for Defense Analysis, IDA Paper P-3792. Retrieved from https://apps.dtic.mil/dtic/tr/fulltext/u2/a468859.pdf
25 Riley, D. (2020, March 16). *Ransomware targets hospitals and medical providers during coronavirus pandemic*. Silicon Angle. Retrieved from https://siliconangle.com/2020/03/16/ransomware-targets-hospitals-medical-providers-coronavirus-pandemic/
26 Riley. (2020). *Ransomware targets hospitals and medical providers during coronavirus pandemic*.
27 Winder, D. (2020, March 6). Beware of this new Windows 10 ransomware threat hiding in plain sight. *Forbes*. Retrieved from https://www.forbes.com/sites/daveywinder/2020/03/05/beware-of-this-new-windows-10-ransomware-threat-hiding-in-plain-sight/#4d7bef232958
28 Computing for IT Leaders. (2020). *Spanish hospitals targeted with coronavirus-themed phishing lures in Netwalker ransomware attacks*.

29 Ross, L. (2020). *Health care cybersecurity: New front in coronavirus battle*. SmartBrief. Retrieved from https://www.smartbrief.com/original/2020/04/health-care-cybersecurity-new-front-coronavirus-battle

30 Bergal, J. (2020). *Hospital hackers seize upon coronavirus pandemic*. GCN. https://gcn.com/articles/2020/04/17/health-care-ransomware-targets.aspx

31 Ross. (2020). *Health care cybersecurity*.

32 Lindsay, J.R. (2013). Stuxnet and the limits of cyber warfare. *Security Studies, 22*(3), 365–404.

33 Clark, D.D., & Landau, S. (2011). Untangling attribution. *Harvard National Security Journal, 2*(2), 25–40.

34 Sanger, D.E., & Perlroth, N. (2014, October 30). New Russian boldness revives a cold war tradition: Testing the other side. *The New York Times*. Retrieved from http://www.nytimes.com/2014/10/31/world/europe/new-russian-boldness-revives-a-cold-war-tradition-testing-the-other-side-.html?_r=0

35 *Critical infrastructure cyberattacks a greater concern than enterprise data breaches*. (2020). Security Magazine. Retrieved from https://www.securitymagazine.com/articles/91992-critical-infrastructure-cyberattacks-a-greater-concern-than-enterprise-data-breaches

36 PwC. (November, 2016). *Five megatrends and their implications for global defense & security*. Retrieved from https://www.pwc.com/gx/en/government-public-services/assets/five-megatrends-implications.pdf

37 Finkle, J. (2014). *Exclusive: FBI warns healthcare sector vulnerable to cyber attacks*. Reuters. https://www.reuters.com/article/instant-article/idUKBREA3M1Q920140423

38 Stein, S., & Jacobs, J. (2020, March 16). *Cyber-attack hits US health agency amid covid-19 outbreak*. Bloomberg. Retrieved from https://www.bloomberg.com/news/articles/2020-03-16/u-s-health-agency-suffers-cyber-attack-during-covid-19-response

39 *US Health and Human Services Department suffers cyberattack*. (2020b). Security Magazine. Retrieved from https://www.securitymagazine.com/articles/91909-us-health-and-human-services-department-suffers-cyberattack

40 Stubbs, J., & Satter, R. (2020) *Vietnam-linked hackers targeted Chinese government over coronavirus response: Researchers*. Reuters. Retrieved from https://www.reuters.com/article/us-health-coronavirus-cyber-vietnam/vietnam-linked-hackers-targeted-chinese-government-over-coronavirus-response-researchers-idUSKCN2241C8

41 Burgess, M. (2020, March 22). Hackers are targeting hospitals crippled by coronavirus. *Wired*. Retrieved from https://www.wired.co.uk/article/coronavirus-hackers-cybercrime-phishing

42 Bergal, J. (2020, April 17). *Hospital hackers seize upon coronavirus pandemic*. GCN. Retrieved from https://gcn.com/articles/2020/04/17/health-care-ransomware-targets.aspx

43 Government of Canada, Department of Justice. (2015). *Economic and organized crime: Challenges for criminal justice*. Retrieved from http://www.justice.gc.ca/eng/rp-pr/csj-sjc/jsp-sjp/rp02_12-dr02_12/p3.html

44 Schwartz, M.J. (2011). 111 arrested in identity theft probe. *InformationWeek*. Retrieved from http://www.informationweek.com/news/security/attacks/231900438

45 *Inside the business model for botnets*. (2018, May 14). MIT Technology Review. Retrieved from https://www.technologyreview.com/s/611123/inside-the-business-model-for-botnets/

46 Wei, W. (2018, March 26). *Leader of hacking group who stole $1 billion from banks arrested in Spain*. The Hacker News. Retrieved from https://thehackernews.com/2018/03/carbanak-russian-hacker.html

47 Кіберполіція викрила українського хакера у взламі комп'ютерів світових банків та готелів. (2018). Retrieved from https://cyberpolice.gov.ua/news/kiberpoliczziya-vykryla-ukrayinskogo-xakera-u-vzlami-kompyuteriv-svitovyx-bankiv-ta-goteliv-4470/

48 Gordon, S., & Ford, R. (2006). On the definition and classification of cybercrime. *Journal in Computer Virology, 2*, 13–20.

49 *CBS News*. (2017, May 16). North Korean hackers behind global cyberattack? Retrieved from cbsn.ws/2qnYHjt

50 Berr, J. (2017, May 16). "WannaCry" ransomware attack losses could reach $4 billion. *CBS News*. Retrieved from cbsn.ws/2rluoXx

51 Shukla, S. (2017, August 3). *WannaCry outbreak: Ransom payments reach $64,000, around 48,000 attack attempts in India*. Moneycontrol. Retrieved from bit.ly/2pQaio5

52 Chemicals and defence firms targeted by hacking attack. (2011, October 31). *BBC News*. Retrieved from http://www.bbc.co.uk/news/technology-15529930

53 Hamid, T. (2011). Smart devices prey for cybercrime. *MEED: Middle East Economic Digest, 55*(2), 25–6.

54 Cramer-Flood, E. (2020, June 1). *Urbanization is driving a wave of new internet users in China*. Insider Intelligence. Retrieved from https://www.emarketer.com/content/urbanization-is-driving-a-wave-of-new-internet-users-in-china

55 Russell, J. (2017, December 19). *US government says North Korea was behind massive WannaCry cyber attack*. TechCrunch. Retrieved from https://techcrunch.com/2017/12/18/us-government-north-korea-wannacry/

56 Chan, E., & Kim, S. (2017, May 23). *Cybersleuths unearth more clues linking WannaCry to North Korea*. Bloomberg News. Retrieved from bloom.bg/2v9Hbzt

57 *Hackers attack 20 million accounts on Alibaba's Taobao shopping site*. (2016, February 4). Reuters. Retrieved from https://www.reuters.com/article/us-alibaba-cyber-idUSKCN0VD14X

58 Blitz, J. (2011, November 1). Security: A huge challenge from China, Russia and organised crime. *Financial Times*. Retrieved from http://www.ft.com/intl/cms/s/0/b43488b0-fe2a-11e0-a1eb-00144feabdc0.html#axzz1dnezI1eF

59 Schneier, B. (2012). Securing medical research: A cybersecurity point of view. *Science, 336*(6088), 1527–9. doi: 10.1126/science.1224321

60 Marketwatch. (2016, September 15). *China's scammers hit victims to the tune of millions as telecoms stand by*. Retrieved from https://www.marketwatch.com/story/chinas-scammers-hit-victims-to-the-tune-of-millions-as-telecoms-stand-by-2016–09–15

61 Blitz. (2011). Security.

62 *China seeks to combat hi-tech crimewave.* (2011, December 30). *BBC News*. Retrieved from http://www.bbc.co.uk/news/technology-16357238

63 *Update 1 – Mengniu website hacked after milk scandal – reports*. (2011, December 29). Reuters. Retrieved from https://www.reuters.com/article/mengniu-hack/update-1-mengniu-website-hacked-after-milk-scandal-reports-idINL3E7NT3B220111229

64 Foremski, T. (2018, May 25). *Ad fraud is out of control – billions lost by media industry, says a new report*. ZDNet. Retrieved from https://www.zdnet.com/article/ad-fraud-is-out-of-control-billions-lost-by-media-industry-says-a-new-report/

65 de Kloet, J. (2002, September 2). Digitisation and its Asian discontents: The internet, politics and hacking in China and Indonesia. *First Monday, 7*(9). Retrieved from http://firstmonday.org/issues/issue7_9/kloet/index.html

66 Glenny, M. (2018, February 2). Partners in crime: Why mafia groups and cybercriminals are joining forces. *The Globe and Mail*. Retrieved from https://www.theglobeandmail.com/opinion/partners-in-crime-why-mafia-groups-and-cybercriminals-are-joining-forces/article37831032/

67 Asiaone. (2011). *Most shady online bank accounts bear Chinese names*. The Yomiuri Shimbun/Asia News Network. Retrieved from https://www.asiaone.com/News/Latest%2BNews/Science%2Band%2BTech/Story/A1Story20111216-316522.html

68 Fletcher, O. (2009). China game boss sniped rivals, took down internet. *PCWorld*. Retrieved from http://www.pcworld.com/businesscenter/article/171018/china_game_boss_sniped_rivals_took_down_internet.html

69 Guillén, M.F., & Suárez, S.L. (2005). Explaining the global digital divide: Economic, political and sociological drivers of cross-national internet use. *Social Forces, 84*(2), 681–708.

70 O'Neill, P. (2017, December 6). *Ethiopia using Israeli spyware to spy on dissidents, journalists*. Cyberscoop. Retrieved from https://www.cyberscoop.com/ethiopia-cyberbit-citizen-lab-elbit-systems/

71 Collier, K. (2020, February 27). Half of US local government offices haven't upgraded their ransomware defenses since 2019's online crime spree, IBM poll says. *Fortune*. Retrieved from https://fortune.com/2020/02/27/local-government-ransomware-preparedness/

72 Emsisoft. (2016). *The state of ransomware in the US: Report and statistics 2019*. Retrieved from https://blog.emsisoft.com/en/34822/the-state-of-ransomware-in-the-us-report-and-statistics-2019/

73 Simonite, T. (2013). *Protecting power grids from hackers is a huge challenge*. MIT Technology Review. Retrieved from https://www.technologyreview.com/2013/02/27/179662/protecting-power-grids-from-hackers-is-a-huge-challenge/

74 Hern, A. (2017, June 13). Industroyer' virus could bring down power networks, researchers warn. *The Guardian*. Retrieved from www.theguardian.com/technology/2017/jun/13/industroyer-malware-virus-bring-down-power-networks-infrastructure-wannacry-ransomware-nhs

75 The European Union Agency for Cybersecurity. (2012, December 19). *Appropriate security measures for smart grids*. Retrieved from https://www.enisa.europa.eu/publications/appropriate-security-measures-for-smart-grids

76 Juniper Research. (2017). *Cybercrime & the internet of threats 2018*. Retrieved from https://www.juniperresearch.com/whitepapers/cybercrime-the-internet-of-threats-2018

77 Sterling, B. (2017, July 19). Global cybercrime. Costs a trillion dollars. Maybe 3. *Wired*. Retrieved from https://www.wired.com/beyond-the-beyond/2017/07/global-cybercrime-costs-trillion-dollars-maybe-3/

78 *Gartner forecasts worldwide security spending will reach $96 billion in 2018, up 8 percent from 2017*. (2017). Gartner. Retrieved from https://www.gartner.com/en/newsroom/press-releases/2017-12-07-gartner-forecasts-worldwide-security-spending-will-reach-96-billion-in-2018

79 Del Matto, D. (2017, December 5). *Changing cybersecurity regulations that global financial services firms need to know about*. CSO. Retrieved from https://www.csoonline.com/article/3239681/security/changing-cybersecurity-regulations-that-global-financial-services-firms-need-to-know-about.html

80 Melendez, S. (2020, March 12). *This free service shows who has your data – and helps you delete it*. Fast Company. Retrieved from https://www.fastcompany.com/90476294/this-free-service-shows-who-has-your-data-and-helps-you-delete-i

81 Nadeau, M. (2018). *General Data Protection Regulation (GDPR) requirements, deadlines and facts.* CSO. Retrieved from https://www.csoonline.com/article/3202771/data-protection/general-data-protection-regulation-gdpr-requirements-deadlines-and-facts.html
82 Lackie, L. (2016, May 11). *High-profile data breaches affecting consumer trust in big brands.* ITProPortal. Retrieved from http://www.itproportal.com/2016/05/11/high-profile-data-breaches-affecting-consumer-trust-in-big-brands/
83 Lackie. (2016). *High-profile data breaches affecting consumer trust in big brands.*
84 McGinnis, D. (2018, December 20). What is the fourth industrial revolution? SalesForce. Retrieved from https://www.salesforce.com/blog/2018/12/what-is-the-fourth-industrial-revolution-4IR.html
85 *Gartner forecasts worldwide security spending will reach $96 billion in 2018, up 8 percent from 2017.* (2017). Gartner.
86 PwC. (2014). *Cybersecurity: Keeping content, brands and people safe online is everyone's business – and a strategic imperative.* Retrieved from https://www.pwc.com/gx/en/global-entertainment-media-outlook/hot-topics/assets/pwc-cybersecurity.pdf
87 National Conference of State Legislatures. (2020). *Budgeting for cybersecurity.* Retrieved from https://www.ncsl.org/documents/taskforces/Budgeting_For_Cybersecurity_32041.pdf
88 Ralph, O. (2019, November 19). Data hacks and big fines drive cyber insurance growth. *Financial Times.* Retrieved from https://www.ft.com/content/751946b2-fb0a-11e9-a354-36acbbb0d9b6
89 *Cowbell cyber adds social engineering coverage, opens platform to non-insureds.* (2020, April 20). Insurance Journal. Retrieved from https://www.insurancejournal.com/news/national/2020/04/20/565391.htm
90 PYMNTS.com. (2019). *Robocalls go unpunished, exposing regulatory gaps.* Retrieved from https://www.pymnts.com/legal/2019/robocalls-regulatory-gaps-fcc/
91 TechInsurance. (n.d.). *First-party vs. third-party cyber liability insurance.* Retrieved from https://www.techinsurance.com/resources/first-party-vs-third-party-cyber-liability-insurance
92 Insureon. (n.d.). *Third-party cyber liability insurance.* Retrieved from https://www.insureon.com/insurance-glossary/cyber-liability-third-party
93 Foremski. (2018). *Ad fraud is out of control.*
94 Cybersecurity & Infrastructure Security Agency. (n.d.). *Critical infrastructure sectors.* Retrieved from www.dhs.gov/critical-infrastructure-sectors
95 Mitchell, C. (2014, September 29). Cyberwar: Not if. Not when. Now. *Washington Examiner.* Retrieved from washingtonexaminer.com/cyberwar-not-if.-not-when.-now./article/2553470
96 Homeland Security Today. (2019). *54 percent in utility sector expect cyber attack on critical infrastructure in next year.* Retrieved from https://www.hstoday.us/subject-matter-areas/cybersecurity/54-percent-in-utility-sector-expect-cyber-attack-on-critical-infrastructure-in-next-year/
97 Schwartz, M. (2020, January 10). *Hackers increasingly probe North American power grid.* Bank Info Security. Retrieved from https://www.bankinfosecurity.com/hackers-increasingly-probe-north-american-power-grid-a-13596
98 Martin, A. (2020, January 6). *What damage could Iran do with a cyber attack?* Sky News. Retrieved from https://news.sky.com/story/what-damage-could-iran-do-with-a-cyber-attack-11902133
99 Khandelwal, S. (2014). *Dragonfly Russian hackers target 1000 western energy firms.* The Hacker News. Retrieved from thehackernews.com/2014/07/dragonfly-russian-hackers-scada-havex.html

100 Perlroth, N. (2014, June 30). Russian hackers targeting oil and gas companies. *The New York Times*. Retrieved from www.nytimes.com/2014/07/01/technology/energy-sector-faces-attacks-from-hackers-in-russia.html

101 Finkle, J. (2014). *US government asks firms to check networks after 'Energetic Bear' attacks*. Reuters. Retrieved from www.reuters.com/article/us-cybersecurity-energeticbear/u-s-government-asks-firms-to-check-networks-after-energetic-bear-attacks-idUSKBN0F722V20140702

102 Goodin, D. (2016). *First known hacker-caused power outage signals troubling escalation*. Ars Technica. Retrieved from arstechnica.com/information-technology/2016/01/first-known-hacker-caused-power-outage-signals-troubling-escalation/

103 Zetter, K. (2017). Everything we know about Ukraine's power plant hack. *Wired*. Retrieved from www.wired.com/2016/01/everything-we-know-about-ukraines-power-plant-hack/

104 Zetter. (2017). Everything we know about Ukraine's power plant hack.

105 Greenberg, A. (2017). Hackers gain direct access to US power grid controls. *Wired*. Retrieved from www.wired.com/story/hackers-gain-switch-flipping-access-to-us-power-systems

106 "Wannacry" ransomware: Bengal power distribution company hit by cyberattack, say officials. (2017). *Hindustan Times*. Retrieved from www.hindustantimes.com/india-news/wannacry-ransomware-bengal-power-distribution-company-hit-by-cyberattack-say-officials/story-biqMQN5cPKng36cIyho2oJ.html

107 Greenberg. (2017). Hackers gain direct access to US power grid controls.

108 Greenberg. (2017). Hackers gain direct access to US power grid controls.

109 Barrett, B. (2019). Security news this week: An unprecedented cyberattack hit US power utilities. *Wired*. Retrieved from https://www.wired.com/story/power-grid-cyberattack-facebook-phone-numbers-security-news/

110 *Hackers leave question of how safe is US power grid*. (2019). Arizona Capitol Times. Retrieved from https://azcapitoltimes.com/news/2019/11/13/hackers-leave-question-of-how-safe-is-u-s-power-grid/

111 *Hackers leave question of how safe is US power grid*. (2019). Arizona Capitol Times.

112 ESET. (2017). *Industroyer: Biggest malware threat to critical infrastructure since Stuxnet*. Retrieved from www.eset.com/lv--/industroyer/

113 Osborne, C. (2018). *Industroyer: An in-depth look at the culprit behind Ukraine's power grid blackout*. ZDNet. Retrieved from https://www.zdnet.com/article/industroyer-an-in-depth-look-at-the-culprit-behind-ukraines-power-grid-blackout/

114 Lyngaas, S. (2019, October 30). *How would MITRE's popular cyberattack framework apply to industrial control systems?* CyberScoop. Retrieved from https://www.cyberscoop.com/mitre-attack-framework-industrial-control-systems/

115 Greenberg. (2017). Hackers gain direct access to US power grid controls.

116 Hern. (2017). Industroyer' virus could bring down power networks, researchers warn.

117 Walton, R. (2019, August 15). *Simultaneous hack of EV chargers could cause Manhattan blackout, NYU researchers find*. Utility Dive. Retrieved from https://www.utilitydive.com/news/simultaneous-hack-of-ev-chargers-could-cause-manhattan-blackout-nyu-resear/560974/

118 Henze, V. (2018, May 21). *E-buses to surge even faster than EVs as conventional vehicles fade*. Bloomberg NEF. Retrieved from https://about.bnef.com/blog/e-buses-surge-even-faster-evs-conventional-vehicles-fade/

119 Carnegie Mellon University. (2012, May 31). *Electricity subsector cybersecurity capability maturity model (ES-C2M2)*. Retrieved from energy.gov/sites/prod/files/Electricity%20Subsector%20

Cybersecurity%20Capabilities%20Maturity%20Model%20%28ES-C2M2%29%20-%20May%20 2012.pdf

120 Shallenberger, K. (2017, May 12). *Trump's cybersecurity executive order calls for power grid assessment.* Utility Dive. Retrieved from www.utilitydive.com/news/trumps-cybersecurity-executive-order -calls-for-power-grid-assessment/442560

121 Groll, E. (2016). Preventing a blackout by taking the power grid offline. *Foreign Policy*. Retrieved from foreignpolicy.com/2016/06/10/preventing-a-blackout-by-taking-the-power-grid-offline

122 Securing Energy Infrastructure Act. S80. Retrieved from www.congress.gov/bill/115th-congress /senate-bill/79

123 Securing the Electric Grid to Protect Military Readiness Act of 2017. H.R.3855 (2017–18). Retrieved from www.congress.gov/bill/115th-congress/house-bill/3855

124 Schwartz. (2020, January 10). *Hackers increasingly probe North American power grid.*

125 Walton, R. (2019). *US power grid attack points surge with proliferating DERs: A hacker "will eventually get in."* Utility Dive. Retrieved from https://www.utilitydive.com/news/security-and -distributed-resources-an-attacker-will-eventually-get-in-s/565966/

126 Cole, B. (2012). *NERC CIP (critical infrastructure protection)*. TechTarget. Retrieved from https:// searchcompliance.techtarget.com/definition/NERC-CIP-critical-infrastructure-protection

127 Nikolewski, R. (2020, January 7). Former California grid director says the state's power system is ready to take on any hacks from Iran. *The San Diego Union-Tribune*. Retrieved from https://www .sandiegouniontribune.com/business/energy-green/story/2020-01-07/california-power-grid -operators-set-to-fend-off-hacks-from-iran-analyst-says

Chapter 2

1 Schwartz, K. (2019, May 2). *Data privacy and data security: What's the difference?* ITPro Today. Retrieved from https://www.itprotoday.com/security/data-privacy-and-data-security-what-s -difference

2 Zimmerman, M., & Bradley, B. (2019, Spring). Intrinsic vs. extrinsic value. In Edward N. Zalta (Ed.), *The Stanford encyclopedia of philosophy*. Retrieved from https://plato.stanford.edu/entries /value-intrinsic-extrinsic/#WhaSorThiHavIntVal

3 Institute of Medicine (US) Committee on Health Research and the Privacy of Health Information: The HIPAA Privacy Rule, Nass, S.J., Levit, L.A., & Gostin, L.O. (Eds.). (2009). *Beyond the HIPAA Privacy Rule: Enhancing privacy, improving health through research*. Academies Press. Retrieved from https://www.ncbi.nlm.nih.gov/books/NBK9579/

4 Symanovich, S. (2020, January 18). *Privacy vs. security: What's the difference?* NortonLifeLock. Retrieved from https://us.norton.com/internetsecurity-privacy-privacy-vs-security-whats-the -difference.html

5 Schwartz. (2019). *Data privacy and data security.*

6 Gostin, L.O. (1995). Health information privacy. *Cornell Law Review, 80*(3), 451–528.

7 Symanovich. (2020). *Privacy vs. security.*

8 Westin, A. (1967). *Privacy and freedom*. Atheneum.

9 Gostin. (2002). Health information privacy.

10 Kenedi, C., Appel, J., & Friedman, S. (2019, April). Medical privacy versus public safety in aviation. *Journal of the American Academy of Psychiatry and the Law.* Retrieved from http://jaapl.org/content /early/2019/04/18/JAAPL.003839-19

11 Knowles, A., & McMahon, M. (1995). Expectations and preferences regarding confidentiality in the psychologist–client relationship. *Australian Psychologist*, *30*(3), 175–8.

12 Kenedi et al. (2019). Medical privacy versus public safety in aviation.

13 Dixon, P. (2006). *A brief introduction to fair information practices*. World Privacy Forum. Retrieved from https://www.worldprivacyforum.org/2008/01/report-a-brief-introduction-to-fair-information-practices/

14 Hoanca, B. (2017, June 29). *If privacy is dead, what can we do instead?* Retrieved from https://technologyandsociety.org/if-privacy-is-dead-what-can-we-do-instead/

15 Gartenberg, C. (2019). *Google reveals just how much of the world it's mapped with street view and earth*. The Verge. Retrieved from https://www.theverge.com/2019/12/13/21020814/google-world-mapped-street-view-earth-square-miles

16 *How a Google street view image of your house predicts your risk of a car accident*. (2019). MIT Technology Review. Retrieved from https://www.technologyreview.com/s/613432/how-a-google-street-view-image-of-your-house-predicts-your-risk-of-a-car-accident/

17 Streitfeld, D. (2013, March 13). Google concedes that drive-by prying violated privacy. *The New York Times*. Retrieved from https://www.nytimes.com/2013/03/13/technology/google-pays-fine-over-street-view-privacy-breach.html

18 Copeland, R. (2019, November 11). Google's "Project Nightingale" gathers personal health data on millions of Americans. *The Wall Street Journal*. Retrieved from https://www.wsj.com/articles/google-s-secret-project-nightingale-gathers-personal-health-data-on-millions-of-americans-11573496790

19 Shaukat, T. (2019). *Our partnership with Ascension*. Google Cloud. Retrieved from https://cloud.google.com/blog/topics/inside-google-cloud/our-partnership-with-ascension

20 Brill, J. (2013). Demanding transparency from data brokers. *The Washington Post*. Retrieved from http://www.washingtonpost.com/opinions/demanding-transparency-from-data-brokers/2013/08/15/00609680-0382-11e3-9259-e2aafe5a5f84_story.html

21 Copeland. (2019). Google's "Project Nightingale" gathers personal health data on millions of Americans.

22 Chan, T. (2018). *Facebook gave use data to 60 companies including Apple, Amazon, and Samsung*. Business Insider. Retrieved from https://www.businessinsider.com/facebook-gave-device-makers-apple-and-samsung-user-data-2018-6

23 Vengattil, M., & Dave, P. (2019, May 5). *Facebook "labels" posts by hand, posing privacy questions*. Reuters. Retrieved from https://www.reuters.com/article/us-facebook-ai/facebook-labels-posts-by-hand-posing-privacy-questions-idUSKCN1SC01T

24 Hern, A. (2018). Privacy policies of tech giants "still not GDPR-compliant." *The Guardian*. Retrieved from https://www.theguardian.com/technology/2018/jul/05/privacy-policies-facebook-amazon-google-not-gdpr-compliant

25 Daniels, J., Gitlitz, M., & Blank Rome LLP. (2018). *On the road to reconciling GDPR and blockchain*. JDSupra. Retrieved from https://www.jdsupra.com/legalnews/on-the-road-to-reconciling-gdpr-and-33122/

26 Knight, P. (2019, November 8). *Can GDPR handle blockchain's privacy problem?* ITProPortal. Retrieved from https://www.itproportal.com/features/can-gdpr-handle-blockchains-privacy-problem/

27 *Article 5 GDPR. Principles relating to processing of personal data*. (n.d.). Intersoft Consulting. Retrieved from https://gdpr-info.eu/art-5-gdpr/

28 Meyer, D. (2018, May 29). *GDPR attacks: First Google, Facebook, now activists go after Apple, Amazon, LinkedIn*. ZDNet. Retrieved from https://www.zdnet.com/article/gdpr-attacks-first-google-facebook-now-activists-go-after-apple-amazon-linkedin/
29 Ikeda, S. (2020). GDPR fines top $126 million with over 160,000 data breaches reported. *CPO Magazine*. Retrieved from https://www.cpomagazine.com/data-protection/gdpr-fines-top-126-million-with-over-160000-data-breaches-reported/
30 Hodge, N. (2020). *Study expects GDPR fines to rise in 2020*. Compliance Week. Retrieved from https://www.complianceweek.com/data-privacy/study-expects-gdpr-fines-to-rise-in-2020/28325.article
31 O'Brien. C. (2020, June 24). *EU report finds GDPR enforcement inadequate in its first 2 years*. VentureBeat. Retrieved from https://venturebeat.com/2020/06/24/eu-report-finds-gdpr-enforcement-inadequate-in-its-first-2-years/
32 Sullivan, J., & Hautala, L. (2018). *What the GDPR means for Facebook, the EU and you*. CNet. Retrieved from https://www.cnet.com/how-to/what-gdpr-means-for-facebook-google-the-eu-us-and-you/
33 Cookiebot. (n.d.). *Cookie policy: Generate a CCPA and GDPR compliant cookie policy for your website*. Retrieved from https://www.cookiebot.com/en/cookie-policy/
34 Onorato, A. (2018, May 25). *Cookies and consent: How GDPR impacts online tracking*. DMNews. Retrieved from https://www.dmnews.com/retail/article/13034543/cookies-and-consent-how-gdpr-impacts-online-tracking
35 Lomas, N. (2018, May 25). *Facebook, Google face first GDPR complaints over "forced consent."* TechCrunch. Retrieved from https://techcrunch.com/2018/05/25/facebook-google-face-first-gdpr-complaints-over-forced-consent/
36 Lubowicka, K. (2017). *How will GDPR affect your web analytics tracking?* Piwikpro. Retrieved from https://piwik.pro/blog/how-will-gdpr-affect-your-web-analytics-tracking/
37 Harding, E., & Hengeli, S. (2019, August 9). New developments in EU for cookies and online tracking. *The National Law Review*. Retrieved from https://www.natlawreview.com/article/new-developments-eu-cookies-and-online-tracking
38 Harding & Hengeli. (2019). New developments in EU for cookies and online tracking.
39 Harding & Hengeli. (2019). New developments in EU for cookies and online tracking.
40 Berzinya, A. (2018, May 30). *GDPR's imminent effect on how businesses collect consumers' location data*. Business.com. Retrieved from https://www.business.com/articles/gdpr-and-geolocation-data/
41 Firfiray, S. (2018). *Microchip implants are threatening workers' rights*. The Conversation. Retrieved from https://theconversation.com/microchip-implants-are-threatening-workers-rights-107221
42 European Commission. (n.d.). *Adequacy decisions: How the EU determines if a non-EU country has an adequate level of data protection*. Retrieved from https://ec.europa.eu/info/law/law-topic/data-protection/international-dimension-data-protection/adequacy-decisions_en
43 *Adequacy decisions*. (n.d.). European Commission. https://tinyurl.com/y5c9gexd
44 Barry, S. (2019, July 18). *GDPR and advertising – the impact so far*. ADWorld. Retrieved from https://www.adworld.ie/2019/07/18/gdpr-and-advertising-the-impact-so-far/
45 Sullivan & Hautala. (2018). *What the GDPR means for Facebook, the EU and you*.
46 Stone, J. (2019, May 21). *Real-time bidding, a thriving ad targeting technique, is becoming a GDPR dilemma*. Cyberscoop. Retrieved from https://tinyurl.com/y63g6gph
47 Porter, J. (2019). *German state bans Office 365 in schools, citing privacy concerns*. The Verge. Retrieved from https://www.theverge.com/2019/7/15/20694797/hesse-german-state-gdpr-office-365-schools-illegal-data-protection

48 Surur. (2019, July 30). *The Netherlands bans Microsoft's mobile Office apps as non-GDPR compliant.* MSPowerUser. Retrieved from https://mspoweruser.com/the-netherlands-bans-microsofts-mobile-office-apps-as-non-gdpr-compliant/
49 Porter. (2019). *German state bans Office 365 in schools, citing privacy concerns.*
50 Surur. (2019). *The Netherlands bans Microsoft's mobile Office apps as non-GDPR compliant.*
51 Afifi-Sabet, K. (2018). *Microsoft under GDPR microscope for Office 365 and OneDrive.* Microsoft News. Retrieved from https://www.alphr.com/microsoft/1010196/microsoft-under-gdpr-microscope-for-office-365-and-onedrive/
52 Meyer, D. (2018, October 12). Twitter under formal investigation for how it tracks users in the GDPR era. *Fortune.* Retrieved from http://fortune.com/2018/10/12/twitter-gdpr-investigation-tco-tracking/
53 Ritzer, C., & Filkina, N. (2019). *First multi-million GDPR fine in Germany: €14.5 million for not having a proper data retention schedule in place.* Data Protection Report. Norton Rose Fulbright. Retrieved from https://www.dataprotectionreport.com/2019/11/first-multi-million-gdpr-fine-in-germany-e14-5-million-for-not-having-a-proper-data-retention-schedule-in-place/
54 McCluskey, C., Newman, K., & Scott, G. (2019, November 8). *German privacy regulator joins the club, issuing hefty fine against real estate company.* JDSupra. Retrieved from https://www.jdsupra.com/legalnews/german-privacy-regulator-joins-the-club-52484/
55 Hunton Andrews Kurth LLP. (2019, November 6). *Berlin commissioner issues fine to Deutsche Wohnen SE.* Privacy & Information Security Law Blog. Retrieved from from https://www.huntonprivacyblog.com/2019/11/06/berlin-commissioner-issues-fine-to-deutsche-wohnen-se/
56 Pollmann, M. (2014, September 11). *Privacy, security & the geography of data protection.* Dark Reading. Retrieved from http://www.darkreading.com/cloud/privacy-security-and-the-geography-of-data-protection-/a/d-id/1315480?_mc=RSS_DR_EDT
57 King, R. (2013, February 7). New EU cyber security directive to impact US companies. *The Wall Street Journal.* Retrieved from http://tinyurl.com/cjujzgl
58 Woods, J. (n.d.). *Federal trade commission's privacy and data security enforcement under Section 5.* American Bar Association. Retrieved from http://www.americanbar.org/groups/young_lawyers/publications/the_101_201_practice_series/federal_trade_commissions_privacy.html
59 Martinez, R. (2014). *Reasonable doubt: Data privacy, cybersecurity, and the FTC.* JDSupra. Retrieved from http://www.jdsupra.com/legalnews/reasonable-doubt-data-privacy-cybersec-06717/
60 Dearie, K.J. (2020, March 26). Comparing the CCPA and the GDPR. *CPO Magazine.* Retrieved from https://www.cpomagazine.com/data-protection/comparing-the-ccpa-and-the-gdpr/
61 European Commission. (n.d.). *Adequacy decisions.*
62 Pearce, S., & Hastings, P. (2019, December 10). *Data privacy in Asia Pacific: A fragmented landscape.* Regulation Asia. Retrieved from https://www.regulationasia.com/data-privacy-in-asia-pacific-a-fragmented-landscape/
63 DLA Piper. (n.d.). *Data protection laws of the world: South Korea.* Retrieved from https://www.dlapiperdataprotection.com/index.html?t=law&c=KR
64 Chan, H. (2020, March 25). *Pervasive personal data collection at the heart of South Korea's COVID-19 success may not translate.* Thomson Reuters. Retrieved from https://blogs.thomsonreuters.com/answerson/south-korea-covid-19-data-privacy/
65 Feng, E. (2020, January 5). *In China, a new call to protect data privacy.* NPR. Retrieved from https://www.npr.org/2020/01/05/793014617/in-china-a-new-call-to-protect-data-privacy

66 Chakraborty, N. (2020, February 9). *Will govt exemption in new data protection bill affect you?* Mint. Retrieved from https://www.livemint.com/money/personal-finance/will-govt-exemption-in-new-data-protection-bill-affect-you-11581246354859.html

67 Rai, S. (2020, February 13). 400 million social media users are set to lose their anonymity in India. *The Economic Times*. Retrieved from https://economictimes.indiatimes.com/tech/internet/400-million-social-media-users-are-set-to-lose-their-anonymity-in-india/articleshow/74110920.cms

68 MPs want more checks on govt. (2020, December 1). *Hindustan Times*. Retrieved from https://www.hindustantimes.com/india-news/mps-want-more-checks-on-govt/story-QEJ8kKjkAekKv6DIUeTa9J.html

69 Orlansky, D. (2019, September 16). Argentina: Personal data protection law – impact on labor issues. *Global Compliance News*. Baker McKenzie. Retrieved from https://globalcompliancenews.com/argentina-personal-data-protection-law-impact-labor-issues-20190903/

70 Egan, M. (2019, February 12). *Data privacy reform gains momentum in Latin America*. IDB. Retrieved from https://blogs.iadb.org/conocimiento-abierto/en/data-privacy-reform-gains-momentum-in-latin-america/

71 Dahir, A. (2018, May 8). *Africa isn't ready to protect its citizens personal data even as EU champions digital privacy*. Quartz Africa. Retrieved from https://qz.com/africa/1271756/africa-isnt-ready-to-protect-its-citizens-personal-data-even-as-eu-champions-digital-privacy/

72 African Union. (n.d.). *List of countries which have signed, ratified/acceded to the African Union Convention on Cybersecurity and Personal Data Protection*. Retrieved from https://au.int/sites/default/files/treaties/29560-sl-african%20union%20convention%20on%20cyber%20security%20and%20personal%20data%20protection.pdf

73 Obulutsa, G. & Miriri, D. (2019, November 8). *Kenya passes data protection law crucial for tech investments*. Reuters. Retrieved from https://www.reuters.com/article/us-kenya-dataprotection/kenya-passes-data-protection-law-crucial-for-tech-investments-idUSKBN1XI1O1

74 Kazeem, Y. (2019, November 12). *Kenya is stepping up its citizens' digital security with a new EU-inspired data protection law*. Quartz Africa. Retrieved from https://qz.com/africa/1746202/kenya-has-passed-new-data-protection-laws-in-compliance-with-gdpr/

75 Frizell, S. (2018, November 8). *How Kenya's new data privacy bill could hurt its economy*. Council on Foreign Relations. Retrieved from https://www.cfr.org/blog/how-kenyas-new-data-privacy-bill-could-hurt-its-economy

76 Jervell, E.E., & Gummer, C. (2014). *U.S. web firms' expansion leaves Germans conflicted*. Retrieved from http://online.wsj.com/articles/amazon-to-launch-web-hosting-in-germany-1414075728

77 Yang, Y. (2018, October 1). China's data privacy outcry fuels case for tighter rules. *Financial Times*. https://www.ft.com/content/fdeaf22a-c09a-11e8-95b1-d36dfef1b89a

78 Ferreira, L. (2019, September 19). *Cybersecurity in China is a business worth $8.9 billion*. CGTN. Retrieved from https://news.cgtn.com/news/2019-09-19/Cybersecurity-in-China-is-a-business-worth-8-9-billion-K6RRVvwmgU/index.html

79 FireEye. (2019, November 11). *FireEye's cyber trendscape 2020*. Retrieved from https://www.fireeye.com/content/dam/fireeye-www/summit/cds-2019/presentations/cds19-executive-s05-fireeye-cyber-trendscape-2020.pdf

80 International Association of Privacy Professionals. (2019, January 22). *CNIL levies $57M fine on Google for GDPR violations*. Retrieved from https://iapp.org/news/a/cnil-levies-57m-fine-on-google-for-gdpr-violations/

81 Fazzini, K. (2019, July 10). *Europe's huge privacy fines against Marriott and British Airways are a warning for Google and Facebook*. CNBC. Retrieved from https://www.cnbc.com/2019/07/10/gdpr-fines-vs-marriott-british-air-are-a-warning-for-google-facebook.html
82 Meyer, D. (2019, January 21). *GDPR: Google hit with €50 million fine by French data protection watchdog*. ZDNet. Retrieved from https://www.zdnet.com/article/gdpr-google-hit-with-eur50-million-fine-by-french-data-protection-watchdog/
83 Merriman, C. (2018). *Google accused of breaching GDPR with "misleading" location tracking*. The Inquirer. Retrieved January 5, 2019 from https://www.theinquirer.net/inquirer/news/3067108/google-location-tracking-gdpr-breach
84 Liao, S. (2018, August 13). *Google still tracks you through the web if you turn off location history*. The Verge. Retrieved from https://www.theverge.com/2018/8/13/17684660/google-turn-off-location-history-data
85 Lomas. (2018). *Facebook, Google face first GDPR complaints over "forced consent."*
86 Brook, C. (2018, December 5). *Is Google violating GDPR by tracking EU users?* Digital Guardian. Retrieved from https://digitalguardian.com/blog/google-violating-gdpr-tracking-eu-users
87 Lomas, N. (2018). *Google faces GDPR complaint over "deceptive" location tracking*. Retrieved from https://techcrunch.com/2018/11/27/google-faces-gdpr-complaint-over-deceptive-location-tracking/
88 Moody, G. (2020, February 7). *Why the EU's General Data Protection Regulation (GDPR) risks turning into a paper tiger.* Privacy News Online. Retrieved from https://www.privateinternetaccess.com/blog/2020/02/why-the-eus-general-data-protection-regulation-gdpr-risks-turning-into-a-paper-tiger/
89 Lindsey, N. (2019, June 6). Google faces new European investigation over potential GDPR violation. *CPO Magazine*. Retrieved from https://www.cpomagazine.com/news/google-faces-new-european-investigation-over-potential-gdpr-violation/
90 Sterling, G. (2019, September 4). *Complaint alleges that Google is circumventing GDPR with RTB personal data sharing*. Marketing Land. Retrieved from https://marketingland.com/complaint-alleges-that-google-is-circumventing-gdpr-with-rtb-personal-data-sharing-266637
91 Lomas. (2018). *Facebook, Google face first GDPR complaints over "forced consent."*
92 *BBC News*. (2020, January 15). Cookies crumbling as Google phases them out. Retrieved from https://www.bbc.com/news/technology-51106526
93 Slefo, G. (2020, January 14). *Behind Google's decision to remove third-party cookies from Chrome*. Ad Age. Retrieved from https://adage.com/article/digital/behind-googles-decision-remove-third-party-cookies-chrome/2227126

Chapter 3

1 Council of Europe. (2004, July 1). Details of Treaty No. 185. Retrieved from http://www.coe.int/en/web/conventions/full-list/-/conventions/treaty/185
2 Volz, D., & Schwartz, F. (2019, November 5). Russia steps up efforts to shield its hackers from extradition to US. *The Wall Street Journal*. Retrieved from https://www.wsj.com/articles/russia-steps-up-efforts-to-shield-its-hackers-from-extradition-to-u-s-11572949802
3 Federal Bureau of Investigation. (2012, January 17). *Manhattan US Attorney and FBI assistant director in charge announce extradition of Russian citizen to face charges for international cyber crimes*. Retrieved from http://www.fbi.gov/newyork/press-releases/2012/manhattan-u.s.-attorney-and

-fbi-assistant-director-in-charge-announce-extradition-of-russian-citizen-to-face-charges-for-international-cyber-crimes

4 CBS. (2015). *These cybercrime statistics will make you think twice about your password: Where is the CSI's cyber team when you need them?* Retrieved January 5, 2016 from http://www.cbs.com/shows/csi-cyber/news/1003888/these-cybercrime-statistics-will-make-you-think-twice-about-your-password-where-s-the-csi-cyber-team-when-you-need-them-/

5 Fletcher, O. (2011, October 5). Patriotic Chinese hacking group reboots. *The Wall Street Journal.* Retrieved from http://blogs.wsj.com/chinarealtime/2011/10/05/patriotic-chinese-hacking-group-reboots/?mg=id-wsj

6 Voice of Russia. (2011). *Real punishment for virtual criminals.* Retrieved January 5, 2012 from http://english.ruvr.ru/2011/01/31/42167678.html

7 Pontell, H., Calavita, K., & Tillman, R. (1994). Corporate crime and criminal justice system capacity: Government response to financial institution fraud. *Justice Quarterly, 11*(3), 385–410.

8 Tillman, R., Calavita, K., & Pontell, H. (1996). Criminalizing white-collar misconduct: Determinants of prosecution in savings and loan fraud cases. *Crime, Law and Social Change, 26*(1), 53–76.

9 Benson, M., Cullen, F., & Maakestad, W. (1990). Local prosecutors and corporate crime. *Crime and Delinquency, 36*, 356–72.

10 CBS. (2015). *These cybercrime statistics will make you think twice about your password.*

11 Travis, A. (2016). Cybercrime figures prompt police call for awareness campaign. *The Guardian.* Retrieved from https://www.theguardian.com/uk-news/2016/jul/21/crime-rate-online-offences-cybercrime-ons-figures

12 Amburn, B. (2009, October 23). Caught in the net: Australian teens. *Foreign Policy.* Retrieved from https://foreignpolicy.com/2009/10/23/caught-in-the-net-australian-teens/

13 Cross, C. (2013). Nobody's holding a gun to your head…: Examining current discourses surrounding victims of online fraud. In J. Tauri & K. Richards (Eds.), *Crime, justice and social democracy: Proceedings of the 2nd international conference, 2013, Volume 1.* Crime and Justice Research Centre, Queensland University of Technology, Australia, pp. 25–32. Retrieved from https://eprints.qut.edu.au/61011/

14 Cross. (2013). Nobody's holding a gun to your head.

15 Tandon, N. (2007). Secondary victimization of children by the media: An analysis of perceptions of victims and journalists. *International Journal of Criminal Justice Sciences, 2*(2), 119–35.

16 Tandon. (2007). Secondary victimization of children by the media.

17 Anand, J. (2011). Cybercrime up by 700% in capital. *Hindustan Times.* Retrieved from https://www.hindustantimes.com/delhi/cyber-crime-up-by-700-in-capital/story-orj190HlR7GZShpzRkSfUL.html

18 Onyshkiv, Y., & Bondarev, A. (2012, March 8). Ukraine thrives as cybercrime haven. *Kyiv Post.* Kadorr Group. Retrieved from http://www.kyivpost.com/news/nation/detail/123965/

19 Claburn, T. (2009). Facebook wins $711 million from spammer. *InformationWeek.* Retrieved January 5, 2016 from http:// www.informationweek.com/news/global-io/security/showArticle.jhtml?articleID=221400140

20 Warren, P. (2007, November 15). Hunt for Russia's web criminals. *The Guardian.* Retrieved from http://www.guardian.co.uk/technology/2007/nov/15/news.crime

21 Walker, C. (2004, June). *Russian Mafia extorts gambling websites.* Retrieved from http://www.americanmafia.com/Feature_Articles_270.html

22 Kole, W. J. (2003, November 2). Romania emerges as major center of computer crime. *The Washington Post.* Retrieved from https://www.washingtonpost.com/archive/politics/2003/11/02/romania-emerges-as-major-center-of-computer-crime/13fe4d2c-13de-4bdc-b143-8b64461aedbc/

23 Neves, Y. (2019, June 4). *Millions stolen by hackers shows vulnerability of Mexico's banks.* InSight Crime. Retrieved from https://www.insightcrime.org/news/brief/hackers-steal-millions-in-mexicos-largest-ever-cyber-theft/

24 Torres, A. (2019, May 17). Mexican authorities arrest hacker gang that stole millions and bought a fleet of luxury cars, including Aston Martin, Lamborghini and Ferrari. *Daily Mail.* Retrieved from https://www.dailymail.co.uk/news/article-7041623/Hackers-average-stole-5-million-banks-Mexico-bought-McLarens-Aston-Martins-Ferrari.html

25 Glassdoor.com. (2020). *IT specialist salaries in Mexico.* Retrieved from https://www.glassdoor.com/Salaries/mexico-it-specialist-salary-SRCH_IL.0,6_IN169_KO7,20.htm

26 Neves. (2019). *Millions stolen by hackers shows vulnerability of Mexico's banks.*

27 van Koppen, P.J., & Jansen, R.W.J. (1998). The road to the robbery: Travel patterns in commercial robberies. *The British Journal of Criminology, 38*(2), 230–46.

28 Bell, R.E. (2002). The prosecution of computer crime. *Journal of Financial Crime, 9*(4), 308–25.

29 Swinhoe, D. (2019, May 30). *Why businesses don't report cybercrimes to law enforcement.* CSO. Retrieved from https://www.csoonline.com/article/3398700/why-businesses-don-t-report-cybercrimes-to-law-enforcement.html

30 Maheshwari, R. (2020, March 12). *1 in 10 Indian adolescents faces cyberbullying, half don't report: Study.* India Spend. Retrieved from https://www.indiaspend.com/1-in-10-indian-adolescents-faces-cyberbullying-half-dont-report-study/

31 The United States Attorney's Office, Western District of Washington. (2015, February 10). *Financial fraud crime victims.* United States Department of Justice. Retrieved from https://www.justice.gov/usao-wdwa/victim-witness/victim-info/financial-fraud

32 BatBlue. (2015). *FBI says banks pay hackers ransom to halt attacks.* Retrieved January 5, 2016 from www.batblue.com/fbi-says-banks-pay-hackers-ransom-to-halt-attacks

33 Kitten, T. (2015, August 24). *DDoS attacks against banks increasing.* Bank Info Security. Retrieved from http://www.bankinfosecurity.com/ddos-a-8497

34 Brandom, R. (2016, April 26). *The DDoS attack that cried wolf.* The Verge. Retrieved from http://www.theverge.com/2016/4/26/11512032/ddos-ransom-armada-collective-denial-of-service-threat

35 Dawar, I. (2014, January 20). *Shoplifting at £511m high as 90% of thefts are never reported.* Express. Retrieved from http://www.express.co.uk/news/uk/454886/Shoplifting-at-511m-high-as-90-of-thefts-are-never-reported

36 Dawar. (2014). *Shoplifting at £511m high as 90% of thefts are never reported.*

37 Clark, J.R., & Davis, W.L. (1995). A human capital perspective on criminal careers. *Journal of Applied Business Research, 11*(3), 58–64.

38 Blau. (2004, May 28). Viruses: From Russia, with love? *PCWorld.* Retrieved from http://www.pcworld.com/article/116304/article.html

39 Galeotti, M. (2011, November 21). Why are Russians excellent cybercriminals? Retrieved January 5, 2012 from http://themoscownews.com/siloviks_scoundrels/20111121/189221309.html

40 Lovejoy, B. (2020, April 14). *Unsurprisingly, Apple customers are the most common phishing target.* 9To5Mac. Retrieved from https://9to5mac.com/2020/04/14/most-common-phishing-target/

41 Lindenberg, S. (2001). Intrinsic motivation in a new light. *Kyklos, 54*(2/3), 317–42.

42 McBride, J. (2020, April 22). *Bill Gates hacked? Coronavirus conspiracy theories rage*. Heavy. Retrieved from https://heavy.com/news/2020/04/bill-gates-hacked-coronavirus-wuhan-lab/

43 Serracino-Inglott, P. (2013). Is it OK to be an Anonymous? *Ethics & Global Politics, 6*(4), 22527. doi: 10.3402/egp.v6i4.22527

44 Palmer, D. (2016, September 8). *The cost of ransomware attacks: $1 billion this year*. ZDNet. Retrieved from http://www.zdnet.com/article/the-cost-of-ransomware-attacks-1-billion-this-year/

45 A payload is "the component of a computer virus that executes a malicious activity." Techopedia. (n.d.). Retrieved from https://www.techopedia.com/definition/5381/payload

46 CTB Locker is mainly delivered through spam. For instance, the email message may say that it needs the receiver's urgent attention. When the email is accessed, the victim is asked to download a zip file. If the zip file attached to the email is accessed, the data on the victim's system will be encrypted. In order to receive a decryption key, the victim is asked for a ransom. (See https://heimdalsecurity.com/blog/ctb-locker-ransomware/)

47 Peters, S. (2015, June 9). *Cybercrime can give attackers 1,425% return on investment*. Dark Reading. Retrieved from http://www.darkreading.com/analytics/cybercrime-can-give-attackers-1425--return-on-investment/d/d-id/1320756

48 Liebowitz, M. (2011). *How to become a cybercriminal for only $7.22*. Retrieved from https://www.nbcnews.com/id/wbna44627713

49 Schwartz, M. (2020, April 15). *Emotet, Ryuk, TrickBot: "Loader-Ransomware-Banker Trifecta."* Data Breach Today. Retrieved from https://www.databreachtoday.com/emotet-ryuk-trickbot-loader-ransomware-banker-trifecta-a-14126?rf=2020-04-16_ENEWS_ACQ_DBT__Slot1_ART14126&mkt_tok=eyJpIjoiT0RReU56aGtPV000WVRobSIsInQiOiIyMUcwbklOSlBYQmcwWHprVVlENVo4cWVneUxOR0pXTVVwNTF6c1JqbjE4V0VYVzgzbWlLZCsxeUpJZnhzeE5TRkQ2ZXBxdUxiMXFSZVdKbGpxTTZUbFwvbGZcL1Fsd2pnaEtvSmJUdDJZZERNZGJcL2tDMDNhV0RrZStuXC9CVXlpZEkifQ%3D%3D

50 Schwartz, M. (2020, March 6). *From cybercrime zero to "hero" – now faster than ever*. Data Breach Today. Retrieved from https://www.databreachtoday.com/from-cybercrime-zero-to-hero-now-faster-than-ever-a-13865

51 Leyden, J. (2010, April 12). *Russian trade body aims to fight cybercrime: Russia no safe haven for spammers and cybercriminals*. The Register. Retrieved from http://www.theregister.co.uk/2010/04/12/russia_cybercrime_feature/

52 Au, A. (2018, April 22). *A former hacker shares the most twisted things he did*. Vice. Retrieved from https://www.vice.com/en_uk/article/xw7km4/a-former-hacker-shares-the-most-twisted-things-he-did

53 Drake, M. (2015, April 4). *One computer hacker a month convicted of cyber crime out of 100,000 incidents a year*. Mirror. Retrieved from http://www.mirror.co.uk/news/uk-news/one-computer-hacker-month-convicted-5461766

54 Gopalakrishnan, C. (2019). *Cyber-crime thrives on legal inefficiency & business leaders turning a blind eye*. SC Media. Retrieved January 5, 2020 from https://www.scmagazineuk.com/cyber-crime-thrives-legal-inefficiency-business-leaders-turning-blind-eye/article/1586202

55 Wylie, I. (2007, December 26). Romania home base for eBay scammers. *Los Angeles Times*, C1.

56 Zulu, B. (2008, October 25). Microsoft combats cybercrime in Nigeria. *PCWorld*. Retrieved from http://www.pcworld.com/businesscenter/article/152784/microsoft_combats_cybercrime_in_nigeria.html

57 Wolverton, T. (2015, February 13). Silicon valley: Obama calls on corporations to work with government to prevent cyberattacks. *The Mercury News*. Retrieved from www.mercurynews.com/business/ci_27520838/obama-issues-cybersecurity-order-at-open-summit

58 Kuchler, H. (2014, September 24). US financial industry launches platform to thwart cyber attacks. *Financial Times*. Retrieved January 23, 2015 from www.ft.com/content/080092b2-437a-11e4-8a43-00144feabdc0

59 Financial Services Information Sharing and Analysis Center. (n.d.). *Intelligence sharing critical cyber-intelligence among members*. Retrieved from https://www.fsisac.com/what-we-do/intelligence

60 Salkever, A. (2003, October 14). *"Phishing" is foul on the Net*. ZDNet. Retrieved from https://www.zdnet.com/article/phishing-is-foul-on-the-net/

61 Yadron, D. (2014, January 6). Target hackers wrote partly in Russian, displayed high skill, report finds. *The Wall Street Journal*. Retrieved from http://online.wsj.com/news/articles/SB10001424052702304419104579324902602426862

62 Vijayan, J. (2014, December 9). Target ruling raises stakes for cybersecurity vigilance. *Christian Science Monitor*. Retrieved from http://www.csmonitor.com/World/Passcode/2014/1209/Target-ruling-raises-stakes-for-cybersecurity-vigilance

63 Schwartz, M.J. (2014, March 14). *Target ignored data breach alarms*. Dark Reading. Retrieved from https://www.darkreading.com/attacks-and-breaches/target-ignored-data-breach-alarms/d/d-id/1127712

64 Blau. (2004). Viruses: From Russia, with love?

65 Paget, F. (2010). *McAfee helps FTC, FBI in case against "Scareware" outfit*. McAfee. Retrieved January 26, 2011 from http://blogs.mcafee.com/mcafee-labs/mcafee-helps-ftc-fbi-in-case-against-scareware-outfit.

66 Finkle, J. (2010, March 24). *Inside a global cybercrime ring*. Reuters. Retrieved from http://www.reuters.com/article/ idUSTRE62N29T20100324

67 Giles, J. (2010). Scareware: The inside story. *New Scientist, 205*(2753), 38–41.

68 Finkle. (2010). *Inside a global cybercrime ring*.

69 Finkle. (2010). *Inside a global cybercrime ring*.

70 Marson, J. (2010). Small victory in the fight against global cybercrime. *Time*. Retrieved from http://content.time.com/time/business/article/0,8599,1998055,00.html

71 Acohido, B. (2010, June 7). "Scareware" ads proliferate across Internet. *USA Today*, B1.

72 Finkle. (2010). *Inside a global cybercrime ring*

73 Finkle. (2010). *Inside a global cybercrime ring*.

74 Giles, J. (2010). Scareware: The inside story.

75 Federal Bureau of Investigation. (2012). *Manhattan US Attorney and FBI assistant director in charge announce extradition of Russian citizen to face charges for international cyber crimes*.

76 Mullins, R. (2010). *Did the government sit on the "scareware" case too long? Critic says authorities took too long to break cyber ring*. Network World. Retrieved from http://www.networkworld.com/community/blog/did-government-sit-scareware-case-too-long

77 Finkle. (2010). *Inside a global cybercrime ring*.

78 Paget. (2010). *McAfee helps FTC, FBI in case against "Scareware" outfit*.

Chapter 4

1 Masters, J. (2011). *Confronting the cyber threat*. Council on Foreign Relations. Retrieved from https://www.cfr.org/backgrounder/confronting-cyber-threat

2 Foremski, T. (2018, May 25). *Ad fraud is out of control – billions lost by media industry, says a new report*. ZDNet. Retrieved from https://www.zdnet.com/article/ad-fraud-is-out-of-control-billions-lost-by-media-industry-says-a-new-report/

3 Kshetri, N. (2009). Positive externality, increasing returns and the rise in cybercrimes. *Communications of the ACM, 52*(12), 141–4.

4 Kshetri, N. (2006). The simple economics of cybercrimes. *IEEE Security and Privacy, 4*(1), 33–9.

5 Scott, R. (2001) *Institutions and organizations*. Sage.

6 Scott. (2001). *Institutions and organizations*.

7 Gaviria, A. (2000). Increasing returns and the evolution of violent crime: The case of Colombia. *Journal of Development Economics, 61*(1), 1.

8 Kshetri, N. (2016, February 16). *Why the IRS was just hacked – again – and what the feds can do about it*. The Conversation. Retrieved from https://theconversation.com/why-the-irs-was-just-hacked-again-and-what-the-feds-can-do-about-it-54524

9 Zapotosky, M. (2016, December 8). Why law enforcement can't get ahead of Pizzagate and other online conspiracy theories. *The Washington Post*. Retrieved from https://www.washingtonpost.com/world/national-security/why-law-enforcement-cant-get-ahead-of-pizzagate-and-other-online-conspiracy-theories/2016/12/08/2960e7ea-bbf5-11e6-ac85-094a21c44abc_story.html

10 Kaste, M. (January 16, 2020). *Coaxing cops to tackle cybercrime? There's an app for that*. NPR. Retrieved from https://www.npr.org/2020/01/15/796252827/coaxing-cops-to-tackle-cybercrime-theres-an-app-for-that

11 Goldman, J. (2013, October 1). *R.T. Jones Capital Equities Management admits security breach*. eSecurity Planet. Retrieved from https://www.esecurityplanet.com/trends/cybersecurity-employment-2021/r

12 Ferguson, S. (2018). *Over 300K cybersecurity jobs remain open in the US, study finds*. Dark Reading. Retrieved from https://www.darkreading.com/operational-security/training/over-300k-cybersecurity-jobs-remain-open-in-the-us-study-finds/a/d-id/743973

13 Aggarwal, V. (2009, August 3). *Cyber crime's rampant*. Express Computer. Retrieved January 5, 2010 from http://www.expresscomputeronline.com/20090803/market01.shtml

14 *Latin America's cyber gangs*. (2011). Southernpulse. Retrieved January 5, 2012 from http://www.southernpulse.com/_webapp_3923733/Latin_America%E2%80%99s_cyber_gangs

15 Blitz, J. (2011, November 1). Security: A huge challenge from China, Russia and organized crime. *Financial Times*. Retrieved from http://www.ft.com/intl/cms/s/0/b43488b0-fe2a-11e0-a1eb-00144feabdc0.html#axzz1dnezI1eF

16 Brazil, 7th most violent country in the world, had 1.1 million murders between 1980 and 2011. (2017, December 6). *Huffington Post*. https://www.huffpost.com/entry/brazil-most-violent-country-murders_n_3618704

17 Jair Bolsonaro will not defeat crime in Brazil by tolerating militias. (2019, May 30). *The Economist*. Retrieved from https://www.economist.com/leaders/2019/05/30/jair-bolsonaro-will-not-defeat-crime-in-brazil-by-tolerating-militias

18 Theriault, C. (2011). *Brazil's cybercrime evolution – it doesn't look pretty*. Naked Security by Sophos. Retrieved from http://nakedsecurity.sophos.com/2011/10/05/brazils-cybercrime-evolution-it-doesnt-look-pretty/

19 Sah, R. (1991). Social osmosis and patterns of crime. *Journal of Political Economy, 99*(6), 169–217.

20 Kshetri. (2006). The simple economics of cybercrimes.

21 Singh, S.R. (2014). India's only cyber appellate tribunal defunct since 2011. *Hindustan Times*. Retrieved from https://www.hindustantimes.com/india/india-s-only-cyber-appellate-tribunal-defunct-since-2011/story-208HGrEN7hXrABg7lAb69N.html

22 Sarda, K. (2018, December 10). Tribunal for cyber crimes headless. *The New Indian Express*. Retrieved from https://www.newindianexpress.com/nation/2018/dec/10/tribunal-for-cyber-crimes-headless-1909362.html

23 EU-OCS. (2019, December 11). *Russian hackers behind "most dangerous cybercrime group in the world" named by US, UK authorities*. Retrieved from https://eu-ocs.com/russian-hackers-behind-most-dangerous-cybercrime-group-in-the-world-named-by-us-uk-authorities/

24 GlobeNewswire. (2020). *March 2020's most wanted malware: Dridex banking Trojan ranks on top malware list for first time*. Retrieved from https://www.globenewswire.com/news-release/2020/04/09/2014156/0/en/March-2020-s-Most-Wanted-Malware-Dridex-Banking-Trojan-Ranks-On-Top-Malware-List-For-First-Time.html

25 National Crime Agency. (2019). *International law enforcement operation exposes the world's most harmful cyber crime group*. Retrieved from https://public-newsroom-nca-01.azurewebsites.net/news/international-law-enforcement-operation-exposes-the-worlds-most-harmful-cyber-crime-group

26 National Crime Agency. (2019). *International law enforcement operation exposes the world's most harmful cyber crime group*.

27 Chimtom, N.K. (2016, April 6). Cameroon's dilemma in fighting cybercrime. *African Independent*. Retrieved from https://www.africanindy.com/business/cameroons-dilemma-in-fighting-cybercrime-5073265

28 Statistics Netherlands (CBS). (2017, September 27). *Three-quarters of cybercrime cases not reported*. Retrieved from https://www.cbs.nl/en-gb/news/2017/39/three-quarters-of-cybercrime-cases-not-reported

29 Kshetri, N. (2005, May/June). Hacking the odds. *Foreign Policy*, 93.

30 Baker, S. (2004, August 9). Gambling sites, this is a holdup. *Business Week*. Retrieved from https://www.bloomberg.com/news/articles/2004-08-08/gambling-sites-this-is-a-holdup

31 Baker. (2004). Gambling sites, this is a holdup.

32 Warren, P. (2007, November 15). Hunt for Russia's web criminals. The Russian Business Network – which some blame for 60% of all Internet crime – appears to have gone to ground. *The Guardian*. Retrieved from http://www.guardian.co.uk/technology/2007/nov/15/news.crime

33 Rogers, E.M. (1995). *Diffusion of innovation* (4th ed.). Free Press.

34 PhishLabs. (2017). *PhishLabs 2017 phishing trends & intelligence report*. Retrieved from https://www.phishlabs.com/phishlabs-2017-phishing-trends-intelligence-report-hacking-the-human/

35 Marquette, S. (2019, July 22). *What is formjacking, and how can you protect your business?* Business News Daily. Retrieved from https://www.businessnewsdaily.com/15219-what-is-formjacking.html

36 Chandrayan, S. (2018, December 5). *Formjacking: Targeting popular stores near you*. Broadcom. Retrieved from https://www.symantec.com/blogs/threat-intelligence/formjacking-targeting-popular-stores

37 Symantec. (2019). *ISTR 24: Symantec's annual threat report reveals more ambitious and destructive attacks*. Retrieved from https://symantec-enterprise-blogs.security.com/blogs/threat-intelligence/istr-24-cyber-security-threat-landscape

38 Shukla, S. (2017, August 4). *Bitcoin a favourite of ransomwares; cyber extortion to shoot up if govts don't act soon*. Money Control. Retrieved from bit.ly/2qiMSrr

39 McAfee. (2017). *McAfee labs threat report*. Retrieved from https://www.mcafee.com/enterprise/en-us/assets/reports/rp-quarterly-threats-dec-2017.pdf

40 Kaspersky. (2019, August 19). *Ransomware modifications double year-on-year in Q2 2019*. Retrieved from https://www.kaspersky.com/about/press-releases/2019_ransomware-modifications-double-year-on-year-in-q2-2019

41 Barth, B. (2020, January 19). *Travelex recovering from ransomware, but more firms at risk of VPN exploit*. SC Media. Retrieved from https://www.scmagazine.com/home/security-news/ransomware/travelex-recovering-from-ransomware-but-more-firms-at-risk-of-vpn-exploit/

42 Asokan, A. (2020, April 10). *Travelex paid $2.3 million to ransomware gang: Report*. Bank Info Security. Retrieved from https://www.bankinfosecurity.com/travelex-paid-23-million-to-ransomware-attackers-report-a-14094

43 Tidy, J. (2020. January 7). Travelex being held to ransom by hackers. *BBC News*. Retrieved from https://www.bbc.com/news/business-51017852

44 Tidy. (2020). *Travelex being held to ransom by hackers.*

45 Usborne, S. (2017, May 15). Digital gold: Why hackers love bitcoin. *The Guardian*. Retrieved from bit.ly/2qlaMW2

46 Converted according to the rate on April 27, 2020, at https://coinmarketcap.com/

47 Swing, D. (2020, April 8). *Learning from a ransomware attack.* Data Centre Dynamics. Retrieved from https://www.datacenterdynamics.com/en/analysis/learning-ransomware-attack/

48 Schwartz, M. (2020, April 7). *No COVID-19 respite: Ransomware keeps pummeling healthcare*. Bank Info Security. Retrieved from https://www.bankinfosecurity.com/no-covid-19-respite-ransomware-keeps-pummeling-healthcare-a-14072

49 *WHO attacks linked to Iran? Mandrake active in Australia. Vollgar vs. MS-SQL servers. COVID-19 and the cyber sector*. (2020, April 2). The Cyber Wire. Retrieved May 5, 2020 from https://thecyberwire.com/newsletters/daily-briefing/9/64

50 Zmudzinski, A. (2020, April 13). *Sodinokibi crypto ransomware switches from bitcoin to monero to hide money trail.* Cointelegraph. Retrieved from https://cointelegraph.com/news/sodinokibi-crypto-ransomware-switches-from-bitcoin-to-monero-to-hide-money-trail

51 Hughes, H. (2020, March 27). *Monero (XMR) quietly gains 99.5% as bitcoin price consolidates.* Cointelegraph. Retrieved from https://cointelegraph.com/news/monero-xmr-quietly-gains-995-as-bitcoin-price-consolidates

52 Global Agenda Council on the Future of Software & Society. (2015, September). *Deep shift: Technology tipping points and societal impact survey report*. World Economic Forum. Retrieved from bit.ly/1KlN8ZR

53 Huckstep, R. (2016, January 14). *What does the future hold for blockchain and insurance?* Daily Fintech. Retrieved from bit.ly/2dRLBkz

54 *Connected Toys Market*. (n.d.). Markets and Markets. Retrieved from https://www.marketsandmarkets.com/Market-Reports/connected-toys-market-38031230.html

55 Wong, G., Chu, K., Osawa, J. (2015, March 2). Inside Alibaba, the sharp-elbowed world of Chinese e-commerce. *The Wall Street Journal*. Retrieved from http://www.wsj.com/articles/inside-alibaba-the-sharp-elbowed-world-of-chinese-e-commerce-1425332447

56 Oxenham, S. (2019, May 28). *"I was a Macedonian fake news writer."* BBC. Retrieved from https://www.bbc.com/future/article/20190528-i-was-a-macedonian-fake-news-writer

57 The interview is available at https://www.youtube.com/watch?v=tNsbKtckzcM

58 Bosilkovski, I. (2019, April 16). Checking in with the Macedonian fake news strategist. *Columbia Journalism Review*. Retrieved from https://www.cjr.org/politics/checking-in-with-the-macedonian-fake-news-strategist.php
59 Geier, E. (2011, December 2). *Hiring hackers and buying malware is easy*. eSecurity Planet. Retrieved from https://www.esecurityplanet.com/hackers/hiring-hackers-and-buying-malware-is-easy.html
60 Becker, G.S. (1995). The economics of crime. *Cross Sections, Federal Reserve Bank of Richmond, 12*(Fall), 8–15.
61 Republic of North Macedonia State Statistical Office. (n.d.). Retrieved from http://www.stat.gov.mk/KlucniIndikatori_en.aspx
62 Subramanian, S. (2017, February 15). Inside the Macedonian fake-news complex. *Wired*. Retrieved from https://www.wired.com/2017/02/veles-macedonia-fake-news/
63 Subramanian. (2017). Inside the Macedonian fake-news complex.
64 Savage, M. (2019). Cybercrime for dummies: Cracking internet passwords is as easy as 123456. *The Guardian*. Retrieved from https://www.theguardian.com/technology/2019/apr/21/cybercrime-hacking-internet-account-passwords
65 Silverman, C. (2016, November 16). *This analysis shows how viral fake election news stories outperformed real news on Facebook*. BuzzFeed News. Retrieved from https://www.buzzfeed.com/craigsilverman/viral-fake-election-news-outperformed-real-news-on-facebook?utm_term=.cxVmeoezNy#.xkL35B5RXK
66 Silverman, C., Singer-Vine, J., & Vo, L.T. (2017, April 4). *In spite of the crackdown, fake news publishers are still earning money from major ad networks*. BuzzFeed News. Retrieved from https://www.buzzfeed.com/craigsilverman/fake-news-real-ads?utm_term=.giD6pjV96d#.yxX5P2075A
67 *See your child's device use with activity reports*. (n.d.). Microsoft. Retrieved from https://support.microsoft.com/en-us/help/12441/microsoft-account-see-child-device-activity
68 Biersdorfer, J.D. (2018, March 2). Tools to keep your children in line when they're online. *The New York Times*. Retrieved from https://www.nytimes.com/2018/03/02/technology/personaltech/setting-up-parental-controls-on-pcs-and-macs.html
69 Takahashi, D. (2019, November 7). *Roblox: 60% of teens rarely discuss inappropriate online behavior with parents*. VentureBeat. Retrieved from https://venturebeat.com/2019/11/07/roblox-60-of-teens-rarely-discuss-inappropriate-online-behavior-with-parents/
70 Hwang, L. (2019, March 12). *Analyze this: Most teens have been cyberbullied*. Science News for Students. Retrieved from https://www.sciencenewsforstudents.org/article/analyze-most-teens-have-been-cyberbullied
71 Mizrahi, A. (2016, August 6). *33% of UK firms are buying bitcoin in anticipation of cyber attacks*. Finance Magnates. Retrieved from bit.ly/2v7t4wb
72 Simonite, T. (2016, June 7). *Companies are stockpiling bitcoin to pay off cybercriminals*. MIT Technology Review. Retrieved from bit.ly/1YeQSpU
73 Méndez, R. (2017, October 11). *Let's focus on cybersecurity for small businesses*. Federal Trade Commission. Consumer Information. Retrieved from https://www.consumer.ftc.gov/blog/2017/10/lets-focus-cybersecurity-small-businesses
74 Firstpost. (2015, December 10). *Nasscom, Mumbai police join hands for Cyber Awareness Month to educate on online security*. Retrieved from http://www.firstpost.com/business/nasscom-mumbai-police-join-hands-for-cyber-awareness-month-to-educate-on-online-security-2540602.html
75 No More Ransom. (n.d.). Retrieved from https://www.nomoreransom.org/

76 Korolov, M. (2017, March 1). *Global cybercrime prosecution a patchwork of alliances*. CSO. Retrieved from https://www.csoonline.com/article/3174439/security/global-cybercrime-prosecution-a-patchwork-of-alliances.html
77 Krebsonsecurity. (2016, October 16). *Are the days of "booter" services numbered?* Retrieved from https://krebsonsecurity.com/tag/bulletproof-hosting/
78 Leyden, J. (2017, October 5). *Bulletproof hosts stay online by operating out of disputed backwaters*. The Register. Retrieved from https://www.theregister.co.uk/2017/10/05/bulletproof_hosting/
79 Thomas, E. (2019, August 8). Inside the bulletproof hosting providers that keep the world's worst websites in business. *ABC News*. Retrieved from https://www.abc.net.au/news/2019-08-09/shining-light-on-the-bulletproof-web-hosts-lurking-in-the-sha/11396986
80 Goldman, D. (2011, July 27). *The cyber Mafia has already hacked you*. CNN Money. Retrieved from https://money.cnn.com/2011/07/27/technology/organized_cybercrime/index.htm
81 Thomas. (2019). Inside the bulletproof hosting providers that keep the world's worst websites in business.
82 Leyden. (2017). *Bulletproof hosts stay online by operating out of disputed backwaters*.
83 Warren. (2007). Hunt for Russia's web criminals.
84 Leyden. (2017). *Bulletproof hosts stay online by operating out of disputed backwaters*.
85 Krebs, B. (2007). Taking on the Russian business network. *The Washington Post*. Retrieved from http://voices.washingtonpost.com/securityfix/2007/10/taking_on_the_russian_business.html
86 Krebs. (2007). Taking on the Russian business network.
87 Kendall, N. (2009, August 1). What the cybercrime fraudsters get up to. *The Times*. Retrieved from https://www.thetimes.co.uk/article/what-the-cybercrime-fraudsters-get-up-to-nlpwp7kqjpx
88 ARN. (2007, September 18). *Symantec reports cyber criminals are becoming increasingly professional*. Retrieved from https://www.arnnet.com.au/article/193423/symantec_reports_cyber_criminals_becoming_increasingly_professional/?fp=2&fpid=1347590468
89 Warren. (2007). Hunt for Russia's web criminals.
90 Leyden. (2017). *Bulletproof hosts stay online by operating out of disputed backwaters*.
91 Leyden, J. (2009, October 22). *FBI and SOCA plot cybercrime smackdown*. The Register. Retrieved from http:// www.theregister.co.uk/2009/10/22/soca_fbi_cybercrime_strategy/
92 *The Economist*. (2007, August 30). A walk on the dark side. Retrieved from https://www.economist.com/unknown/2007/08/30/a-walk-on-the-dark-side
93 Espiner, T. (2007). *Cracking open the cybercrime economy*. CNet. Retrieved from https://www.cnet.com/news/cracking-open-the-cybercrime-economy-1/
94 Blakely, R., Richards, J., & Halpin, T. (2007, November 10). Cybergang raises fear of new crime wave. *The Times* (London), 13.
95 Lauder, M. (2018, July 7). *"Wolves of the Russian spring": Examination of night wolves as proxy for Russian government – analysis*. Eurasia Review News & Analysis. Retrieved from https://www.eurasiareview.com/07072018-wolves-of-the-russian-spring-examination-of-night-wolves-as-proxy-for-russian-government-analysis/
96 Soshnikov, A. (2016, September 20). Bears with keyboards: Russian hackers snoop on West. *BBC News*. Retrieved from https://www.bbc.com/news/world-europe-37409456
97 Kabay, M.E., & Guinen, B. (2011, March 28). *The Russian Cybermafia: RBN & the RBS WorldPay attack*. Network World. Retrieved from https://www.networkworld.com/article/2201011/the-russian-cybermafia--rbn---the-rbs-worldpay-attack.html

98 Gorman, S., & Perez, E. (2009, December 22). FBI probes hack at Citibank. *The Wall Street Journal*. Retrieved from https://www.wsj.com/articles/SB126145280820801177

99 Masters, G. (2009, December 22). *Citibank refutes reported hack by Russian gang*. SC Media. Retrieved from https://www.scmagazine.com/home/security-news/citibank-refutes-reported-hack-by-russian-gang/

100 Leyden. (2017). *Bulletproof hosts stay online by operating out of disputed backwaters.*

101 *What Is A Fast Flux?* (n.d.). Infoblox. Retrieved from https://www.infoblox.com/glossary/fast-flux/

102 Mahjoub, D. (2016, May 16). *Black Hat 2016 preview: Fast Flux with SSL, a unique and popular Bulletproof Hosting option for cyber criminals*. Cisco. Retrieved from https://umbrella.cisco.com/blog/black-hat-2016-fast-flux-ssl-unique-popular-bulletproof-hosting-option-cyber-criminals

Chapter 5

1 Lawal, L. (2006, June 1). *Online scams create "Yahoo! millionaires": In Lagos, where scamming is an art, the quickest way to wealth for the cyber-generation runs through a computer screen.* CNN Money. Retrieved from http://money.cnn.com/magazines/fortune/fortune_archive/2006/05/29/8378124/

2 Kshetri, N. (2013). *Cybercrime and cybersecurity in the Global South*. Palgrave Macmillan.

3 Jandt, F.E., & Tanno, D.V.H. (2001). Decoding domination, encoding self-determination: Intercultural communication research processes. *Journal of Communications, 12*(3), 119–35.

4 Arabianbusiness. (2010). *Saudi Arabia's Hai'a to set up cybercrime unit*. Retrieved from http://www.arabianbusiness.com/saudi-arabia-s-hai-a-set-up-cyber-crime-unit-342746.html

5 North, D.C. (1990). *Institutions, institutional change and economic performance*. Harvard University Press.

6 Scott, W.R. (2001). *Institutions and organizations* (2nd ed.). Sage.

7 Scott, W.R., Ruef, M., Mendel, P.J., & Caronna, C.A. (2000). *Institutional change and healthcare organizations: From professional dominance to managed care*. University of Chicago Press.

8 Scott. (2001). *Institutions and organizations.*

9 Scott. (2001). *Institutions and organizations.*

10 Schjølberg, S. (2008, December). *The history of global harmonization on cybercrime legislation – the road to Geneva*. Retrieved from https://www.cybercrimelaw.net/documents/cybercrime_history.pdf

11 Tickner, J.A. (1995). Revisioning security. In K. Booth and S. Smith (Eds.), *International Relations Theory Today* (pp. 175–97). Polity.

12 Ball, N. (1988). *Security and economy in the third world*. Adamantine Press.

13 Rayman, N. (2014, August 7). The world's top 5 cybercrime hotspots. *Time*. Retrieved from http://time.com/3087768/the-worlds-5-cybercrime-hotspots/

14 Asia-Pacific Center for Security Studies. (2001, January 10–11). *Report from the conference on: Domestic determinants of security: Security institutions and policy-making processes in the Asia-Pacific region*. Retrieved from http://www.apcss.org/Publications/Report_DomesticDeterminantsOfSecurity.html

15 Scott. (1995). *Institutions and organizations.*

16 Serracino-Inglott. (2013). Is it OK to be an Anonymous?

17 Fox-Brewster, T. (2014, May 11). *Did China order hackers to cripple the Hong Kong protest?* Vice. Retrieved from https://www.vice.com/en_us/article/539wnz/inside-the-unending-cyber-siege-of-hong-kong

18 Moran, N., Scott, M., & Oppenheim, M. (2014, November 3). *Operation poisoned handover: Unveiling ties between APT activity in Hong Kong's pro-democracy movement.* FireEye. Retrieved from https://www.fireeye.com/blog/threat-research/2014/11/operation-poisoned-handover-unveiling-ties-between-apt-activity-in-hong-kongs-pro-democracy-movement.html

19 Fox-Brewster. (2014). *Did China order hackers to cripple the Hong Kong protest?*

20 *About us.* (n.d.). NASSCOM. Retrieved from https://www.nasscom.in/

21 Jandt & Tanno. (2001). Decoding domination, encoding self-determination.

22 Hoare, C.H. (1991). Psychosocial identity development and cultural others. *Journal of Counseling & Development*, *70*(1), 45–53.

23 Denning, D.E. (2001). Chapter eight: Activism, hacktivism, and cyberterrorism: The internet as a tool for influencing foreign policy. In J. Arquilla and D. Ronfeldt (Eds.), *Networks and netwars: The future of terror, crime, and militancy.* Rand Corporation Monograph/Report MR1382. Retrieved from http://www.rand.org/pubs/monograph_reports/MR1382/index.html

24 Kshetri, N. (2005). Pattern of global cyber war and crime: A conceptual framework. *Journal of International Management, 11*(4), 541–62.

25 Bridis, T. (2001, June 27). E-espionage rekindles cold-war tensions – US tries to identify hackers; millions of documents are stolen. *The Wall Street Journal*, A18.

26 *ABC News*. (2008, January 28). Hackers hit Scientology with online attack. Retrieved from https://abcnews.go.com/Technology/GadgetGuide/story?id=4194143

27 Shepardson. D. (2018, May 15). *Homeland Security unveils new cyber security strategy amid threats.* Reuters. Retrieved from https://www.reuters.com/article/us-usa-cyber/homeland-security-unveils-new-cyber-security-strategy-amid-threats-idUSKCN1IG2L9

28 Sands, G. (2016). What to know about the worldwide hacker group "Anonymous." *ABC News.* Retrieved from https://abcnews.go.com/US/worldwide-hacker-group-anonymous/story?id=37761302

29 Shubert, A. (2003, February 6). *Taking a swipe at cyber card fraud.* CNN. Retrieved from http://www.cnn.com/2003/WORLD/asiapcf/southeast/02/06/indonesia.fraud/

30 Antariksa (2001, July). I am a thief, not a hacker: Indonesia's electronic underground. *Latitudes Magazine*, 12–17.

31 Busy signals. (2005, September 8). *The Economist, 376*(8443), 66.

32 Fest, G. (2005). Offshoring: Feds take fresh look at India BPOs; major theft has raised more than a few eyebrows. *Bank Technology News, 18*(9), 1.

33 Kaka, N.F. (2006, May). Running a customer service center in India: An interview with the head of operations for Dell India. *Mckinsey Quarterly.* Retrieved from http://www.ceoexpress.com/asp/mckinseyalls4.asp?id=m0382

34 Sonn, J. (2020, March 19). *Coronavirus: South Korea's success in controlling disease is due to its acceptance of surveillance.* The Conversation. Retrieved from https://theconversation.com/coronavirus-south-koreas-success-in-controlling-disease-is-due-to-its-acceptance-of-surveillance-134068

35 Joshi, S. (2013, June 19). *An IT superpower, India has just 556 cyber security experts.* The Hindu. Retrieved from http://www.thehindu.com/news/national/an-it-superpower-india-has-just-556-cyber-security-experts/article4827644.ece

36 newpaper.asia1.com. (2004, August 3). *Hackers – the new breed of gangsters.* Retrieved May 4, 2005 from http://newpaper.asia1.com.sg/top/story/0,4136,69503-1-1098892740,00.html

37 Walker. (2004). *Russian Mafia extorts gambling websites.*

38 McCarthy, N. (2017). The countries with the most STEM graduates [Infographic]. *Forbes.* Retrieved from https://www.forbes.com/sites/niallmccarthy/2017/02/02/the-countries-with-the-most-stem-graduates-infographic/#7257558c268a

39 Xinhua. (2019, September 9). *China to lead global cybersecurity market growth in next 5 years.* XinhuaNet. Retrieved from http://www.xinhuanet.com/english/2019-09/09/c_138377152.htm

40 *Research: Russian cybersecurity market could grow by 10% in 2019.* (2019, July 19). Tech Driven Information. Retrieved from https://blockchainjournal.news/es/research-russian-cybersecurity-market-could-grow-by-10-in-2019/

41 Dignan, L. (2019, March 20). *Global security spending to top $103 billion in 2019, says IDC.* ZDNet. Retrieved from https://www.zdnet.com/article/global-security-spending-to-top-103-billion-in-2019-says-idc/

42 IDC. (2019). *New IDC spending guide sees commercial sector investments leading worldwide ICT spending to $4.8 trillion in 2023.* Businesswire. Retrieved from https://www.businesswire.com/news/home/20190815005073/en/New-IDC-Spending-Guide-Sees-Commercial-Sector-Investments-Leading-Worldwide-ICT-Spending-to-4.8-Trillion-in-2023

43 ET Bureau. (2020, January 7). ICT spending in India will reach $144 billion in 2023: GlobalData. *The Economic Times.* Retrieved from https://economictimes.indiatimes.com/tech/ites/ict-spending-in-india-will-reach-144-billion-in-2023-globaldata/articleshow/73138501.cms

44 IDC. (2020). *ICT spending growth in Central and Eastern Europe in 2020 expected to fall short of previous year, according to IDC.* Retrieved from https://www.idc.com/getdoc.jsp?containerId=prEUR245975320

45 Statista Research Department. (2020). *Global technology market spending by region from 2014 to 2019.* Statista. Retrieved from https://www.statista.com/statistics/886413/tech-spending-worldwide-by-region/

46 Moran, T., Kapetaneas, J., & Effron, L. (2019, January 3). A journey through "Hackerville," Romanian city with a reputation as a criminal hacker breeding ground. *ABC News.* Retrieved from https://abcnews.go.com/International/journey-hackerville-romanian-city-reputation-criminal-hacker-breeding/story?id=60123285

47 Bitdefender. (2018, September 11). *Bitdefender on growth trajectory with strategic investments in second half 2018.* PRNewswire. Retrieved from https://www.prnewswire.com/news-releases/bitdefender-on-growth-trajectory-with-strategic-investments-in-second-half-2018-808024865.html

48 Bitdefender. (2018). *Bitdefender on growth trajectory with strategic investments in second half 2018.*

49 Barberá, M. (2020, March 27). *Romania: From "hackerville" to cybersecurity powerhouse.* BalkanInsight. Retrieved from https://balkaninsight.com/2020/03/27/romania-from-hackerville-to-cybersecurity-powerhouse/

50 Granato, A., & Polacek, A. (2019). *The growth and challenges of cyber insurance.* Federal Reserve Bank of Chicago. Retrieved from https://www.chicagofed.org/publications/chicago-fed-letter/2019/426

51 Granato & Polacek. (2019). *The growth and challenges of cyber insurance.*

52 Mondelez Intl. Inc. v. Zurich Am. Ins. Co., No. 2018-L-11008, 2018 WL 4941760 (Ill. Cir. Ct., Cook Cty., complaint filed October 10, 2018).

53 McCarthy, K. (2019). *Cyber-insurance shock: Zurich refuses to foot NotPetya ransomware clean-up bill – and claims it's "an act of war."* The Register. Retrieved from https://www.theregister.co.uk/2019/01/11/notpetya_insurance_claim/

54 Ferland, J. (2019). Cyber insurance – what coverage in case of an alleged act of war? Questions raised by the Mondelez v. Zurich case. *Computer Law & Society Review, 35*, 369–76.

55 Standard & Poor's Financial Services. (2021, March 2). *Cyber risk in a new era: Let's not be quiet about insurers' exposure to silent cyber*. Retrieved from https://www.allnews.ch/ckfinder/userfiles/files/20210302_SP_Global_Ratings_Insurers_Cyber_Risk_Silent_Exposure.pdf

56 Aon Benfield. (2017). *Global cyber market overview: Uncovering the hidden opportunities*. Retrieved from https://www.aon.com/inpoint/bin/pdfs/white-papers/Cyber.pdf

57 Below W., & Wolfrom, L. (2018, February 21). *The cyber insurance market: Responding to a risk with few boundaries*. The Forum Network. Retrieved from https://www.oecd-forum.org/users/85359-william-below-and-leigh-wolfrom/posts/30529-the-cyber-insurance-market-responding-to-a-risk-with-few-boundaries

58 Ralph. (2019). Data hacks and big fines drive cyber insurance growth.

59 Insikt Group. (2019, August 21). *Return to normalcy: False flags and the decline of international hacktivism*. RecordedFuture. Retrieved from https://www.recordedfuture.com/international-hacktivism-analysis/

60 Renaud, K. (2016). How smaller businesses struggle with security advice. *Computer Fraud and Security, 8*, 10–18. Retrieved from https://doi.org/10.1016/S1361-3723(16)30062-8

61 Co, J. (2017, July 31). *Small and medium-sized businesses are not investing in cyber protection despite spate of attacks*. Independent. Retrieved from https://tinyurl.com/y8b8pb9n

62 Hinkelman, M. (2014, October 15). For small biz, cybercrime can be deadly. *The Philadelphia Inquirer*. Retrieved from https://www.inquirer.com/philly/business/20141015_For_small_biz__cybercrime_can_be_deadly.html

63 Tmcnet News. (2011, November 11). *GFI labs reports on cybercriminals exploiting search engine ads and user inexperience*. Retrieved from http://www.tmcnet.com/usubmit/2011/11/11/5922728.htm

64 Anderson, E. (2015, February 24). *SMEs failing to guard against cyberattacks, government warns*. The Telegraph. Retrieved from https://tinyurl.com/prwcvnx

65 Perera, D., Mershon, E., Warmbrodt, Z., Alexander, A., & Waterman, S. (2014, October 23). *N.Y. financial chief eyes cybersecurity*. Politico. Retrieved from https://tinyurl.com/ycaaunnr

66 *New ₺5000 government grant for small businesses to boost cyber security*. (2015, July 16). Gov.uk. Retrieved form https://tinyurl.com/pv3t5kg

67 Leydon, J. (2015, November 3). *UK SMEs with weak security risk procurement exclusion – survey*. The Register. Retrieved from http://www.theregister.co.uk/2015/11/03/uk_sme_weak_security_procurement_exclusion/

68 *US organizations not battle ready in war against cybercrime*. (2014, May 29). TechMonitor. Retrieved from https://tinyurl.com/y9kjckg7

69 Co. (2017). Small and medium-sized businesses are not investing in cyber protection despite spate of attacks.

70 Barth, B. (2019, October 7). *Data on 92M Brazilians found for sale on underground forums*. SCMedia. Retrieved from https://www.scmagazine.com/home/security-news/data-breach/data-on-92m-brazilians-found-for-sale-on-underground-forums/

71 Mari, A. (2018, January 24). *More than half of connected Brazilians suffered cyberattacks*. ZD Net. Retrieved from https://www.zdnet.com/article/more-than-half-of-connected-brazilians-suffered-cyberattacks/

72 BNAmericas. (2019, September 6). *Brazil watch: Cable costs, cyber losses*. Retrieved from https://www.bnamericas.com/en/news/brazil-watch-cable-costs-cyber-losses

73 Mari, A. (2019, June 27). *Brazil leads in ransomware attacks*. ZD Net. Retrieved from https://www.zdnet.com/article/brazil-leads-in-ransomware-attacks/

74 BNAmericas. (2020, January 6). *Why is Brazil so vulnerable to cyber attacks?* Retrieved from https://www.bnamericas.com/en/features/why-is-brazil-so-vulnerable-to-cyber-attacks

75 Accenture Security. (2019). *The cost of cybercrime*. Accenture. Retrieved from https://www.accenture.com/_acnmedia/pdf-96/accenture-2019-cost-of-cybercrime-study-final.pdf

76 Muggah, R., & Thompson, N. (2018, April 10). *Brazil's critical infrastructure faces a growing risk of cyberattacks*. Council on Foreign Relations. Retrieved from https://www.cfr.org/blog/brazils-critical-infrastructure-faces-growing-risk-cyberattacks

77 Schreiber, C. (2018, September 10). *Cybersecurity challenges for Latin America*. Grupo de Estudios en Seguirdad Internacional. Retrieved from https://www.seguridadinternacional.es/?q=es/content/cybersecurity-challenges-latin-america

78 Mari, A. (2020, February 11). *Brazil launches cybersecurity strategy*. ZDNet. Retrieved from https://www.zdnet.com/article/brazil-launches-cybersecurity-strategy/

79 Mari, A. (2019, November 8). *Brazilian government announces creation of AI lab network*. ZDNet. Retrieved from https://www.zdnet.com/article/brazilian-government-announces-creation-of-ai-lab-network/

80 Henriques, B. (2020, January 12). *Brazil is emerging as a world-class AI innovation hub*. VentureBeat. Retrieved from https://venturebeat.com/2020/01/12/brazil-is-emerging-as-a-world-class-ai-innovation-hub/

81 Compliance Point, & Dumiak, M. (2020, September 21). *Brazil's LGPD data privacy regulation now in effect*. JDSupra. Retrieved from https://www.jdsupra.com/legalnews/brazil-s-lgpd-data-privacy-regulation-63190/

82 Baptistaluz. (2017). *Proteção de dados a legislação vigente no Brasil* [White-paper]. Retrieved from http://baptistaluz.com.br/wp-content/uploads/2017/11/Privacy-Hub-Leis-Setoriais.pdf

83 Monteiro, R. (2018, August 15). *The new Brazilian General Data Protection Law – a detailed analysis*. International Association of Privacy Professionals. Retrieved from https://iapp.org/news/a/the-new-brazilian-general-data-protection-law-a-detailed-analysis/

84 Woods, D. (2018, November 27). Brazil's new president and the changing cyber risk landscape. *Forbes*. Retrieved from https://www.forbes.com/sites/riskmap/2018/11/27/brazils-new-president-and-the-changing-cyber-risk-landscape/#59f548e05453

85 Ikeda, S. (2019, October 10). Citizen data of 92 million Brazilians offered for sale on underground forum. *CPO Magazine*. Retrieved from https://www.cpomagazine.com/cyber-security/citizen-data-of-92-million-brazilians-offered-for-sale-on-underground-forum/

86 Kujawski, F.F., Rodrigues Brancher, P.M., & Sombra, T.L. (2019). *Data protection and privacy in Brazil*. Lexology. Retrieved from https://www.lexology.com/library/detail.aspx?g=980b7a87-a569-499d-b631-88595d8c1927

87 Muggah, R., & Francisco, P. (2019, December 5). *Brazil's data protection paradox*. Council on Foreign Relations. Retrieved from https://www.cfr.org/blog/brazils-data-protection-paradox

88 Muggah & Francisco. (2019). *Brazil's data protection paradox.*
89 Blickensderfer, S., Swanson, J., & Rego, A.C. Jr. (2019, May 2). Brazil's new data protection law: An overview and four key takeaways for U.S. companies. *The National Law Review, 11*(111). Retrieved from https://www.natlawreview.com/article/brazil-s-new-data-protection-law-overview-and-four-key-takeaways-us-companies
90 Baig, A. (2020, March 19). *Understanding data encryption reuqirements for GDPR, CCPA, LGPD & HIPAA.* Hashedout. Retrieved from https://www.thesslstore.com/blog/understanding-data-encryption-requirements-for-gdpr-ccpa-lgpd-hipaa/
91 Blickensderfer et al. (2019). Brazil's new data protection law.

Chapter 6

1 The White House. (2015). *Five things to know: The administration's priorities on cybersecurity.* Retrieved from https://obamawhitehouse.archives.gov/issues/foreign-policy/cybersecurity
2 Cronk, T.M. (2018, September 21). *White House releases first national cyber strategy in 15 years.* US Department of Defense. https://www.defense.gov/Explore/News/Article/Article/1641969/white-house-releases-first-national-cyber-strategy-in-15-years/
3 Kshetri, N. (2014). Japan's changing cybersecurity landscape. *IEEE Computer, 47*(1), 83–6.
4 Glosserman, B. (2017, May 18). *The ARF moves forward on cybersecurity.* CSS Blog Network. Retrieved from https://isnblog.ethz.ch/security/the-arf-moves-forward-on-cybersecurity
5 Steed, D. (2020, March 9). *The future of values in cyber security strategies.* CSS Blog Network. Retrieved from https://isnblog.ethz.ch/cyber/the-future-of-values-in-cyber-security-strategies
6 National Cyber Security Centre Netherlands (CSC-NL). (2018, April 20). *National cyber security agenda: A cyber secure Netherlands.* Retrieved from https://english.ncsc.nl/publications/publications/2019/juni/01/national-cyber-security-agenda
7 International Telecommunication Union. (n.d.). *National strategies.* Retrieved from https://www.itu.int/en/ITU-D/Cybersecurity/Pages/cybersecurity-national-strategies.aspx
8 ITU. (2018). *Global cybersecurity index 2018.* ITU Publications. Retrieved from https://www.itu.int/dms_pub/itu-d/opb/str/D-STR-GCI.01-2018-PDF-E.pdf
9 Heinl, C. (2018). *Can ASEAN continue to improve cybersecurity in the region and beyond?* Council on Foreign Relations. Retrieved from https://www.cfr.org/blog/can-asean-continue-improve-cybersecurity-region-and-beyond
10 Bracewell LLP. (2014, October 30). *10 million more reasons to pay attention.* Lexology. Retrieved from https://www.lexology.com/library/detail.aspx?g=5875291a-e556-4b75-9ded-3a317bce4671
11 *Oversight of executive order 13636 and development of the cybersecurity framework.* (2013, July 18). National Institute of Standards and Technology. Retrieved from https://www.nist.gov/speech-testimony/oversight-executive-order-13636-and-development-cybersecurity-framework
12 Gillis, A. (n.d.). *NIST cybersecurity framework.* Tech Target SearchSecurity. Retrieved from https://searchsecurity.techtarget.com/definition/NIST-Cybersecurity-Framework?src=6041122&asrc=EM_ERU_124194809&utm_content=eru-rd2-rcpE&utm_medium=EM&utm_source=ERU&utm_campaign=20200228_ERU%20Transmission%20for%2002/28/2020%20(UserUniverse:%20698513
13 *NIST implementation tiers translated into plain English.* (2020, January 27). Praxiom Research Group. Retrieved from https://www.praxiom.com/nist-cybersecurity-tiers.htm

14 *Cybersecurity framework FAQs framework components*. (2015, September 30). National Institute of Standards and Technology. Retrieved from https://www.nist.gov/cyberframework/cybersecurity-framework-faqs-framework-components
15 Murillo, H.K. (2017, May 11). *A summary of the cybersecurity executive order*. Lawfare. Retrieved from https://www.lawfareblog.com/summary-cybersecurity-executive-order
16 Sternstein, A. (2014, September 8). *WH official: Cyber coverage will be a basic insurance policy by 2020*. Nextgov. Retrieved from http://www.nextgov.com/cybersecurity/2014/09/wh-official-cyber-coverage-will-be-basic-insurance-policy-2020/93503/
17 Miliard, M. (2014). *FDA offers final guidance for medical device cybersecurity*. Retrieved from https://www.healthcareitnews.com/news/fda-offers-final-guidance-medical-device-cybersecurity
18 Peterson, A. (2014, September 23). The FDA wants to talk about medical device cybersecurity. *The Washington Post*. Retrieved from http://www.washingtonpost.com/blogs/the-switch/wp/2014/09/23/the-fda-wants-to-talk-about-medical-device-cybersecurity/
19 Finkle, V., & Johnson, C. (2014). Ask regulators for details on cybersecurity measures. *American Banker, 179*(203).
20 Finkle & Johnson. (2014). Ask regulators for details on cybersecurity measures.
21 Corbin, K. (2014, October 15). *SEC examiners zeroing in on cybersecurity*. Retrieved from https://www.financial-planning.com/news/sec-examiners-zeroing-in-on-cybersecurity
22 *Cybersecurity checklist*. (n.d.). FINRA. Retrieved from https://www.finra.org/compliance-tools/cybersecurity-checklist
23 Arampatzis, A. (2018, October 18). *US national cyber strategy: What you need to know*. Tripwire. Retrieved from https://www.tripwire.com/state-of-security/government/us-cyber-strategy/
24 Arampatzis. (2018). *US national cyber strategy*.
25 Arampatzis. (2018). *US national cyber strategy*.
26 European Commission, High Representative of the European Union for Foreign Affairs and Security Policy. (2013, February 7). *Cybersecurity strategy of the European Union: An open, safe and secure cyberspace*. Retrieved from http://tinyurl.com/cdejw3a
27 European Commission. (2013). *Cybersecurity strategy of the European Union*.
28 Parliament of Romania, Chamber of Deputies, Committee for information technologies and communications. (2012, September 24). *The reform of the EU data protection framework – Building trust in a digital and global world*. Retrieved from http://tinyurl.com/bf5u6bd
29 Stassen, M. (2019). *The EU Cybersecurity Act: Addressing the risks of a connected Europe*. Crowell & Moring. Retrieved from https://www.retailconsumerproductslaw.com/2019/08/the-eu-cybersecurity-act-addressing-the-risks-of-a-connected-europe/
30 Day, J., Haas, O., Hladjk, J., Little, J., von Diemar, U. (2019, July 1). *The EU Cybersecurity Act is now applicable*. JDSupra. Retrieved from https://www.jdsupra.com/legalnews/the-eu-cybersecurity-act-is-now-68811/
31 *The EU Cybersecurity Act*. (2021, March 26). European Commission. Retrieved from https://ec.europa.eu/digital-single-market/en/eu-cybersecurity-act
32 *The EU Cybersecurity Act*. (2021).
33 Caron, C., & Morise, A. (2020, January 14). *European Union: The New EU Cybersecurity Act: One step closer to a more secure future*. Mondaq. Retrieved from https://www.mondaq.com/france/security/883102/the-new-eu-cybersecurity-act-one-step-closer-to-a-more-secure-future
34 *The EU Cybersecurity Act*. (2021).

35 Nicolás, E. (2020, January 23). *EU to unveil 5G "toolbox" to tackle security threats*. EU Observer. Retrieved from https://euobserver.com/science/147238
36 *Secure 5G networks: Questions and answers on the EU toolbox*. (2020, January 29). European Commission. Retrieved from https://ec.europa.eu/commission/presscorner/detail/en/QANDA_20_127
37 Hunton Andrews Kurth. (2011, February 16). *Update: Privacy and the protection of personal information in China*. Retrieved from www.huntonprivacyblog.com/2011/02/articles/update-privacy-and-the-protection-of-personal-information-in-china
38 Maranto, L. (2020, June 25). *Who benefits from China's cybersecurity laws?* Center for Strategic & International Studies. Retrieved from https://www.csis.org/blogs/new-perspectives-asia/who-benefits-chinas-cybersecurity-laws#:~:text=In%20June%202017%2C%20the%20China,for%20China's%20present%20day%20guidelines.&text=The%20law%20requires%20that%20data,to%20government%2Dconducted%20security%20checks
39 Cimpanu, C. (2019, February 9). *China's cybersecurity law update lets state agencies "pen-test" local companies*. ZDNet. Retrieved from https://www.zdnet.com/article/chinas-cybersecurity-law-update-lets-state-agencies-pen-test-local-companies/
40 Chandrasekaran, S. (2019, October 31). *China's cybersecurity future and its impact on US business*. Jolt Digest. Retrieved from https://jolt.law.harvard.edu/digest/chinas-cybersecurity-future-and-its-impact-on-u-s-business#:~:text=In%202017%2C%20the%20Chinese%20government,more%20directly%20monitor%20corporate%20networks.&text=MLPS%202.0%20allows%20the%20government,at%20any%20time%20without%20permission
41 Swinhoe, D. (2019, October 28). *China's MLPS 2.0: Data grab or legitimate attempt to improve domestic cybersecurity?* CSO. Retrieved from https://www.csoonline.com/article/3448578/chinas-mlps-20-data-grab-or-legitimate-attempt-to-improve-domestic-cybersecurity.html
42 Zhang, T. (2020, October 21). *How to manage IT compliance requirements in China*. Dezan Shira & Associates. Retrieved from https://www.china-briefing.com/news/it-compliance-in-china-how-to-manage-requirements/
43 Uchill, J. (2019, October 17). *China's upgraded cybersecurity law could take a toll*. Axios. Retrieved from https://www.axios.com/china-cybersecurity-law-upgrade-938714cd-de65-4a71-a7c2-03da33c51bc9.html
44 Uchill. (2019). China's upgraded cybersecurity law could take a toll.
45 Ji, H., & Fang, J. (2017, January 24). *Costs and unanswered questions of China's new cybersecurity regime*. International Association of Privacy Professionals. Retrieved from https://iapp.org/news/a/costs-and-unanswered-questions-of-chinas-new-cybersecurity-regime/
46 Beedham, M. (2019, April 2). *Alibaba, Baidu, and Tencent are among the first 197 regulated blockchain firms in China*. The Next Web. Retrieved from https://thenextweb.com/hardfork/2019/04/02/china-cyberspace-admin-regulates-blockchain-firms/
47 *Cyber administration releases first list of registered blockchain service providers*. (2019, April 2). Global Times. Retrieved from http://www.globaltimes.cn/content/1144347.shtml
48 Barboza, D., & Markoff, J. (2011, December 5). Power in numbers: China aims for high-tech primacy. *The New York Times*. Retrieved from http://www.nytimes.com/2011/12/06/science/china-scrambles-for-high-tech-dominance.html?pagewanted=all&_r=0
49 *The Economist*. (2013). Masters of the cyber-universe. Retrieved from https://www.economist.com/special-report/2013/04/06/masters-of-the-cyber-universe

50 Levite, A., & Jinghua, L. (2019, January 24). *Chinese-American relations in cyberspace: Toward collaboration or confrontation?* Carnegie Endowment for International Peace. Retrieved from https://carnegieendowment.org/2019/01/24/chinese-american-relations-in-cyberspace-toward-collaboration-or-confrontation-pub-78213

51 Aneja, A. (2019, June 29). *At G20 trilateral, China proposes 5G partnership with India, Russia.* The Hindu. Retrieved from https://www.thehindu.com/news/international/at-g-20-trilateral-china-proposes-5g-partnership-with-india-russia/article28226759.ece

52 Betts, W. (1999). Third party mediation: An obstacle to peace in Nagorno Karabakh. *SAIS Review, 19*(2), 161–83.

53 Zartman, I.W., & Touval, S. (1996). International mediation in the post-cold war era. In C.A. Chester, F.O. Hampson, & P. Aall (Eds.), *Managing global chaos: Sources of and responses to international conflict* (pp. 445–61). United States Institute of Peace.

54 Sanger, D.E. (2014, May 20). Fine line seen in US spying on companies. *The New York Times*. Retrieved from http://www.nytimes.com/2014/05/21/business/us-snooping-on-companies-cited-by-china.html?_r=0

55 Guillén, M.F., & Suárez, S.L. (2005). Explaining the global digital divide: Economic, political and sociological drivers of cross-national internet use. *Social Forces, 84*(2), 681–708.

56 King, G., Pan, J., & Roberts, M. (2013). How censorship in China allows government criticism but silences collective expression. *American Political Science Review, 107*(2), 1–18.

57 FlorCruz, M. (2015, April 28). *China to use big data to rate citizens in new "social credit system."* International Business Times. Retrieved from http://www.ibtimes.com/china-use-big-data-rate-citizens-new-social-credit-system-1898711

58 *China rates its own citizens – including online behavior*. (2015, April 25). deVolksrant. Retrieved from http://www.volkskrant.nl/buitenland/china-rates-its-own-citizens-including-online-behaviour~a3979668/

59 Kshetri, N. (2020). China's social credit system: Data, algorithms and implications. *IEEE IT Professional, 22*(2), 14–18.

60 Bandurski, D., Goldkorn, J., Creemers, R., Qiang, X., & Ng, J. (2014, December 3). *Can China conquer the internet?* A ChinaFile Conversation. Retrieved from http://www.chinafile.com/conversation/can-china-conquer-internet

61 Kitaev, E.G. (2013, January 8). *China adopts privacy legislation strengthening online personal data protection*. Data Counsel. Retrieved from http://www.dataprivacymonitor.com/online-privacy/china-adopts-privacy-legislation-strengthening-online-personal-data-protection/

62 Devlin, K. (2019, July 1). Cybersecurity threat looms large in Japan. *The Japan Times*. Retrieved from https://www.japantimes.co.jp/opinion/2019/07/01/commentary/japan-commentary/cybersecurity-threat-looms-large-japan/#.XpDzn8hKjIU

63 McCurry, J. (2018, November 15). System error: Japan cybersecurity minister admits he has never used a computer. *The Guardian*. Retrieved from https://www.theguardian.com/world/2018/nov/15/japan-cyber-security-ministernever-used-computer-yoshitaka-sakurada

64 Tidey, A. (2018, November 16). *Japan's new minister of cybersecurity has never used a computer – ever*. Euronews. Retrieved from https://www.euronews.com/2018/11/16/japan-s-new-minister-of-cybersecurity-has-never-used-a-computer-ever

65 Arie, H. (2017, January 17). *Japan's approach to tackling cybersecurity challenges*. Japan Industry News. Retrieved from https://www.japanindustrynews.com/2017/01/japans-approach-tackling-cybersecurity-challenges/

66 Matsubara, M. (2017, November 15). *Japan's new cybersecurity strategies have the right priorities in mind*. Palo Alto Netowrks. Retrieved from https://researchcenter.paloaltonetworks.com/2017/11/cso-japans-new-cybersecurity-strategies-right-priorities-mind/

67 The National Institute for Defense Studies. (2012). *East Asian strategic review 2012: Chapter 7: Japan: Toward the establishment of a dynamic defense force*. The National Institute for Defense Studies.

68 Kshetri, N. (2013). The expert opinion: An interview with Dr. Taro Komukai, executive director and senior consultant, Infocom Research, Japan. *Journal of Global Information Technology Management*, *16*(4), 68–71.

69 Akimoto, D. (2020, March 18). *Japan's emerging "multi-domain defense force."* The Diplomat. Retrieved from https://thediplomat.com/2020/03/japans-emerging-multi-domain-defense-force/

70 *Japan-United States of America relations: Japan-U.S. security consultative committee*. (2019, April 19). Ministry of Foreign Affairs of Japan. Retrieved from https://www.mofa.go.jp/na/fa/page3e_001008.html

71 Cheah, G.G. (2011, October). *Protection of personal financial information in China*. Retrieved May 5, 2012 from http://www.nortonrosefulbright.com/knowledge/publications/56148/protection-of-personal-financial-information-in-china

72 Cheah. (2011). *Protection of personal financial information in China*.

73 Ferris, R.J., Lacktman, N.M., & Tianran, Y. (2014). *Realizing the potential of telemedicine in China, part 2: Data privacy and security*. Health Care Law Today. Lexology. Retrieved from http://www.lexology.com/library/detail.aspx?g=c8c9aa69-4efd-4686-a524-f52b98d98fc8

74 Zhu, G. (1997). The right to privacy: An emerging right in Chinese law. *Statute Law Review 18*(3), 208–14.

75 O'Donoghue, C., & Kimoto, T. (2013). *Japan promotes the use of big data*. Lexology. Retrieved from http://tinyurl.com/kevjxa6

76 Chavez-Dreyfuss, G. (2019, January 17). *Venture capital funding of cybersecurity firms hit record high in 2018: Report*. Reuters. Retrieved from https://www.reuters.com/article/us-usa-cyber-investment/venture-capital-funding-of-cybersecurity-firms-hit-record-high-in-2018-report-idUSKCN1PB163

77 Leitersdorf, Y., & Schreiber, O. (2019, January 6). *A look back at the Israeli cyber security industry in 2018*. TechCrunch. Retrieved from https://techcrunch.com/2019/01/06/a-look-back-at-the-israeli-cyber-security-industry-in/

78 *Cybersecurity startup exits in Israel total to US$11.3 billion: IVC report*. (2020, January 21). CISOmag. Retrieved from https://www.cisomag.com/cybersecurity-startup-exits-in-israel-total-to-us11-3-billion-ivc-report/

79 In a venture capital deal, a startup creates large ownership chunks and sells to a few investors through limited partnership. The VC investors are limited partners who do not take part in managing the business. Partnerships consist of a pool of several similar enterprises.

80 Kram, J., & Voci, V. (2019, June 28). *US-Israel cybersecurity collaborative: A roadmap for global private, public partnership*. US Chamber of Commerce. Retrieved from https://www.uschamber.com/series/above-the-fold/us-israel-cybersecurity-collaborative-roadmap-global-private-public

81 Friedman, S. (2017, May 16). *What's next for NIST cybersecurity framework?* GCN. Retrieved from https://gcn.com/articles/2017/05/16/nist-cybersecurity-framework.aspx

82 Housen-Couriel, D. (2018, July 2). *A look at Israel's new draft cybersecurity law*. Council on Foreign Relations. Retrieved from https://www.cfr.org/blog/look-israels-new-draft-cybersecurity-law

83 Dvorin, T. (2014, September 21). *Israel launches national cyber-defense authority*. Arutz Sheva 7: Israel National News. Retrieved from www.israelnationalnews.com/News/News.aspx/185349#.VCm0pVfNJmc

84 Ramnani, M. (2019, June 28). *Look out for Israeli technology as the country is turning into a cybersecurity hub*. Enterprise Talk. Retrieved from https://enterprisetalk.com/featured/look-out-for-israeli-technology-as-the-country-is-turning-into-a-cybersecurity-hub/
85 *What India can learn from Israel, EU, Singapore and US cybersecurity strategy?* (2019, December 20). Retrieved from https://medium.com/@cyberdiplomacy/what-india-can-learn-from-israel-eu-singapore-and-us-cybersecurity-strategy-ba177832fca3
86 Solomon, S. (2019, June 25). *Israel cybersecurity sector hamstrung by shortage of labor, report says*. The Times of Israel. Retrieved from https://www.timesofisrael.com/israel-cybersecurity-sector-hamstrung-by-shortage-of-labor-report-says/
87 Sun, L. (2018, January 26). *How to invest in Israel's top cybersecurity stocks*. The Motley Fool. Retrieved from https://www.fool.com/investing/2018/01/26/how-to-invest-in-israels-top-cybersecurity-stocks.aspx
88 Ramnani. (2019). Look out for Israeli technology as the country is turning into a cybersecurity hub.
89 Solomon, S. (2018, April 15). *Israel wins second-largest number of cybersecurity deals globally*. The Times of Israel. Retrieved from www.timesofisrael.com/israel-nabs-second-largest-number-of-cybersecurity-deals-globally/
90 Miller, Z. (2014, January 29). 7 Israeli security companies who could help Putin protect the Sochi Olympics. *Forbes*. Retrieved from www.forbes.com/sites/zackmiller/2014/01/29/7-israeli-security-companies-who-could-help-putin-protect-the-sochi-olympics/
91 Leichman, A.K. (2013, November 3). *Israel is the go-to address for cyber-security*. Israel 21c. Retrieved from www.israel21c.org/technology/israel-is-the-go-to-address-for-cyber-security/
92 Solomon. (2019). *Israel cybersecurity sector hamstrung by shortage of labor, report says*.
93 Ramnani. (2019). Look out for Israeli technology as the country is turning into a cybersecurity hub.
94 Moskowitz, J. (2014, December 5). Cybersecurity unit drives Israeli Internet economy. *The Christian Science Monitor*. Retrieved from www.csmonitor.com/World/Passcode/2014/1205/Cybersecurity-unit-drives-Israeli-Internet-economy
95 Barnes, J. (2014, February 23). Israel utilises its cyber security expertise. *Financial Times*. Retrieved from www.ft.com/intl/cms/s/0/8b6e572c-97e7-11e3-8dc3-00144feab7de.html#axzz3EnpyGV6S

Chapter 7

1 Meltzer, J., & Kerry, C. (2019, September 18). *Cybersecurity and digital trade: Getting it right*. Brookings. https://www.brookings.edu/research/cybersecurity-and-digital-trade-getting-it-right/
2 European Union. (2016, July 19). *Directive (EU) 2016/1148 of the European Parliament and of the Council of 6 July 2016 concerning measures for a high common level of security of network and information systems across the Union*. Retrieved from https://eur-lex.europa.eu/legal-content/EN/TXT/?toc=OJ:L:2016:194:TOC&uri=uriserv:OJ.L_.2016.194.01.0001.01.ENG
3 Meltzer & Kerry. (2019). *Cybersecurity and digital trade.*
4 Ruhl, C., Hollis, D., Hoffman, W., & Maurer, T. (2020, February 26). *Cyberspace and geopolitics: Assessing global cybersecurity norm processes at a crossroads.* Carnegie Endowment for International Peace. Retrieved from https://carnegieendowment.org/2020/02/26/cyberspace-and-geopolitics-assessing-global-cybersecurity-norm-processes-at-crossroads-pub-81110
5 Radu, S. (2019, February 1). *China, Russia biggest cyber offenders*. US News. Retrieved from https://www.usnews.com/news/best-countries/articles/2019-02-01/china-and-russia-biggest-cyber-offenders-since-2006-report-shows

6 Kshetri, N. (2014). Cyberwarfare in the Korean Peninsula: Asymmetries and strategic responses. *East Asia, 31*, 183–201.
7 Swan, B. (2020, April 21). *State report: Russian, Chinese and Iranian disinformation narratives echo one another.* Politico. Retrieved from https://www.politico.com/news/2020/04/21/russia-china-iran-disinformation-coronavirus-state-department-193107
8 Brewster, T. (2020, April 15). Iran-linked group caught spreading COVID-19 "disinformation" on Facebook and Instagram. *Forbes*. Retrieved from https://www.forbes.com/sites/thomasbrewster/2020/04/15/iran-linked-group-caught-spreading-covid-19-disinformation-on-facebook-and-instagram/#5c7a22a21f21
9 Swan. (2020). *State report.*
10 Bilyeau, N. (2020, April 20). *Cybersabotage during crisis earns cash for N Korea, Russia, Iran: Experts.* The Crime Report. Retrieved from https://thecrimereport.org/2020/04/20/cybersabotage-during-crisis-earns-cash-for-n-korea-russia-iran-experts/
11 Brantly, A. (2020, April 8). *The cybersecurity of health.* Council on Foreign Relations. Retrieved from https://www.cfr.org/blog/cybersecurity-health
12 Rahn, W. (2019, August 5). *North Korea cyberattacks generate $2 billion for weapons program.* DW. Retrieved from https://www.dw.com/en/north-korea-cyberattacks-generate-2-billion-for-weapons-program/a-49903673
13 Goldsmith, J. (2011). *Cybersecurity treaties: A skeptical view.* Retrieved from https://media.hoover.org/sites/default/files/documents/FutureChallenges_Goldsmith.pdf
14 Owens, W.A., Dam, K.D., & Lin, H.S. (Eds.). (2009). *Technology, policy, law, and ethics regarding US acquisition and use of cyberattack capabilities.* National Academies Press.
15 Pi, Y.I. (2011). *New China criminal legislations in the progress of harmonization of criminal legislation against cybercrime.* Council of Europe. Retrieved May 5, 2012 from http://www.coe.int/t/dghl/cooperation/economiccrime/cybercrime/documents/countryprofiles/Cyber_cp_china_Pi_Yong_Dec11.pdf
16 Zhuangzhi, S. (2018, June 5). *A new SCO "milestone" in the making.* China Daily. Retrieved from http://www.chinadaily.com.cn/a/201806/05/WS5b15d53aa31001b82571e1f7.html
17 cyberBRICS. (n.d.). Retrieved from https://cyberbrics.info/
18 Belli, L. (2019, November 18). *BRICS countries to build digital sovereignty.* openDemocracy. Retrieved from https://www.opendemocracy.net/en/hri-2/brics-countries-build-digital-sovereignty/
19 The North Atlantic Treaty Organization. (2008, May 14). *NATO opens new centre of excellence on cyber defence.* Retrieved from https://www.nato.int/docu/update/2008/05-may/e0514a.html
20 Korzak, E. (2017, July 31). *UN GGE on cybersecurity: The end of an era?* The Diplomat. Retrieved from https://thediplomat.com/2017/07/un-gge-on-cybersecurity-have-china-and-russia-just-made-cyberspace-less-safe/
21 Iasiello, E. (2019, December 5). *OEWG or GGE – which has the best shot of succeeding?* TechNative. Retrieved from https://www.technative.io/oewg-or-gge-which-has-the-best-shot-of-succeeding/
22 *Declaration by Miguel Rodríguez, representative of Cuba, at the final session of group of governmental experts on developments in the field of information and telecommunications in the context of international security. New York, June 23, 2017.* (2017, June 23). Retrieved from https://www.justsecurity.org/wp-content/uploads/2017/06/Cuban-Expert-Declaration.pdf
23 *UN GGE and OEWG.* (n.d.). Digital Watch. Retrieved from https://dig.watch/processes/un-gge
24 Korzak, E. (2020). *What's ahead in the cyber norms debate?* Lawfare. Retrieved from https://www.lawfareblog.com/whats-ahead-cyber-norms-debate

25 Open-ended working group on developments in the field of informatin and telecommunications in the context of international security. (2021, March 10). *Final substantive report* [Conference room paper]. United Nations General Assembly. Retrieved from https://front.un-arm.org/wp-content/uploads/2021/03/Final-report-A-AC.290-2021-CRP.2.pdf

26 Hakmeh, J. (2018, October 10). *Cyberattack revelations appear to undercut Russia's UN efforts*. Chatham House. Retrieved from https://www.chathamhouse.org/expert/comment/cyberattack-revelations-appear-undercut-russia-un

27 Stolton, S. (2020, January 23). *UN backing of controversial cybercrime treaty raises suspicions*. Euractiv. Retrieved from https://www.euractiv.com/section/digital/news/un-backing-of-controversial-cybercrime-treaty-raises-suspicions/

28 *Open letter to UN General Assembly: Proposed international convention on cybercrime poses a threat to human rights online*. (n.d.). Retrieved from https://www.apc.org/sites/default/files/Open_letter_re_UNGA_cybercrime_resolution_0.pdf

29 Hakmeh, J., & Peters, A. (2020, January 13). *A new UN cybercrime treaty? The way forward for supporters of an open, free, and secure internet*. Council on Foreign Relations. Retrieved from https://www.cfr.org/blog/new-un-cybercrime-treaty-way-forward-supporters-open-free-and-secure-internet

30 Prince, T. (2019, December 20). *US concerned Russia-backed UN resolution will hurt online freedom*. Radio Free Europe. Retrieved from https://www.rferl.org/a/us-russia-internet-un/30335318.html

31 Goldsmith, J. (2011). *Cybersecurity treaties*.

32 Kizekova, A. (2012, February 27). *The Shanghai Cooperation Organization: Challenges in cyberspace – analysis*. Eurasia Review. Retrieved from http://www.eurasiareview.com/27022012-the-shanghai-cooperation-organisation-challenges-in-cyberspace-analysis/

33 *Obama administration reportedly considering fines, trade penalties for cybertheft*. (2013, February 20). FoxNews. Retrieved from http://www.foxnews.com/politics/2013/02/20/obama-administration-developing-penalties-for-cybertheft/

34 Kshetri, N. (2013). Cybercrimes in the former Soviet Union and Central and Eastern Europe: Current status and key drivers. *Crime, Law and Social Change, 60*(1), 39–65.

35 Department of Economic and Social Affairs, United Nations. (n.d.). *Internet Governance Forum: We must act now to tackle the threats of cyberspace*. Retrieved from https://www.un.org/es/desa/internet-governance-forum-we-must-act-now-tackle-threats-cyberspace

36 Azar, E.E., & Moon, C. (1998). *National security in the third world: The management of internal and external threats*. Edward Elgar Publishing.

37 Information colonialism. (1999, February 18–24). *Beijing Scene, 7*(5). Retrieved from http://www.beijingscene.com/v07i005/inshort.html

38 Kizekova. (2012). *The Shanghai Cooperation Organization*.

39 Kenya defence Twitter account hacked. (2014, July 21). *BBC News*. Retrieved from http://www.bbc.com/news/world-africa-28398976

40 Ball. (1988). *Security and economy in the third world*.

41 Brenner, J.F. (2013). Eyes wide shut: The growing threat of cyber attacks on industrial control systems. *Bulletin of the Atomic Scientists, 69*(5), 15–20.

42 Gross, G. (2013, March 20). *Experts: Iran and North Korea are looming cyber-threats to US*. Computerworld. Retrieved from https://www.computerworld.com/article/2495664/experts--iran-and-north-korea-are-looming-cyberthreats-to-u-s-.html

43 Waterman, S. (2012, August 23). North Korean jamming of GPS shows system's weakness. *The Washington Times*. Retrieved from www.washingtontimes.com/news/2012/aug/23/north-korean-jamming-gps-shows-systems-weakness

44 Masters, J. (2011, May 23). *Confronting the cyber threat*. Council on Foreign Relations. Retrieved from https://www.cfr.org/backgrounder/confronting-cyber-threat
45 Lewis, J.A., Cha, V., Jun, J., LaFoy, S., & Sohn, E. (2015, December 30). *North Korea's cyber operations: Strategy and responses*. Center for Strategic and International Studies, Office of the Korea Chair. Retrieved from https://www.csis.org/analysis/north-korea%E2%80%99s-cyber-operations
46 Odobescu, V. (2014, January 13). US data thefts turn spotlight on Romania. *USA Today*. Retrieved from http://www.usatoday.com/story/news/world/2014/01/13/credit-card-hacking-romania/4456491/
47 Wylie. (2007). Romania home base for eBay scammers.
48 Lipson, C. (1991). Why are some international agreements informal? *International Organization, 45*(4), 495–538.
49 Bach, D., & Newman, A.L. (2010). Transgovernmental networks and domestic policy convergence: Evidence from insider trading regulation. *International Organization, 64*(3), 505–28.
50 *The FBI around the world*. (2018, July 11). Department of State. Retrieved from https://2017-2021.state.gov/the-fbi-around-the-world/index.html
51 *FBI to station cybercrime expert in Estonia*. (2009, May 11). NBC News Digital. Retrieved from https://www.nbcnews.com/id/wbna30683801
52 Kshetri. (2013). *Cybercrime and cybersecurity in the Global South*.
53 Kshetri. (2013). *Cybercrime and cybersecurity in the Global South*.
54 Marusic, S. (2020, April 15). *NATO to help North Macedonia combat fake news about virus*. Balkan Insight. Retrieved from https://balkaninsight.com/2020/04/15/nato-to-help-north-macedonia-combat-fake-news-about-virus/
55 *Joint ministerial statement of the ASEAN-Japan ministerial policy meeting on cybersecurity cooperation*. (2013, September 13). Association of Southeast Asian Nations. Retrieved from https://asean.org/joint-ministerial-statement-of-the-asean-japan-ministerial-policy-meeting-on-cybersecurity-cooperation/
56 Heinl. (2018). *Can ASEAN continue to improve cybersecurity in the region and beyond?*
57 *ASEAN regional forum (ARF)*. (n.d.). Australian Government Department of Foreign Affairs and Trade; Retrieved from https://www.dfat.gov.au/international-relations/regional-architecture/Pages/asean-regional-forum-arf
58 McGraw, G. (2012). *Proactive defense prudent alternative to cyberwarfare*. Search Security Tech Target. Retrieved May 5, 2013 from http://searchsecurity.techtarget.com/news/2240169976/Gary-McGraw-Proactive-defense-prudent-alternative-to-cyberwarfare
59 Schafer, S. (2006, January 16). A piracy culture: Beijing continues to defy US and European efforts to stop IP theft. *Newsweek* (International Edition). Retrieved May 5, 2006 from http://www.msnbc.msn.com/id/10756810/site/newsweek/
60 Riley, M. (2012, January 10). SEC push may yield new disclosures of company cyber attacks. *Business Week*. Retrieved from https://www.bloomberg.com/news/articles/2012-01-10/sec-push-may-yield-new-disclosures-of-cyber-attacks-on-companies
61 Mozur, P. (2014, May 29). China questions evidence used in US cybercrimes indictment. *The Wall Street Journal*. Retrieved from http://online.wsj.com/articles/chinas-defense-ministry-questions-evidence-used-by-u-s-to-indict-five-peoples-liberation-army-officers-for-cybercrimes-1401359575
62 Bennett, C. (2017, January 28). *Trump talks tough on China's hacking. Why not Russia's?* Politico. Retrieved from https://www.politico.com/story/2017/01/trump-russia-china-hacking-234301

63 Yingzi, T., & Limin, C. (2012, October 9). *Telecom giants hit back at allegations*. ChinaDaily. Retrieved from http://usa.chinadaily.com.cn/china/2012-10/09/content_15802157.htm
64 Li, X. (2012, June 5). *China is victim of hacking attacks*. English.people.cn. Retrieved from http://english.peopledaily.com.cn/90883/8271052.html
65 Ming, J.W. (2013). *Commentary: Washington owes world explanations over troubling spying accusations*. Xinhuanet. Retrieved from http://usa.chinadaily.com.cn/world/2013-06/23/content_16647575.htm
66 Khan, M. (2014, October 9). *China: US is fabricating cyber attack claims*. International Business Times. Retrieved from http://www.ibtimes.co.uk/china-us-fabricating-cyber-attack-claims-1469302
67 *The Guardian*. (2014). China says US must change "mistaken policies" before deal on cyber security. Retrieved May 5, 2015 from http://www.theguardian.com/world/2014/oct/19/china-cyber-security-cooperation-problematic-mistaken-us-policies
68 Lam, O. (2014, May 28). *China to perform security inspections for tech products*. Advox. Retrieved from http://advocacy.globalvoicesonline.org/2014/05/28/china-to-perform-security-inspections-for-tech-products/
69 Cronan, B. (2014, September 22). China tightens censorship on Google. *The Christian Science Monitor*. Retrieved from https://www.csmonitor.com/Technology/2014/0922/China-tightens-censorship-on-Google
70 Bennett. (2017). *Trump talks tough on China's hacking*.
71 Bennett. (2017). *Trump talks tough on China's hacking*.
72 Department of Defense. (2018). *Cyber strategy*. Retrieved from https://media.defense.gov/2018/Sep/18/2002041658/-1/-1/1/cyber_strategy_summary_final.pdf
73 Jinghua, L. (2018, October 19). *A Chinese perspective on the Pentagon's cyber strategy: From "active cyber defense" to "defending forward."* Lawfare. Retrieved from https://www.lawfareblog.com/chinese-perspective-pentagons-cyber-strategy-active-cyber-defense-defending-forward
74 Buchanan, B., & Williams, R. (2018, October 24). *A deepening US–China cybersecurity dilemma*. Lawfare. Retrieved from https://www.lawfareblog.com/deepening-us-china-cybersecurity-dilemma
75 *US accuses China of violating bilateral anti-hacking deal*. (2018, November 8). Reuters. Retrieved from https://www.reuters.com/article/us-usa-china-cyber/u-s-accuses-china-of-violating-bilateral-anti-hacking-deal-idUSKCN1NE02E
76 *US accuses China of violating bilateral anti-hacking deal*. (2018). Reuters.
77 Reuters. (2020, March 4). *Chinese cybersecurity firm Qihoo 360 accuses CIA of 11-year-long hacking campaign*. South China Morning Post. Retrieved from https://www.scmp.com/news/china/diplomacy/article/3064853/chinese-cybersecurity-firm-qihoo-360-accuses-cia-11-year-long
78 *Japan bans Huawei and ZTE networking hardware*. (2019). IndustryHerald24. Retrieved May 5, 2020 from https://www.industryherald24.com/japan-bans-huawei-and-zte-networking-hardware/
79 *Entity list*. (n.d.). Bureau of Industry and Security. Retrieved from https://www.bis.doc.gov/index.php/policy-guidance/lists-of-parties-of-concern/entity-list
80 Leonard, J., & King, I. (2019, October 24). Five months after Huawei export ban, US companies are confused. *Los Angeles Times*. Retrieved from https://www.latimes.com/business/story/2019-10-24/huawei-export-ban-us-companies-confusion
81 Pham, S. (2019, October 30). *Huawei and ZTE could lose what little business they have in the United States*. CNN. Retrieved from https://www.cnn.com/2019/10/29/tech/fcc-huawei-5g-ajit-pai/index.html

82 Leonard & King. (2019). Five months after Huawei export ban, US companies are confused.
83 Tao, M. (2019, October 29). *ZTE steps up 5G network gear deployment overseas as capital spending in China is set to decline.* South China Morning Post. Retrieved from https://www.scmp.com/tech/gear/article/3035254/zte-steps-5g-network-gear-deployment-overseas-capital-spending-china-set
84 Peel, M. (2019, January 1). EU eyes tougher scrutiny of China cyber security risks. *Financial Times.* Retrieved from https://www.ft.com/content/3d13c208-0545-11e9-99df-6183d3002ee1
85 *Czech cyber watchdog calls Huawei, ZTE products a security threat.* (2018, December 17). Reuters. Retrieved from https://www.reuters.com/article/us-czech-huawei/czech-cyber-watchdog-calls-huawei-zte-products-a-security-threat-idUSKBN1OG1Z3

Chapter 8

1 Yadron. (2014). Target hackers wrote partly in Russian, displayed high skill, report finds.
2 O'Connor, C. (2014, May 5). Target CEO Gregg Steinhafel resigns in data breach fallout. *Forbes.* Retrieved from http://www.forbes.com/sites/clareoconnor/2014/05/05/target-ceo-gregg-steinhafel-resigns-in-wake-of-data-breach-fallout/
3 Norton, S. (2014, June 10). SEC official: Corporate boards need to oversee cybersecurity risk. *The Wall Street Journal.* Retrieved from https://www.wsj.com/articles/BL-CIOB-4703
4 Stanford, D.D. (2014, October 10). *Kmart says card data stolen in latest retail cyber hack.* Bloomberg. Retrieved from http://www.bloomberg.com/news/2014-10-10/sears-s-kmart-says-hackers-stole-payment-card-data-in-attack.html
5 Creswell, J., & Perlroth, N. (2014, September 19). Ex-employee say Home Depot left data vulnerable. *The New York Times.* Retrieved from http://www.nytimes.com/2014/09/20/business/ex-employees-say-home-depot-left-data-vulnerable.html
6 Lehman, S. (2014, September 22). How a Times cybersecurity reporter protects her data, and what you can do to protect yours. *The New York Times.* Retrieved from http://www.nytimes.com/times-insider/2014/09/22/how-a-times-cyber-security-reporter-protects-her-data-and-what-you-can-do-to-protect-yours/
7 Galvin, J. (2018, May 7). *60 percent of small businesses fold within 6 months of a cyber attack. Here's how to protect yourself.* Inc. https://www.inc.com/joe-galvin/60-percent-of-small-businesses-fold-within-6-months-of-a-cyber-attack-heres-how-to-protect-yourself.html Retrieved from https://www.inc.com/joe-galvin/60-percent-of-small-businesses-fold-within-6-months-of-a-cyber-attack-heres-how-to-protect-yourself.html
8 Lardner, R. (2012, July 6). *Cybercrime: Companies reluctant to report breaches despite SEC guidelines.* Las Vegas Review Journal. Retrieved from https://www.reviewjournal.com/business/cybercrime-companies-reluctant-to-report-breaches-despite-sec-guidelines/
9 Vijayan, J. (2014). Target ruling raises stakes for cybersecurity vigilance. *The Christian Science Monitor.* Retrieved from http://www.csmonitor.com/World/Passcode/2014/1209/Target-ruling-raises-stakes-for-cybersecurity-vigilance
10 Sattar, E. (2020, February 14). Is your cybersecurity workforce ready to win against cybercriminals? *Forbes.* Retrieved from https://www.forbes.com/sites/forbesbusinesscouncil/2020/02/14/is-your-cybersecurity-workforce-ready-to-win-against-cybercriminals/#56e549c13c07
11 *Select Texas House of Representatives interim report 2018: A report to the House of Representatives 86th Texas Legislature.* (2019). House Select Committee on Cybersecurity.

12 *What is cyber security?* (n.d.). Kaspersky. Retrieved from https://www.kaspersky.com/resource-center/definitions/what-is-cyber-security

13 Olcott, J. (2019, February 13). *6 cybersecurity KPI examples for your next report*. BitSight. Retrieved from https://www.bitsight.com/blog/6-cybersecurity-kpis-examples-for-your-next-report

14 Rose, Z. (2020, April 20). *Building effective cybersecurity budgets*. Tripwire. Retrieved from https://www.tripwire.com/state-of-security/featured/building-effective-cybersecurity-budgets/

15 National Conference of State Legislatures. (n.d.). *Budgeting for cybersecurity*. Retrieved from https://www.ncsl.org/documents/taskforces/Budgeting_For_Cybersecurity_32041.pdf

16 Moore, S. (2016, December 9). *Gartner says many organizations falsely equate IT security spending with maturity*. Gartner. Retrieved from https://www.gartner.com/en/newsroom/press-releases/2016-12-09-gartner-says-many-organizations-falsely-equate-it-security-spending-with-maturity

17 Faile, C., & Hockin, N. (2019, May 3). *Just how much are financial institutions spending on cybersecurity? An average of about $2,300 per employee, Deloitte survey finds*. Deloitte. Retrieved from https://www2.deloitte.com/us/en/pages/about-deloitte/articles/press-releases/deloitte-fsisac-survey-press-release.html

18 Gupta, U. (2011, January 13). *On-the-job cybersecurity training*. Bank Info Security. Retrieved from http://www.bankinfosecurity.com/on-the-job-cybersecurity-training-a-3259

19 ENISA. (2012). *Introduction to return on security investment helping CERTs assessing the cost of (lack of) security* [Deliverable – December 2012]. Retrieved from https://www.enisa.europa.eu/publications/introduction-to-return-on-security-investment

20 Bissell, K., LaSalle, R., & dal Cin, P. (2020, January 28). *Lessons from leaders to master cybersecurity execution*. Accenture. Retrieved from https://www.accenture.com/us-en/insights/security/invest-cyber-resilience

21 Finch, B. (2014, December 11). Why cybersecurity must be defined by process, not tech. *The Wall Street Journal*. Retrieved from https://www.wsj.com/articles/BL-CIOB-5905

22 Busetta, A., & Milito, A.M. (2009). Socio-demographic vulnerability: The condition of Italian young people. *Social Indicators Research, 97*(3), 375–96.

23 Nusca, A. (2014, April 25). It's time for corporate boards to tackle cybersecurity. Here's why. *Fortune*. Retrieved from https://fortune.com/2014/04/25/its-time-for-corporate-boards-to-tackle-cybersecurity-heres-why/

24 High, M. (2020, April 18). *Why culture is essential to cybersecurity strategy*. Technology. Retrieved from https://www.gigabitmagazine.com/ai/why-culture-essential-cybersecurity-strategy

25 Colón, M. (2014, September 11). *Retail trade association appoints new VP of cybersecurity*. SCMagazine. Retrieved from https://staging.scmagazine.com/home/security-news/retail-trade-association-appoints-new-vp-of-cybersecurity/

26 Varonis Systems. (2008, June 30). *Ponemon study – survey on the governance of unstructured data*. Retrieved May 5, 2010 from http://www.varonis.com/metadata/ponemon-study/

27 Stoller, D. (2020, February 5). *Cyber insurance purchases will surge with California privacy law*. Bloomberg Law. Retrieved from https://news.bloomberglaw.com/privacy-and-data-security/cyber-insurance-purchases-will-surge-with-california-privacy-law

28 *Cyber re/insurance is already rapidly maturing: PCS*. (2018, February 7). Reinsurance News. Retrieved from https://www.reinsurancene.ws/cyber-reinsurance-already-rapidly-maturing-pcs/

29 Strohm, C. (2014, December 4). Hack that crippled Sony Pictures' computers highlights growing threat of malware as cyber weapons. *Financial Post*. Retrieved from http://business.financialpost

.com/2014/12/04/hack-that-crippled-sony-pictures-computers-highlights-growing-threat-of-malware-as-cyber-weapons/?__lsa=8c2b-c5b1

30 Sondhi, T. (2014, November 3). 5 ways to reinforce your company's cybersecurity program today. *Forbes*. Retrieved from http://www.forbes.com/sites/symantec/2014/11/03/5-ways-to-reinforce-your-companys-cybersecurity-program-today/

31 Srivastava, R., Shervani, T., & Fahey, L. (1998). Market-based assets and shareholder value: A framework for analysis. *Journal of Marketing, 62*(1), 2–18.

32 Harber, C. (2019, October 8). *For cybersecurity to be proactive, terrains must be mapped*. Dark Reading. Retrieved from https://www.darkreading.com/vulnerabilities---threats/for-cybersecurity-to-be-proactive-terrains-must-be-mapped/a/d-id/1335965

33 Lemos, R. (2019, December 12). *Lessons from the NSA: Know your assets*. Dark Reading. Retrieved from https://www.darkreading.com/careers-and-people/lessons-from-the-nsa-know-your-assets/d/d-id/1336596

34 Gibbs, S. (2014, December 12). *FBI: 90% of US companies could be hacked just like Sony*. Insider. Retrieved from http://www.businessinsider.com/fbi-90-of-cyber-security-systems-out-there-would-not-have-been-able-to-block-the-sony-hackers-2014-12

35 Sheridan, T. (2014, September 19). *The 6 questions for understanding cybersecurity*. Accounting Web. Retrieved from http://www.accountingweb.com/article/6-questions-understanding-cybersecurity/223859

36 Dougherty, C. (2014, September 1). *Cyber attacks on banks color terrorism insurance renewal debate*. Retrieved from https://www.insurancejournal.com/news/national/2014/09/01/339146.htm

37 Kaiman, J. (2013, December 4). Hack Tibet: Welcome to Dharamsala, ground zero in China's cyberwar. *Foreign Policy*. Retrieved from https://foreignpolicy.com/2013/12/04/hack-tibet/

38 Kaiman. (2013). Hack Tibet.

39 Leahy, J. (2013, August 22). *Brazil's Petrobras to invest heavily in data security*. Retrieved from http://www.ft.com/intl/cms/s/0/f3195d0a-2081-11e3-9a9a-00144feab7de.html#axzz3AlFqmIaB

40 O'Neill. (2017). *Ethiopia using Israeli spyware to spy on dissidents, journalists*.

41 Nakashima, E. (2014, May 29). Iranian hackers are targeting U.S. officials through social networks, report says. *The Washington Post*. Retrieved from https://www.washingtonpost.com/world/national-security/iranian-hackers-are-targeting-us-officials-through-social-networks-report-says/2014/05/28/7cb86672-e6ad-11e3-8f90-73e071f3d637_story.html

42 Reuters. (2014, May 29). *Iranian hackers used fake Facebook accounts to spy on US, others*. iSight. Retrieved from https://gadgets.ndtv.com/internet/news/iranian-hackers-used-fake-facebook-accounts-to-spy-on-us-others-isight-532574

43 Rooney, B. (2011, April 19). *Cyber crime now an industry*. Retrieved May 5, 2011 from http://blogs.wsj.com/tech-europe/2011/04/19/cyber-crime-now-an-industry/?mod=google_news_blog

44 Gale, A. (2014, December 7). North Korea denies direct role in hacking Sony pictures. *The Wall Street Journal*. Retrieved from http://www.wsj.com/articles/north-korea-denies-role-in-hacking-sony-pictures-1417931200

45 Strohm. (2014). Hack that crippled Sony Pictures' computers highlights growing threat of malware as cyber weapons.

46 Kshetri. (2016). *Why the IRS was just hacked – again – and what the feds can do about it.*

47 Kshetri. (2005). Pattern of global cyber war and crime.

48 Gabberty, J. (2014, October 29). How banks can step up to bat on cybersecurity. *American Banker*. Retrieved from http://www.americanbanker.com/bankthink/how-banks-can-step-up-to-bat-on-cybersecurity-1070900-1.html

49 Murphy, A., Tucker, H., Coyne, M., & Touryalai, H. (2020, May 13). Global 2000: The world's largest public companies. *Forbes*. Retrieved from https://www.forbes.com/global2000/list/#tab:overall

50 *Update 3 – Hackers target Brazilian statistics agency.* (2011, June 24). Reuters. Retrieved from https://www.reuters.com/article/cybersecurity-brazil-hackers/update-3-hackers-target-brazilian-statistics-agency-idUSN1E75N0IK20110624

51 McDonald, J. (2014, May 20). China suspends cooperation in joint task force over cyberspying charges. *Huffington Post*. Retrieved May 25, 2014 from http://www.huffingtonpost.com/2014/05/20/china-suspends-cooperation-us-_n_5356447.html

52 Purkayastha, P., & Bailey, R. (2014). US control of the internet. *Monthly Review: An Independent Socialist Magazine 66*(3), 103–27.

53 Sanger. (2014). Fine line seen in US spying on companies.

54 Cutler, S. (2012, September 18). *The mirage campaign*. Secureworks. Retrieved from http://www.secureworks.com/cyber-threat-intelligence/threats/the-mirage-campaign/

55 Muggah, R., & Thompson, N. (2018, April 10). *Brazil's critical infrastructure faces a growing risk of cyberattacks*. Council on Foreign Relations. Retrieved from https://www.cfr.org/blog/brazils-critical-infrastructure-faces-growing-risk-cyberattacks

56 Leahy. (2013). *Brazil's Petrobras to invest heavily in data security.*

57 Ozores, P. (2013). *Petrobras on cloud: "We are conservative."* BNAmericas. Retrieved from https://www.bnamericas.com/en/news/ict/petrobras-on-cloud-we-are-conservative?idioma=I&tipoContenido=detalle&pagina=company&idContenido=2471

58 Kovacs, E. (2013, September 20). *Brazil's Petrobras to invest billions in cybersecurity in light of NSA spying allegations*. Softpedia News. Retrieved from http://news.softpedia.com/news/Brazil-s-Petrobras-to-Invest-Billions-in-Cybersecurity-in-Light-of-NSA-Spying-Allegations-384874.shtml

59 Leahy. (2013). *Brazil's Petrobras to invest heavily in data security.*

60 Kovacs. (2013). *Brazil's Petrobras to invest billions in cybersecurity in light of NSA spying allegations.*

61 Associated Press. (2017, May 13). *Global "WannaCry" ransomware cyberattack seeks cash for data.* WREG News Channel 3. Retrieved from http://wreg.com/2017/05/13/global-wannacry-ransomware-cyberattack-seeks-cash-for-data/

Chapter 9

1 Ofonseca. (2012, January 17). *How data breaches harm reputations*. Experian. Retrieved from http://www.experian.com/blogs/data-breach/2012/01/17/how-data-breaches-harm-reputations/

2 Goldman, J. (2016, June 20). *Acer hacked*. eSecurity Planet. Retrieved from http://www.esecurityplanet.com/hackers/acer-hacked.html

3 Warren, T. (2020, April 1). *Zoom faces a privacy and security backlash as it surges in popularity.* The Verge. Retrieved from https://www.theverge.com/2020/4/1/21202584/zoom-security-privacy-issues-video-conferencing-software-coronavirus-demand-response

4 Lyons, K. (2020, January 28). *Zoom vulnerability would have allowed hackers to eavesdrop on calls.* The Verge. Retrieved from https://www.theverge.com/2020/1/28/21082331/zoom-vulnerability-hacker-eavesdrop-security-google-hangouts-skype-checkpoint

5 Wu, D., Savov, V., & Nguyen, L. (2020, April 25). *Zoom backlash widens with Daimler, Ericsson and BofA curbs*. Economic Times Telecom. Retrieved from https://telecom.economictimes.indiatimes.com/news/zoom-backlash-widens-with-daimler-ericsson-and-bofa-curbs/75376355

6 Cox, J. (2020, March 26). *Zoom iOS app sends data to Facebook even if you don't have a Facebook account*. Vice. Retrieved from https://www.vice.com/en_us/article/k7e599/zoom-ios-app-sends-data-to-facebook-even-if-you-dont-have-a-facebook-account
7 Walcott, J. (2020, April 9). Foreign spies are targeting Americans on Zoom and other video chat platforms, U.S. intel officials say. *Time*. Retrieved from https://time.com/5818851/spies-target-americans-zoom-others/
8 Paul, K. (2020, April 2). "Zoom is malware": Why experts worry about the video conferencing platform. *The Guardian*. Retrieved from https://www.theguardian.com/technology/2020/apr/02/zoom-technology-security-coronavirus-video-conferencing
9 Krause, R. (2020, April 8). *Zoom video stock rally cools as Google bans video app for employees*. Investor's Business Daily. Retrieved from https://www.investors.com/news/technology/zoom-stock-google-ban/
10 Ikeda. (2020). GDPR fines top $126 million with over 160,000 data breaches reported.
11 Ikeda. (2020). GDPR fines top $126 million with over 160,000 data breaches reported.
12 Swartz, J. (2020, April 7). *Zoom video lurches from boom to backlash amid privacy issues, "Zoom bombing" attacks*. MarketWatch. Retrieved from https://www.marketwatch.com/story/zoom-video-lurches-from-boom-to-backlash-amid-privacy-issues-2020-04-01
13 Brooks, K. (2020, April 1). Zoom sued for allegedly sharing users' personal data with Facebook. *CBS News*. Retrieved from https://www.cbsnews.com/news/zoom-app-personal-data-selling-facebook-lawsuit-alleges/
14 Rizzi, C. (2020, May 21). *Class action alleges Zoom shares user info with Facebook, other third parties without proper notice [UPDATE]*. Class Action. Retrieved from https://www.classaction.org/blog/class-action-alleges-zoom-shares-user-info-with-facebook-other-third-parties-without-proper-notice
15 *Zoom sued for overstating, not disclosing privacy, security flaws*. (2020, April 8). Reuters. Retrieved from https://www.reuters.com/article/us-zoom-video-commn-privacy-lawsuit/zoom-sued-for-overstating-not-disclosing-privacy-security-flaws-idUSKBN21Q10V
16 Davis, W. (2020, April 13*). Facebook, LinkedIn sued for allegedly 'eavesdropping' on Zoom users*. MediaPost. Retrieved from https://www.mediapost.com/publications/article/349913/facebook-linkedin-sued-for-allegedly-eavesdroppi.html
17 *Sales navigator plans & pricing*. (n.d.). LinkedIn. Retrieved April 2020 from https://business.linkedin.com/sales-solutions/compare-plans
18 Weiss, D. (2020, April 14). *Another lawsuit is filed against Zoom over alleged privacy problems*. ABA Journal. Retrieved from https://www.abajournal.com/news/article/another-lawsuit-is-filed-against-zoom-over-alleged-privacy-problems
19 *Zoom unveils 90-day plan to rebuild reputation*. (2020, April 2). Verdict. Retrieved from https://www.verdict.co.uk/zoom/
20 Ikeda. (2020). GDPR fines top $126 million with over 160,000 data breaches reported.
21 Barr, R. (2020, April 15). *Luta Security, helmed by Katie Moussouris, joins forces with Zoom on bug bounty program*. Zoom Blog. Retrieved from https://blog.zoom.us/wordpress/2020/04/15/luta-security-katie-moussouris-zoom-bug-bounty/
22 Gallagher, K. (2017, March 16). *Ad fraud estimates doubled*. Business Insider. Retrieved from http://www.businessinsider.com/ad-fraud-estimates-doubled-2017-3
23 Johnson, M. (2019, June 5). What marketers need to know about cybersecurity. *CPO Magazine*. Retrieved from https://www.cpomagazine.com/cyber-security/what-marketers-need-to-know-about-cybersecurity/

24 Deloitte. (n.d.). *Why CMOs should care about cyber risk*. Retrieved from https://www2.deloitte.com/content/dam/Deloitte/us/Documents/consumer-business/us-why-consumer-products-cmos-should-care-about-cyber-risk-final.pdf
25 Bocetta, S. (2019, February 11). *Cybersecurity for marketers: Teamwork is key to protect data.* Marketing Land. Retrieved from https://marketingland.com/cybersecurity-for-marketers-teamwork-is-key-to-protect-data-256749
26 Aaker, D. (1991). *Managing brand equity*. The Free Press.
27 Burt, A. (2019, March 4). Cybersecurity is putting customer trust at the center of competition. *Harvard Business Review*. Retrieved from https://hbr.org/2019/03/cybersecurity-is-putting-customer-trust-at-the-center-of-competition
28 *Global survey from ping identity shows consumers are abandoning brands after data breaches*. (2018, November 17). Ping Identity. Retrieved from https://www.pingidentity.com/en/company/press-releases-folder/2018/global-survey-ping-identity-shows-consumers-abandoning-brands-after-data-breaches.html
29 Gemalto. (2018). *Data breaches & customer loyalty report 2018.* Thales. Retrieved May 25, 2019 from https://safenet.gemalto.com/resources/data-breaches-customer-loyalty-2018-report/
30 *How consumers see cybersecurity and privacy risks and what to do about it*. (n.d.). PwC. Retrieved May 25, 2018 from https://www.pwc.com/us/en/services/consulting/library/consumer-intelligence-series/cybersecurity-protect-me.html
31 *Research: 69% of travelers say hotels aren't investing enough in cybersecurity.* (2019, October 25). Hotel Technology News. Retrieved from https://hoteltechnologynews.com/2019/10/research-69-of-travelers-say-hotels-arent-investing-enough-in-cybersecurity/
32 Murray, A. (2017, June 8). Fortune 500 CEOs see A.I. as a big challenge. *Fortune*. Retrieved from https://fortune.com/2017/06/08/fortune-500-ceos-survey-ai/
33 Greenberg, A. (2015, July 21). Hackers remotely kill a jeep on the highway – with me in it. *Wired*. Retrieved from https://www.wired.com/2015/07/hackers-remotely-kill-jeep-highway/
34 *Research: 69% of travelers say hotels aren't investing enough in cybersecurity*. (2019).
35 *Global daily brand tracking data*. (n.d.). YouGov. Retrieved from http://www.brandindex.com/
36 Tadena, N. (2014, December 19). Sony's data hack delivers huge blow to its brand. *The Wall Street Journal*. Retrieved from https://www.wsj.com/articles/BL-269B-2346
37 Abelson, R., & Goldstein, M. (2015, February 5). Anthem hacking points to security vulnerability of health care industry. *The New York Times*. Retrieved from https://tinyurl.com/yacopd2h
38 Wall, J. (2015, May 11). *Anthem's brand suffers small ding from data breach*. Indianapolis Business Journal. Retrieved from https://tinyurl.com/lm5yo7u
39 Spillett, R. (2015, November 6). TalkTalk reveals personal details of 150,000 customers and 15,000 bank accounts were accessed in last month's cyber attack. *Daily Mail*. Retrieved from http://www.dailymail.co.uk/news/article-3306600/Hackers-accessed-details-150-000-TalkTalk-customers.html
40 Sword, A. (2016, January 22). *Rebuilding brand trust: TalkTalk's path back from cyber attack*. TechMonitor. Retrieved from https://tinyurl.com/yclfpajy
41 Peters, J. (2016, November 2). *Yahoo and others face cybercrime-related brand damage*. SurfWatch Labs. Retrieved from https://blog.surfwatchlabs.com/2016/11/02/yahoo-and-others-face-cybercrime-related-brand-damage/
42 Dua, T. (2016, September 23). *Yahoo's data breach further dents its already flailing brand*. Digiday. Retrieved from http://digiday.com/brands/yahoos-data-breach-dents-already-flailing-brand/

43 Lewin, J. (2016, February 1). Cyber attack cost TalkTalk up to £60m and 101k customers. *Financial Times*. Retrieved from https://www.ft.com/content/410f5477-d061-3cd4-a064-5563c91bd7fb
44 Peters, J. (2016, November 2). *Yahoo and others face cybercrime-related brand damage*. SurfWatch Labs. Retrieved from https://tinyurl.com/yc7pdy85
45 Condliffe, J. (2016, December 21). *How Russian hackers stole $5 million a day from US advertisers*. MIT Technology Review. Retrieved from https://www.technologyreview.com/s/603233/how-russian-hackers-stole-5-million-a-day-from-us-advertisers/
46 Saba, J., & Finkle, J. (2015, March 23). *Online ad revenue at risk in war on "click fraud."* Reuters. Retrieved from https://www.reuters.com/article/us-advertising-cyberfraud/online-ad-revenue-at-risk-in-war-on-click-fraud-idUSKBN0MJ0Z820150323
47 Liu, Y. (2018, January 20). *Rampant click fraud on China's internet hurts digital advertisers*. South China Morning Post. Retrieved from http://www.scmp.com/tech/china-tech/article/2129712/rampant-click-fraud-chinas-internet-hurts-digital-advertisers
48 Raval, T. (2019, October 28). Click deception: Why marketers finally need to address the growing ad fraud issue. *Forbes*. Retrieved from https://www.forbes.com/sites/forbestechcouncil/2019/10/28/click-deception-why-marketers-finally-need-to-address-the-growing-ad-fraud-issue/#3561177e50bd
49 Nair-Ghaswalla, A. (February 28, 2020). *Click-bait digital fraud bleeding ad industry*. The Hindu Business Line. Retrieved from https://www.thehindubusinessline.com/info-tech/click-bait-digital-fraud-bleeding-ad-industry/article30934282.ece
50 Saba & Finkle. (2015). *Online ad revenue at risk in war on "click fraud."*
51 Goo, S.K. (2006, October 22). "Click fraud" threatens foundation of web ads Google faces another lawsuit by businesses claiming overcharges. *The Washington Post*. Retrieved from https://www.washingtonpost.com/archive/politics/2006/10/22/click-fraud-threatens-foundation-of-web-ads-span-classbankheadgoogle-faces-another-lawsuit-by-businesses-claiming-overchargesspan/d929eaa7-3187-4323-9c07-39222d5f1255/
52 Koetsier, J. (2017, August 16). Mobile ad fraud in India: Click fraud is 2.4x worse, app install fraud 1.7x higher. *Forbes*. Retrieved from https://www.forbes.com/sites/johnkoetsier/2017/08/16/mobile-ad-fraud-in-india-click-fraud-is-2-4x-worse-app-install-fraud-1-7x-higher/#3b64a39d729b
53 Penenberg, A.L. (2005, August 1). *So many clicks, so few sales*. Inc. Retrieved from https://www.inc.com/magazine/20050801/marketing.html
54 Businessweek.com. (2006, April 24). *Your ad here. And here. And here*. Retrieved May 25, 2008 from http://www.businessweek.com/magazine/content/06_17/b3981046.htm
55 Arthur, C. (2013, August 2). How low-paid workers at "click farms" create appearance of online popularity. *The Guardian*. Retrieved from https://www.theguardian.com/technology/2013/aug/02/click-farms-appearance-online-popularity
56 Associated Press. (2017, June 13). *Thai police raid WeChat "click farm," find 347,200 SIM cards, arrest three Chinese men*. South China Morning Post. Retrieved from http://www.scmp.com/news/asia/southeast-asia/article/2098167/thai-police-raid-wechat-click-farm-find-347200-sim-cards
57 Liu, Y. (2018, January 20). *Rampant click fraud on China's internet hurts digital advertisers*. South China Morning Post. Retrieved from http://www.scmp.com/tech/china-tech/article/2129712/rampant-click-fraud-chinas-internet-hurts-digital-advertisers
58 Businessweek.com. (2006). *Your ad here*.
59 Kaufman, L. (2014, December 9). Study puts a price tag on fake ad clicks. *The New York Times*. https://www.nytimes.com/2014/12/10/business/media/study-puts-a-price-tag-on-digital-ad-click-fraud-.html

60 Leyden, J. (2006, October 6). *Worm automates Google AdSense fraud*. The Register. Retrieved from https://www.theregister.com/2006/10/06/google_adsense_worm/
61 Mello, J.P., Jr. (2006, January 5). *Pornographers turn to click fraud*. E-Commerce Times. Retrieved from http://www.ecommercetimes.com/story/48135.html
62 Hayes, P. (2020, April 13). *Golf equipment retailer sues competitor for ad-click fraud*. Bloomberg Law. Retrieved from https://news.bloomberglaw.com/tech-and-telecom-law/golf-equipment-retailer-sues-competitor-for-ad-click-fraud
63 Kim, L. (2017, July 20). *These are the 10 most expensive Google keywords*. Inc. Retrieved from https://www.inc.com/larry-kim/these-are-the-10-most-expensive-google-keywords.html
64 Cimpanu, C. (2019, August 6). *Facebook files lawsuit against two Android app developers for click fraud*. ZDNet. Retrieved from https://www.zdnet.com/article/facebook-files-lawsuit-against-two-android-app-developers-for-click-fraud/
65 *Alert (TA18-331A)*. (2018, November 27). Cybersecurity & Infrastructure Security Agency. Retrieved from https://www.us-cert.gov/ncas/alerts/TA18-331A
66 Goodfellow, J. (2019, October 24). *Inside ad fraud: What it takes to dismantle a $5.8bn enterprise*. Campaign. Retrieved from https://www.campaignlive.co.uk/article/inside-ad-fraud-takes-dismantle-58bn-enterprise/1663536
67 Beer, J. (2019, May 22). *How the No. 1 most creative person in business this year led the FBI to its biggest ad-fraud bust ever*. Fast Company. Retrieved from https://www.fastcompany.com/90345976/most-creative-people-2019-white-ops-tamer-hassan
68 Hammett, E. (2019, June 18). *Brands and tech giants come together to launch first digital safety alliance*. Marketing Week. Retrieved from https://www.marketingweek.com/brands-and-tech-giants-come-together-to-launch-first-digital-safety-alliance/
69 *Keeping browsing experience in users' hands*. (2015, December 21). Microsoft. Retrieved from https://www.microsoft.com/security/blog/2015/12/21/keeping-browsing-experience-in-users-hands/?source=mmpc
70 Staufenberg, J. (2016, January 28). *TalkTalk cyberattack: Call centre workers arrested in India as part of hacking probe*. Independent. Retrieved from https://tinyurl.com/ycms2n5b
71 Bottom, W.P. (2002). When talk is not cheap: Substantive penance and expressions of intent in rebuilding cooperation. *Organization Science, 13*, 497–513.
72 Korsgaard, M.A. (2002). Trust in the face of conflict: Role of managerial trustworthy behavior and organizational context. *Journal of Applied Psychology, 87*, 312–19.
73 BankThink: Good marketing can help banks survive a cybersecurity backlash. (2015, January 12). *American Banker*. Retrieved from https://tinyurl.com/yd2kgon5
74 Bray, C. (2011, November 9). Seven accused of infecting computers with malware in more than 100 countries. *The Wall Street Journal*. Retrieved from http://online.wsj.com/article/SB10001424052970204358004577028090371514700.html
75 *Estonian court approves extradition of six persons to US for cybercrime*. (2012, February 21). Baltic Business News. Retrieved May 25, 2013 from http://balticbusinessnews.com/article/2012/2/21/estonian-court-approvesextradition-of-six-persons-to-us-for-cybercrime
76 *FBI arrest may lead to Estonia's largest asset seizure*. (2011, November 11). Baltic Business News. Retrieved May 25, 2013 from http://balticbusinessnews.com/article/2011/11/11/fbi-arrest-may-lead-to-estonia-s-largest-asset-seizure
77 Ferguson, P. (2011). *Esthost taken down – biggest cybercriminal takedown in history*. Retrieved May 25, 2013 from http:// blog.trendmicro.com/?p=38093
78 Ferguson. (2011). *Esthost taken down – biggest cybercriminal takedown in history*.

79 *FBI arrest may lead to Estonia's largest asset seizure.* (2011).
80 Sengupta, S., & Jenna, W. (2011, November 9). 7 charged in web scam using ads. *The New York Times.* Retrieved from http:// www.nytimes.com/2011/11/10/technology/us-indicts-7-in-online-ad-fraud-scheme.html
81 Hurtado, P., & Michael, R. (2011, November 9). *Hackers hijack millions of computers in "massive" fraud case.* Business Week. Retrieved from http://www.businessweek.com/news/2011-11-09/hackers-hijack-millions-ofcomputers-in-massive-fraud-case.html
82 Sengupta & Jenna. (2011). 7 charged in web scam using ads.
83 Hurtado & Michael. (2011). *Hackers hijack millions of computers in "massive" fraud case.*
84 Menn, J. (2011, November 9). US uncovers alleged "click fraud" ring. *Financial Times.* Retrieved from https://www.ft.com/content/96c244ae-0b16-11e1-ae56-00144feabdc0
85 Hurtado & Michael. (2011). *Hackers hijack millions of computers in "massive" fraud case.*
86 Biggest cybercriminal takedown in history. (2011, November 9). KrebsonSecurity. Retrieved from https://krebsonsecurity.com/2011/11/malware-click-fraud-kingpins-arrested-in-estonia/
87 Mello, J. (2011, November 12). Spam researchers help bust global cybercrime ring. *PCWorld.* Retrieved from https://www.pcworld.com/article/243748/spam_researchers_help_bust_global_cybercrime_ring.html
88 *Estonian court approves extradition of six persons to US for cybercrime.* (2012). Baltic Business News.
89 Raymond, N. (2016, April 26). *Estonian sentenced in US to seven years in prison for cyber fraud.* Reuters. Retrieved from https://www.reuters.com/article/us-usa-cybersecurity-malware/estonian-sentenced-in-u-s-to-seven-years-in-prison-for-cyber-fraud-idUSKCN0XN2WX

Chapter 10

1 Terrill, C. (2017, January 19). How to tell when your company should hire a CISO. *Forbes.* Retrieved from http://www.forbes.com/sites/christieterrill/2017/01/19/how-to-tell-when-your-company-should-hire-a-ciso/#259c30e4f67a
2 Kshetri, N. (2016). *The quest to cyber superiority: Cybersecurity regulations, frameworks, and strategies of major economies.* Springer-Verlag.
3 Lonergan, K. (2015, January 7). *The world's first CISO explains why technology alone has never beaten cyber crime.* Information Age. Retrieved from http://www.information-age.com/worlds-first-ciso-explains-why-technology-alone-has-never-beaten-cyber-crime-123458809/
4 Gartner predicts 75% of CEOs will be personally liable for cyber-physical security incidents by 2024. (2020, September 1). Gartner. Retrieved from https://www.gartner.com/en/newsroom/press-releases/2020-09-01-gartner-predicts-75--of-ceos-will-be-personally-liabl#:~:text=Soon%2C%20CEOs%20won't%20be,over%20%2450%20billion%20by%202023.&text=Gartner%20clients%20can%20learn%20more,Security%20and%20Risk%20Management%20Programs.%E2%80%9D
5 Nash, K.S. (2015, July 1). Boards struggle with cybersecurity, especially in health care. *The Wall Street Journal.* Retrieved from https://www.wsj.com/articles/BL-CIOB-7483
6 Torsten, G. (2016, April 13). *Cyber security oversight: Why it belongs in the board room.* SecurityWeek. Retrieved from http://www.securityweek.com/cyber-security-oversight-why-it-belongs-board-room
7 Morgan, S. (2016, February 21). Are HR chiefs the biggest cyber threat? *Forbes.* Retrieved from http://www.forbes.com/sites/stevemorgan/#159e2a5b662e

8 Huselid, M.A., Jackson, S.E., & Schuler, R.S. (1997). Technical and strategic human resource management effectiveness as determinants of firm performance. *Academy of Management Journal, 40*(1), 171–88.

9 Muncaster, P. (2020, April 23). *Most remote workers have received no security training for a year*. InfoSecurity Magazine. Retrieved from https://www.infosecurity-magazine.com/news/most-remote-workers-received-no/

10 McKendrick, J. (2019, September 30). So easy, even a CEO can understand it: A call for greater clarity in cybersecurity. *Forbes*. Retrieved from https://www.forbes.com/sites/joemckendrick/2019/09/30/so-easy-even-a-ceo-can-understand-it-a-call-for-greater-clarity-in-cybersecurity/#de5473330cc2

11 Sweeney, B. (2016, September 13). Cybersecurity is every executive's job. *Harvard Business Review*. Retrieved from https://hbr.org/2016/09/cybersecurity-is-every-executives-job

12 Damouni, N. (2014, May 30). *Exclusive: U.S. companies seek cyber experts for top jobs, board seats*. Reuters. http://www.reuters.com/article/us-usa-companies-cybersecurity-exclusive-idUSKBN0EA0BX20140530

13 Hubbard, T. (2016, October 12). Why cyber attacks remain a challenge for U.S. *Forbes*. Retrieved from http://www.forbes.com/sites/kpmg/2016/10/12/why-cyber-attacks-remain-a-challenge-for-u-s/#ca9bfa216de8

14 Balaouras, S. (2020, February 24). *Cybersecurity in 2020: The rise of the CISO*. MIT Technology Review. Retrieved from https://www.technologyreview.com/s/615092/cybersecurity-in-2020-the-rise-of-the-ciso/

15 Torsten. (2016). *Cyber security oversight*.

16 *CISO role still in flux: Despite small gains, CISOs face an uphill battle in the C-suite*. (2015, October 24). Retrieved from https://cloudsecuritysol.com/ciso-role-still-in-flux/

17 Shah, S. (2014, June 27). *Salaries soar for cyber security high-flyers*. Computing. Retrieved from http://www.computing.co.uk/ctg/analysis/2352505/salaries-soar-for-cyber-security-high-flyers

18 Boyer, S. (2019, October 11). *We're all at risk when 65% of stressed-out cybersecurity and IT workers are thinking about quitting, tech exec warns*. CNBC. Retrieved from https://www.cnbc.com/2019/10/11/65percent-of-stressed-out-cybersecurity-it-workers-think-about-quitting.html

19 McKendrick. (2019). So easy, even a CEO can understand it.

20 Shah. (2014). *Salaries soar for cyber security high-flyers*.

21 Dimon, J. (2019, April 4). *Letter to shareholders*. JPMorgan Chase. Retrieved from https://www.jpmorganchase.com/corporate/investor-relations/document/ceo-letter-to-shareholders-2018.pdf

22 Slefo, G. (2019, December 27). *The 10 biggest brand data breaches of the decade*. Ad Age. Retrieved from https://adage.com/article/year-end-lists-2019/10-biggest-brand-data-breaches-decade/2222096

23 Williams, M. (2017, October 4). *Inside the Russian hack of Yahoo: How they did it*. CSO. Retrieved from https://www.csoonline.com/article/3180762/inside-the-russian-hack-of-yahoo-how-they-did-it.html

24 Battat, R. (2017, October 2). *Lessons from the Yahoo hack*. Risk Management. Retrieved from http://www.rmmagazine.com/2017/10/02/lessons-from-the-yahoo-hack/

25 Perlroth, N., & Goel, V. (2016, September 28). Defending against hackers took a back seat at Yahoo, insiders say. *The New York Times*. Retrieved from https://www.nytimes.com/2016/09/29/technology/yahoo-data-breach-hacking.html#:~:text=Defending%20Against%20Hackers%20Took%20a%20Back%20Seat%20at%20Yahoo%2C%20Insiders%20Say,-Alex%20Stamos%2C%20

then&text=SAN%20FRANCISCO%20%E2%80%94%20Six%20years%20ago,technology%20companies%20were%20also%20hit

26 Szoldra, P. (2016, October 4). *A former insider says Marissa Mayer kept secrets from Yahoo's security team more than once*. Business Insider. Retrieved from https://www.businessinsider.com/marissa-mayer-secret-yahoo-security-2016-10

27 Szoldra. (2016). *A former insider says Marissa Mayer kept secrets from Yahoo's security team more than once.*

28 Swisher, K. (2013, January 14). *Yahoo's chief information security officer departs – with more top execs under CEO scrutiny*. AllThingsD. Retrieved from http://allthingsd.com/20130114/yahoos-chief-information-security-officer-departs-with-more-top-execs-under-ceo-scrutiny/

29 Carlson, N. (2013, January 15). *Marissa Mayer is plotting a major purge of executives at Yahoo.* BusinessInsider. Retrieved from https://www.businessinsider.com/marissa-mayer-is-plotting-a-major-purge-of-executives-at-yahoo-2013-1

30 Szoldra, P. (2016, September 28). *Yahoo won't answer the most important question about its massive hack*. BusinessInsider. Retrieved from http://www.businessinsider.com/yahoo-massive-hack-2016-9

31 Menn, J. (2016, October 4). *Yahoo secretly helped the US government scan millions of customer emails*. BusinessInsider. Retrieved from https://www.businessinsider.com/r-exclusive-yahoo-secretly-scanned-customer-emails-for-us-intelligence-sources-2016-10

32 Szoldra. (2016). *Yahoo won't answer the most important question about its massive hack*.

33 Ackerman, R. (2020, February 10). *Develop a serious cybersecurity strategic plan that incorporates CCM.* TechCrunch. Retrieved from https://techcrunch.com/2020/02/10/develop-a-serious-cybersecurity-strategic-plan-that-incorporates-ccm/

34 Johnson, A. (2018, December 12). Verizon signals its Yahoo and AOL divisions are almost worthless. *NBC News*. Retrieved from https://www.nbcnews.com/tech/tech-news/verizon-signals-its-yahoo-aol-divisions-are-almost-worthless-n946846

35 Johnson. (2018). Verizon signals its Yahoo and AOL divisions are almost worthless.

36 Lomas, N. (2017, June 19). *Sources: Yahoo CISO Bob Lord out after AOL-Yahoo merger*. TechCrunch. Retrieved from https://techcrunch.com/2017/06/09/sources-yahoo-ciso-bob-lord-out-after-aol-yahoo-merger/

37 Hart, K. (2017, November 7). *Congress subpoenaed former Yahoo CEO Marissa Mayer to testify on data breach*. Axios. Retrieved from https://www.axios.com/congress-subpoenaed-former-yahoo-ceo-marissa-mayer-to-testify-on-data-breach-2507557870.html

38 Volz, D. (2016, September 24). *Yahoo faces growing scrutiny over when it learned of data breach.* Thomson Reuters Foundation News. Retrieved from http://news.trust.org/item/20160923211729-stk2c/

39 O'Donnell, L. (2019, April 10). *Yahoo offers $117.5m settlement in data breach lawsuit*. ThreatPost. Retrieved from https://threatpost.com/yahoo-offers-117-5m-settlement-in-data-breach-lawsuit/143671/

40 Nichols, S. (2018, March 12). *Yahoo! Can't! Toss! Hacking! Lawsuit!* The Register. Retrieved from https://www.theregister.co.uk/2018/03/12/yahoo_hacking_lawsuit/

41 Cushing, T. (2018, March 14). *Judge says Yahoo still on the hook for multiple claims related to three billion compromised email accounts*. Techdirt. Retrieved from https://www.techdirt.com/articles/20180313/10201339416/judge-says-yahoo-still-hook-multiple-claims-related-to-three-billion-compromised-email-accounts.shtml

42 Cooper, P. (2019, January 29). *Yahoo breach settlement riddled with problems, court says*. Bloomberg Law. Retrieved from https://news.bloomberglaw.com/class-action/yahoo-breach-settlement-riddled-with-problems-court-says

43 O'Flaherty, K. (2019, September 25). Yahoo breach: Find out if you can claim up to $25,000. *Forbes*. Retrieved from https://www.forbes.com/sites/kateoflahertyuk/2019/09/25/yahoo-breach-find-out-if-you-can-claim-up-to-25000/#1297cf919b3f

44 O'Flaherty. (2019). Yahoo breach.

45 Cooper. (2019). *Yahoo breach settlement riddled with problems, court says*.

46 Atkinson, C. (2018, October 23). Verizon's advertising business is getting eclipsed by Amazon. *NBC News*. Retrieved from https://www.nbcnews.com/news/all/verizon-s-ad-business-getting-eclipsed-amazon-n923316

47 *Yahoo and Oath: A failed mega business plan*. (2019, April 12). VisionTimes. Retrieved from https://www.visiontimes.com/2019/04/12/yahoo-and-oath-a-failed-mega-business-plan.html

48 *Yahoo and Oath: A failed mega business plan*. (2019).

49 *Bitglass Fortune 500 cybersecurity report: Leading companies failing to demonstrate commitment to cybersecurity*. (2019, September 30). Bitglass. Retrieved from https://www.bitglass.com/press-releases/bitglass-report-fortune-500-cybersecurity

50 Pritchard, S. (2020, 24 April). *How to become a CISO – your guide to climbing to the top of the enterprise security ladder*. The Daily Swig. Retrieved from https://portswigger.net/daily-swig/how-to-become-a-ciso-your-guide-to-climbing-to-the-top-of-the-enterprise-security-ladder

51 Torsten. (2016). *Cyber security oversight.*

52 *Eight reasons the CISO should report to the CEO and not the CIO.* (2017, January 9). MIS-Asia. Retrieved from https://www.businessinsider.com/marissa-mayer-is-plotting-a-major-purge-of-executives-at-yahoo-2013-1

53 Mortlock, S. (2016, March 2). *The Asian IT job where salaries are up by 30% and banks still want foreign talent.* Efinancialcarreers. Retrieved from http://news.efinancialcareers.com/sg-en/238033/cyber-security-salaries-soar-30-in-singapore-hong-kong

54 Lavelle, J. (2019, January 17). *Gartner survey shows global talent shortage is now the top emerging risk facing organizations*. Gartner. Retrieved from https://www.gartner.com/en/newsroom/press-releases/2019-01-17-gartner-survey-shows-global-talent-shortage-is-now-the-top-emerging-risk-facing-organizations

55 Lavelle. (2019). *Gartner survey shows global talent shortage is now the top emerging risk facing organizations.*

56 Kalra, A.S. (2016, September 29). *Why HR directors' concerns about tech security aren't unfounded.* HumanResources. Retrieved May 25, 2017 from http://www.humanresourcesonline.net/hr-directors-concerns-tech-security-arent-unfounded/

57 Mee, P. (2020, March 2). *We need to start teaching young children about cybersecurity*. World Economic Forum. Retrieved from https://www.weforum.org/agenda/2020/03/we-need-to-start-teaching-young-children-about-cybersecurity/

58 Kelly, E. (2016, November 3). *UK to invest £1.9B to improve cyber defenses*. Science Business. Retrieved from http://sciencebusiness.net/news/79987/UK-to-invest-%C2%A31.9B-to-improve-cyber-defences

59 Magshimim Program. (n.d.). Retrieved from https://www.rashi.org.il/magshimim-cyber-program

60 NICE (National Initiative for Cybersecurity Education). (2015). Retrieved from https://www.nist.gov/itl/applied-cybersecurity/nice

61 Setalvad, A. (2015, March 21). *Demand to fill cybersecurity jobs booming.* SFGate. Retrieved from http://blog.sfgate.com/inthepeninsula/2015/03/31/cybersecurity-jobs-growth/
62 Center for Strategic and International Studies. (2016, July 27). *Hacking the skills shortage* [Event recording]. Retrieved from https://www.csis.org/events/hacking-skills-shortage
63 Risen, T. (2014, May 28). *Companies unprepared as hacking increases.* USNews. Retrieved from http://www.usnews.com/news/articles/2014/05/28/companies-unprepared-as-hacking-increases
64 Kim, P.H., Pinkley, R.L., & Fragale, A.R. (2005). Power dynamics in negotiation. *Academy of Management Review, 30*(4), 799–822.
65 Pandya, D. (2009, November 22). *CISO reporting to board of directors: Myth or for real?* Computer Weekly. Retrieved from http://www.computerweekly.com/news/1375176/CISO-reporting-to-board-of-directors-Myth-or-for-real
66 Morgan, S. (2020, October 13). *Backstory of the world's first chief information security officer.* Cybercrime Magazine. Retrieved from https://cybersecurityventures.com/backstory-of-the-worlds-first-chief-information-security-officer/
67 Harmon, A. (1995, August 19). Hacking theft of $10 million from Citibank revealed. *Los Angeles Times.* Retrieved from http://articles.latimes.com/1995-08-19/business/fi-36656_1_citibank-system
68 Terrill. (2017). How to tell when your company should hire a CISO.
69 Kirk, J. (2016, September 13). *How Hearst's CISO talks security with the board.* Bank Info Security. Retrieved from http://www.bankinfosecurity.com/blogs/tips-for-talking-security-to-board-p-2250
70 Boodaei, M. (2011, July 11). *Mobile malware: Why fraudsters are two steps ahead.* Security Intelligence. Retrieved from https://securityintelligence.com/mobile-malware-why-fraudsters-are-two-steps-ahead/
71 Norton. (2014). SEC official.
72 Tobias, S. (2014, December 31). 2014: The year in cyberattacks. *Newsweek.* Retrieved from http://www.newsweek.com/2014-year-cyber-attacks-295876
73 Gale, A. (2014). North Korea denies direct role in hacking Sony pictures. *The Wall Street Journal.* Retrieved from http://www.wsj.com/articles/north-korea-denies-role-in-hacking-sony-pictures-1417931200
74 SonyHack. (2015, February 12). Ex-Sony chief Amy Pascal acknowledges she was fired. *NBC News.* Retrieved from http://www.nbcnews.com/storyline/sony-hack/ex-sony-chief-amy-pascal-acknowledges-she-was-fired-n305281
75 Chellel, K. (2015, July 7). *A London hedge fund lost $1.2 million in a Friday afternoon phone scam.* Bloomberg. Retrieved from http://www.bloomberg.com/news/articles/2015-07-07/friday-afternoon-scam-cost-hedge-fund-1-2-million-and-cfo-s-job
76 *Putin says DNC hack was public service, Russia didn't do it.* (2016, September 2). Matzav. Retrieved from http://matzav.com/putin-says-dnc-hack-was-public-service-russia-didnt-do-it/
77 Martin, J., & Rapperport, A. (2016, July 24). Debbie Wasserman Schultz to resign D.N.C. post. *The New York Times.* Retrieved from https://www.nytimes.com/2016/07/25/us/politics/debbie-wasserman-schultz-dnc-wikileaks-emails.html?_r=0
78 *Human error to blame for 9 in 10 UK cyber data breaches in 2019.* (2020, February 17). CYBSafe. Retrieved from https://www.cybsafe.com/press-releases/human-error-to-blame-for-9-in-10-uk-cyber-data-breaches-in-2019/

79 Babati, B. (2019). *Phishing attacks and scams in 2019 and beyond*. Hoxhunt. Retrieved from https://www.hoxhunt.com/blog/phishing-attacks-and-scams-in-2019-and-beyond/
80 Bernard, A. (2020, April 8). *Cybersecurity prevention can save your company $682K*. TechRepublic. Retrieved from https://www.techrepublic.com/article/cybersecurity-prevention-can-save-your-company-682k/
81 Sheridan, K. (2020, February 12). *FBI: Business email compromise cost businesses $1.7B in 2019*. DarkReading. Retrieved from https://www.darkreading.com/fbi-business-email-compromise-cost-businesses-$17b-in-2019/d/d-id/1337035
82 *Training employees to be smarter than AI-powered phishing messages*. (2020, April 21). Interesting Engineering. Retrieved from https://interestingengineering.com/training-employees-to-be-smarter-than-ai-powered-phishing-messages
83 Kapner, S. (2011, October 31). Hackers press the "schmooze" button. *The Wall Street Journal*. Retrieved from http://online.wsj.com/article/SB10001424052970203911804576653393584528906.html?mod=WSJ_article_onespot
84 Rozenfeld, M. (2016, October 10). *Company tests how employees handle social engineering attacks*. IEE Spectrum. Retrieved from http://theinstitute.ieee.org/technology-topics/cybersecurity/company-tests-how-employees-handle-social-engineering-attacks
85 *Table 1. Workers who could work at home, did work at home, and were paid for work at home, by selected characteristics, averages for the period 2017–2018*. (2019, September 24). US Bureau of Labor Statistics. Retrieved from https://www.bls.gov/news.release/flex2.t01.htm
86 Soper, T. (2020, March 19). *Microsoft Teams hits 44M daily active users, spiking 37% in one week amid remote work surge*. GeekWire. Retrieved from https://www.geekwire.com/2020/microsoft-teams-hits-44m-users-huge-37-growth-spike-1-week-amid-remote-work-surge/
87 Thompson, D. (2020, March 13). The coronavirus is creating a huge, stressful experiment in working from home. *The Atlantic*. Retrieved from https://www.theatlantic.com/ideas/archive/2020/03/coronavirus-creating-huge-stressful-experiment-working-home/607945/
88 Kelly, J. (2019, September 5). *Consumer vs. enterprise security: There is a difference*. Law Technology Today. Retrieved from https://www.lawtechnologytoday.org/2018/09/consumer-vs-enterprise-security/
89 Boehm, J., Kaplan, J., & Sportsman, N. (2020, March 25). *Cybersecurity's dual mission during the coronavirus crisis*. McKinsey & Company. Retrieved from https://www.mckinsey.com/business-functions/risk/our-insights/cybersecuritys-dual-mission-during-the-coronavirus-crisis?cid=other-eml-alt-mip-mck&hlkid=34fdd9f7b5154c8d88ee3d25ac6cfd3f&hctky=2762145&hdpid=ffacc134-7120-4a94-ba44-bd4b2fe75bcc
90 Gordon, C. (2020, March 16). Here's what you need to know about cybersecurity when working from home. *WSMV News 4*. Retrieved from https://www.wsmv.com/news/here-s-what-you-need-to-know-about-cybersecurity-when/article_df3b5c90-67f9-11ea-959a-0f1c00f9c28d.html
91 Miller, M. (2020, March 14). Hackers find new target as Americans work from home during outbreak. *The Hill*. Retrieved from https://thehill.com/policy/cybersecurity/487542-hackers-find-new-target-as-americans-work-from-home-during-outbreak
92 Murphy, H. (2019, December 8). How remote working increases cyber security risks. *Financial Times*. Retrieved from https://www.ft.com/content/f7127666-0c80-11ea-8fb7-8fcec0c3b0f9
93 *Remote work is the future – but is your organization ready for it?* (n.d). Open VPN. Retrieved from https://openvpn.net/remote-workforce-cybersecurity-quick-poll/
94 Boehm et al. (2020). *Cybersecurity's dual mission during the coronavirus crisis*.

95 Cappelli, P. (2020, March 17). Vague remote work policies won't cut it during the coronavirus pandemic. *Fortune*. Retrieved from https://fortune.com/2020/03/17/coronavirus-remote-work-from-home-policies/

96 Bernick, M. (2020, March 16). Remote work and best practices: The coronavirus workplace series. *Forbes*. Retrieved from https://www.forbes.com/sites/michaelbernick/2020/03/16/remote-work-and-best-practices-the-coronavirus-workplace-series/#f2051e2769c3

97 Sharp, S. (2020, February 06). *Remote workers prime targets for cyber attacks*. TechRadar.pro. Retrieved from https://www.techradar.com/news/remote-workers-prime-targets-for-cyber-attacks

98 *Top tips for cybersecurity when working remotely*. (2020, March 15). European Union Agency for Cybersecurity. Retrieved from https://www.enisa.europa.eu/news/executive-news/top-tips-for-cybersecurity-when-working-remotely

99 *Remote work is the future – but is your organization ready for it?* (n.d). Open VPN.

100 Mendoza, N.F. (2020, March 5). *How to maintain safe cybersecurity practices while transitioning workers from the office to remote workstations*. TechRepublic. Retrieved from https://www.techrepublic.com/article/how-to-maintain-safe-cybersecurity-practices-while-transitioning-workers-from-the-office-to-remote-workstations/

101 Mendoza. (2020). *How to maintain safe cybersecurity practices while transitioning workers from the office to remote workstations.*

102 Miller. (2020). Hackers find new target as Americans work from home during outbreak.

103 Murphy. (2019). How remote working increases cyber security risks.

104 Mendoza. (2020). *How to maintain safe cybersecurity practices while transitioning workers from the office to remote workstations.*

105 Sweeney. (2016). Cybersecurity is every executive's job.

106 Wright, A.D. (2016, August 1). *Study: More cybersecurity workers needed.* SHRM. Retrieved from https://www.shrm.org/resourcesandtools/hr-topics/technology/pages/experts-say-more-cybersecurity-workers-needed.aspx

107 *IBM study finds C-suite and CISOs not aligned on how to combat cybercriminals.* (2016, February 17). IBM. Retrieved from http://www-03.ibm.com/press/us/en/pressrelease/49100.wss

108 Thornton, D. (2016, October 25). *3 biggest cybersecurity concerns of agency CISOs*. FederalNews Radio. Retrieved from http://federalnewsradio.com/cybersecurity-awareness-october-2016/2016/10/3-biggest-cybersecurity-concerns-agency-cisos/

109 Rozenfeld, M. (2016, 10 October). *Company tests how employees handle social engineering attacks: Posing as hackers, researchers identify vulnerabilities at Croatian telecom giant.* IEEE Spectrum. Retrieved May 25, 2017 from http://theinstitute.ieee.org/technology-topics/cybersecurity/company-tests-how-employees-handle-social-engineering-attacks

110 Sweeney. (2016). Cybersecurity is every executive's job.

111 Mueller, B. (2015, October 29). *4 ways HR can help build a cyber-secure culture*. Namely. Retrieved from http://blog.namely.com/blog/2015/10/29/4-ways-hr-can-help-build-a-cyber-secure-culture

112 Thornton. (2016). *3 biggest cybersecurity concerns of agency CISOs.*

113 BBB (Better Business Bureau). (2016, October 25). *The human factor of cybersecurity: Employees can be both first line of defense or biggest vulnerability*. The Oklahoman. Retrieved from http://newsok.com/article/5523925

114 Strohm. (2014). Hack that crippled Sony Pictures' computers highlights growing threat of malware as cyber weapons.

115 Dreyfuss, J. (2019, December 17). *The hacker behind your company's data breach may be sitting right in the next cubicle*. CNBC. Retrieved from https://www.cnbc.com/2019/12/17/hacker-behind-your-companys-data-breach-may-be-in-the-next-cubicle.html

116 Frost & Sullivan. (2013). *Agents of change: Women in the information security profession*. (ISC)^2. Retrieved from https://1c7fab3im83f5gqiow2qqs2k-wpengine.netdna-ssl.com/wp-content/uploads/2019/03/Women-in-the-Information-Security-Profession-GISWS-Subreport.pdf

117 Higgins, K. (2018, February 22). *Best practices for recruiting & retaining women in security*. Dark Reading. Retrieved from https://www.darkreading.com/careers-and-people/best-practices-for-recruiting-and-retaining-women-in-security/d/d-id/1331114

118 *Women in cybersecurity: Underrepresented, untapped potential*. Pwc. Retrieved May 25, 2020 from https://www.pwc.com/us/en/services/consulting/cybersecurity/women-in-cybersecurity.html

119 Bauer, M. (2019, January 17). *How more women on cybersecurity teams can create advantages*. Fifth Domain. Retrieved from https://www.fifthdomain.com/workforce/2019/01/18/how-more-women-on-cybersecurity-teams-can-create-advantages/

120 Poster, W. (2018, March 26). *Cybersecurity needs women*. Nature. Retrieved from https://www.nature.com/articles/d41586-018-03327-w

121 Higgins. (2018). *Best practices for recruiting & retaining women in security*.

122 Poster. (2018). *Cybersecurity needs women*.

123 Frost & Sullivan. (2013). *Agents of change*.

124 Davis, C. (2019, September 7). *Why Mastercard is betting on middle school girls to detect cyberthreats and protect our personal data*. CNBC. Retrieved from https://www.cnbc.com/2019/09/07/mastercard-is-betting-on-middle-school-girls-to-detect-cyberthreats.html

125 *By age 6, gender stereotypes can affect girls' choices*. (2017, January 26). National Science Foundation. Retrieved from https://www.nsf.gov/news/news_summ.jsp?cntn_id=190924&WT.mc_id=USNSF_51&WT.mc_ev=click

126 Peacock, D., & Irons, A. (2017) Gender inequality in cybersecurity: Exploring the gender gap in opportunities and progression. *International Journal of Gender, Science, & Technology, 9*(1). Retrieved from http://genderandset.open.ac.uk/index.php/genderandset/article/view/449

127 Ashford, W. (2017, June 15). *PaloAlto Networks partners with US Girl Scouts on security skills*. Computer Weekly. Retrieved from https://www.computerweekly.com/news/450420822/PaloAlto-Networks-partners-with-US-Girl-Scouts-on-security-skills

128 Maurer, R. (2017, January 10). *Why aren't women working in cybersecurity*? SHRM. Retrieved from https://www.shrm.org/resourcesandtools/hr-topics/talent-acquisition/pages/women-working-cybersecurity-gender-gap.aspx

129 *Economic impact and perceptions around the cybersecurity gender gap*. (2020, March 12). Help Net Security. Retrieved from https://www.helpnetsecurity.com/2020/03/12/cybersecurity-gender-gap/

130 Maurer. (2017). *Why aren't women working in cybersecurity*?

131 Arghuire, I. (2016, June 14). *CISOs risk getting fired over poor reporting*. SecurityWeek. Retrieved from http://www.securityweek.com/cisos-risk-getting-fired-over-poor-reporting-survey

132 Creswell, J., & Perlroth, N. (2014, September 20). Ex-employees say Home Depot left data vulnerable. *The New York Times*. Retrieved from http://www.nytimes.com/2014/09/20/business/ex-employees-say-home-depot-left-data-vulnerable.html

133 *Cost of a retail data breach: $179 million for Home Depot*. (2017, March 14). WebTitan. Retrieved from https://www.webtitan.com/blog/cost-retail-data-breach-179-million-home-depot/

134 Creswell & Perlroth. (2014). Ex-employees say Home Depot left data vulnerable.

135 Creswell & Perlroth. (2014). Ex-employees say Home Depot left data vulnerable.
136 Smith, G. (2014, September 24). Home Depot security team understaffed and overwhelmed for years, insiders say. *Huffington Post.* Retrieved from http://www.huffingtonpost.com/2014/09/24/home-depot-security-team-_n_5874178.html
137 Smith. (2014). Home Depot security team understaffed and overwhelmed for years, insiders say.
138 Smith. (2014). Home Depot security team understaffed and overwhelmed for years, insiders say.
139 Pauli, D. (2014, September 22). *Home Depot ignored staff warnings of security fail laundry list.* The Register. Retrieved from https://www.theregister.co.uk/2014/09/22/home_depot_ignored_staff_warnings_of_security_fail_laundry_list/
140 Reilly, R. (2014, September 18). *After attack and investigation, Home Depot says it has installed new security tools.* VentureBeat. Retrieved from https://venturebeat.com/2014/09/18/56-million-home-depot-customers-victim-in-stealthy-cyber-attack/
141 Reilly. (2014). *After attack and investigation, Home Depot says it has installed new security tools.*
142 Woolley, D. (2015, March 31). *Home Depot hires Jamil Farshchi as first CISO.* FierceRetail. Retrieved from https://www.fierceretail.com/operations/home-depot-hires-jamil-farshchi-as-first-ciso
143 Norton, S. (2015, March 30). Home Depot names Jamil Farshchi chief information security officer. *The Wall Street Journal.* Retrieved from https://www.wsj.com/articles/BL-CIOB-6658
144 Pwc. (2018). *Strengthening digital society against cyber shocks.* Retrieved from https://www.pwc.com/us/en/cybersecurity/assets/pwc-2018-gsiss-strengthening-digital-society-against-cyber-shocks.pdf
145 Pwc. (2018). *Strengthening digital society against cyber shocks.*
146 Uchill, J. (2018, May 22). *Equifax's new head of cyber enjoys direct line to CEO, board.* Axios. Retrieved from https://www.axios.com/equifax-chief-information-security-officer-ceo-board-fc44c2ca-2ba9-4789-b04b-bb253b41ca6c.html

Chapter 11

1 Kemp, S. (2020, January 30). *Digital 2020: 3.8 billion people use social media.* Wearesocial. Retrieved from https://wearesocial.com/blog/2020/01/digital-2020-3-8-billion-people-use-social-media
2 Cheung, M. (2020, March 11). *Social networking is nearly ubiquitous among internet users in China.* eMarketer. Retrieved from https://www.emarketer.com/content/social-networking-is-nearly-ubiquitous-among-internet-users-in-china
3 Dore, B. (2020, April 17). Fake news, real arrests. *Foreign Policy.* Retrieved from https://foreignpolicy.com/2020/04/17/fake-news-real-arrests/
4 Okunoye, B. (2019, January 10). *In Africa, a new tactic to suppress online speech: Taxing social media.* Council on Foreign Relations. Retrieved from https://www.cfr.org/blog/africa-new-tactic-suppress-online-speech-taxing-social-media
5 Gesenhues, A. (October 14, 2019). *Social media ad spend to surpass print for first time.* Marketing Land. Retrieved from https://marketingland.com/social-media-ad-spend-to-surpass-print-for-first-time-268998
6 *PPC News: Pinterest, Twitter, Facebook, Snapchat report Q4 earnings.* (2020, February 7). AORI. Retrieved from https://aori.com/blog/ppc-news-roundup-feb-7-2020
7 Gesenhues. (2019). *Social media ad spend to surpass print for first time.*

8 McCracken, H. (2019, December 24). *5 things Google got right in 2019 – and 5 it got wrong*. Fast Company. Retrieved from https://www.fastcompany.com/90441295/5-things-google-got-right-in-2019-and-5-it-got-wrong

9 Crawford, A. (2016, February 12). Paedophiles use secret Facebook groups to swap images. *BBC News*. Retrieved from http://www.bbc.com/news/uk-35521068

10 Crawford, A. (2017, March 7). Facebook failed to remove sexualized images of children. *BBC News*. Retrieved from http://www.bbc.com/news/technology-39187929

11 Kalra, A., & Varadhan, S. (2019, April 16). *TikTok vanishes from Google, Apple app stores in India after ban*. Reuters. Retrieved from https://www.reuters.com/article/us-tiktok-india-court/google-blocks-chinese-app-tiktok-in-india-after-court-order-idUSKCN1RS1HT

12 Meisenzahl, M. (2020, February 25). *US government agencies are banning TikTok, the social media app teens are obsessed with, over cybersecurity fears – here's the full list*. Business Insider. Retrieved from https://www.businessinsider.com/us-government-agencies-have-banned-tiktok-app-2020-2

13 Meisenzahl. (2020). *US government agencies are banning TikTok*.

14 Ashford, W. (2019, August 30). *Social media and enterprise apps pose big security risks*. Computer Weekly. Retrieved from https://www.computerweekly.com/news/252469873/Social-media-and-enterprise-apps-pose-big-security-risks

15 *2019 data breach investigations report*. (2019). Verizon. Retrieved May 5, 2020 from https://enterprise.verizon.com/resources/reports/dbir/

16 Bradley, T. (2020, February 5). Cybersecurity priorities are a matter of perspective. *Forbes*. Retrieved from https://www.forbes.com/sites/tonybradley/2020/02/05/cybersecurity-priorities-are-a-matter-of-perspective/#5575d6345d17

17 Cubrilovic, N. (2009, June 21). *Facebook click fraud enraging advertisers*. TechCrunch. Retrieved from http://techcrunch.com/2009/06/21/facebook-click-fraud-enraging-advertisers/#ixzz0onAAps6u

18 Lomas, N. (2018, February 12). *Unilever warns social media to clean up "toxic" content*. Tech Crunch. Retrieved from https://techcrunch.com/2018/02/12/unilever-warns-social-media-to-clean-up-toxic-content/

19 Graham, M. (2020, August 6). *Zuckerberg was right: Ad boycotts won't hurt Facebook that much*. CNBC. Retrieved from https://www.cnbc.com/2020/08/04/some-major-companies-will-keep-pausing-facebook-ads-as-boycott-ends.html

20 Palmer, D. (2020, January 28). *CEOs are deleting their social media accounts to protect against hackers*. ZDNet. Retrieved from https://www.zdnet.com/article/ceos-are-deleting-their-social-media-accounts-to-protect-against-hackers/

21 ComPsych Corporation. (2017, March 20). *Digital distraction at work: Almost 20 percent of workers check social media more than 10 times during workday, according to ComPsych*. Business Wire. Retrieved from https://www.businesswire.com/news/home/20170320005837/en/Digital-Distraction-Work-20-Percent-Workers-Check

22 Gervis, Z. (2019, December 6). More than 60% of Americans say they've been a victim of an online scam. *New York Post*. Retrieved from https://nypost.com/2019/12/06/more-than-60-of-americans-say-theyve-been-a-victim-of-an-online-scam/

23 Combs, V. (2020, February 6). *IoT is a gold mine for hackers using fileless malware for cyberattacks*. TechRepublic. Retrieved from https://www.techrepublic.com/article/iot-is-a-gold-mine-for-hackers-using-fileless-malware-for-cyberattacks/

24 Bajak, F. (2019, December 19). *Researcher: Data on 267 million Facebook users exposed*. The Associated Press. Retrieved from https://apnews.com/bdf02dbe7bf266b025b6f1b0ae5860fd

25 An update on our security incident. (2020, July 18). Twitter Inc. Retrieved from https://blog.twitter.com/en_us/topics/company/2020/an-update-on-our-security-incident.html
26 Greenberg, A. (2020, August 18). The attack that broke Twitter is hitting dozens of companies. *Wired*. Retrieved from https://www.wired.com/story/phone-spear-phishing-twitter-crime-wave/
27 Turnbull, M. (2019, December 4). *Africa faces a balancing act on social media regulation*. The Africa Report. Retrieved from https://www.theafricareport.com/20741/africa-faces-a-balancing-act-on-social-media-regulation/
28 Molla, R. (2002, March 12). *How coronavirus took over social media*. Vox. Retrieved from https://www.vox.com/recode/2020/3/12/21175570/coronavirus-covid-19-social-media-twitter-facebook-google
29 Allem, J. (2020, April 6). *Social media fuels wave of coronavirus misinformation as users focus on popularity, not accuracy*. The Conversation. Retrieved from https://theconversation.com/social-media-fuels-wave-of-coronavirus-misinformation-as-users-focus-on-popularity-not-accuracy-135179
30 Dorn, S. (2020, March 21). Don't believe these fake coronavirus stories on social media. *New York Post*. Retrieved from https://nypost.com/2020/03/21/dont-believe-these-fake-coronavirus-stories-on-social-media/
31 *Why the coronavirus has morphed into a social media nightmare*. (2020, March 29). The Straits Times. Retrieved from https://www.straitstimes.com/world/united-states/why-the-coronavirus-has-morphed-into-a-social-media-nightmare
32 Newall, M. (2018). *Cyberbullying: A global advisor survey*. Ipsos. Retrieved from https://www.ipsos.com/sites/default/files/ct/news/documents/2018-06/cyberbullying_june2018.pdf
33 Tanase, S. (2010). When Web 2.0 sneezes … everyone gets sick. *Engineering & Technology, 5*(5), 28–9.
34 Chheda, M. (2020, January 13). *Malicious Android app 'Shopper' takes over Google Play Store: Do not download this app!* International Business Times. Retrieved from https://www.ibtimes.sg/malicious-android-app-shopper-takes-over-google-play-store-do-not-download-this-app-37572
35 *Mystery shopper: Malicious app spreads fake reviews thanks to in-app accessibility service abuse*. (2020, January 9). Kaspersky. Retrieved from https://www.kaspersky.com/about/press-releases/2020_mystery-shopper-malicious-app-spreads-fake-reviews-thanks-to-in-app-accessibility-service-abuse
36 Marks, P. (2010). Social networks must heed the human element. *New Scientist, 206*(2763), 19.
37 *FTC imposes $5 billion penalty and sweeping new privacy restrictions on Facebook*. (2019, July 24). Federal Trade Commission. Retrieved from https://www.ftc.gov/news-events/press-releases/2019/07/ftc-imposes-5-billion-penalty-sweeping-new-privacy-restrictions
38 Rodrigo, C. (2020, March 31). As misinformation surges, coronavirus poses AI challenge. *The Hill*. Retrieved from https://thehill.com/policy/cybersecurity/490257-as-misinformation-surges-coronavirus-poses-ai-challenge
39 Douek, E. (2020, March 25). *COVID-19 and social media content moderation*. Lawfare. Retrieved from https://www.lawfareblog.com/covid-19-and-social-media-content-moderation
40 Statt, N. (2020, March 4). *Facebook's new AI-powered moderation tool helps it catch billions of fake accounts*. The Verge. Retrieved from https://www.theverge.com/2020/3/4/21164695/facebook-ai-moderation-dec-deep-entity-classification-fake-accounts-spam-scams
41 Constine, J. (2020, January 29). *Facebook hits 2.5B users in Q4 but shares sink from slow profits*. TechCrunch. Retrieved from https://techcrunch.com/2020/01/29/facebook-earnings-q4-2019/

42 *Facebook users in India.* (2020, January). NapoleonCat. Retrieved from https://napoleoncat.com/stats/facebook-users-in-india/2020/01

43 Kshetri, N. (2010). *The global cyber-crime industry: Economic, institutional and strategic perspectives.* Springer-Verlag.

44 Javers, E. (2019, November 8). *Here's why LinkedIn is a 'gold mine' for foreign spies digging for corporate and government secrets.* CNBC. Retrieved from https://www.cnbc.com/2019/11/08/linkedin-is-a-gold-mine-for-spies-seeking-corporate-govt-secrets.html

45 Javers. (2019). *Here's why LinkedIn is a 'gold mine' for foreign spies digging for corporate and government secrets.*

46 Bradshaw, S., Neudert, L., & Howard, P. (2018). *Government responses to malicious use of social media.* NATO Stratcom Centre of Excellence. Retrieved from https://www.stratcomcoe.org/government-responses-malicious-use-social-media

47 Dore. (2020) Fake news, real arrests.

48 Silverman, C., Singer-Vine, J., & Vo, L. (2017, April 4). *In spite of the crackdown, fake news publishers are still earning money from major ad networks.* BuzzFeed News. Retrieved from https://www.buzzfeednews.com/article/craigsilverman/fake-news-real-ads

49 Bosilkovski. (2019). Checking in with the Macedonian fake news strategist.

50 Bradshaw et al. (2018). *Government responses to malicious use of social media.*

51 Isi, G. (2018, April 24). *Governments' social media regulation in Africa: How possible?* Afro Hustler. Retrieved from https://www.afrohustler.com/social-media-regulation-africa-possible/

52 *Asia cracks down on coronavirus "fake news."* (2020, April 10). The Straits Times. Retrieved from https://www.straitstimes.com/asia/coronavirus-asia-cracks-down-on-virus-fake-news

53 *Asia cracks down on coronavirus "fake news."* (2020, April 10). The Straits Times.

54 Bradshaw et al. (2018). *Government responses to malicious use of social media.*

55 Mwakideu, C. (2019, May 11). *Some African governments using social media to monitor citizens: Freedom House report.* DW. Retrieved from https://www.dw.com/en/some-african-governments-using-social-media-to-monitor-citizens-freedom-house-report/a-51096624

56 Mwakideu. (2019). *Some African governments using social media to monitor citizens.*

57 Wild, S. (2020, April 3). *Citing virus misinformation, South Africa tests speech limits.* Undark. Retrieved from https://undark.org/2020/04/03/fake-news-south-africa-covid-19/

58 Kazeem, Y. (2016, April 6). *Politics and activism are driving Africa's Twitter conversations to new highs.* Quartz Africa. Retrieved from https://qz.com/africa/654958/politics-and-activism-are-driving-africas-twitter-conversations-to-new-highs/

59 Okunoye. (2019). *In Africa, a new tactic to suppress online speech.*

60 Consumers International. (2019, May). *Social media scams: Understanding the consumer experience to create a safer digital world.* Retrieved from https://www.consumersinternational.org/media/293343/social-media-scams-final-245.pdf

61 Thompson, L.A., Dawson K., & Ferdig, R. (2008). The intersection of online social networking with medical professionalism. *Journal of General Internal Medicine, 23*(7), 954–7.

62 Zimlich, R. (2019, October 27). *Beware the social media trap.* Contemporary Pediatrics. Retrieved from https://www.contemporarypediatrics.com/pediatrics/beware-social-media-trap

63 Fanaroff, J. (2019). *Fired for using Facebook: Law, ethics, and professionalism in social media.* Presented on October 27, 2019 at the 2019 American Academy of Pediatrics (AAP) Annual Conference and Exhibition in New Orleans, Louisiana. Retrieved from https://downloads.aap.org/aap/pdf/comlrm_nce_Sessions_2019.pdf

64 Jimenez, K. (2019, February 7). *Social media, texts have fueled numerous instances of teacher misconduct.* Voice of San Diego. Retrieved from https://www.voiceofsandiego.org/topics/education/social-media-texts-have-fueled-numerous-instances-of-teacher-misconduct/

65 Taylor, C. (2019, July 22). *Millennials posting pet pictures on Instagram are at risk of fraud, Santander warns.* CNBC. Retrieved from https://www.cnbc.com/2019/07/22/millennials-at-risk-of-fraud-over-instagram-pet-pictures-santander.html

66 Kirk, J. (2019, June 14). *Instagram shows kids' contact details in plain sight.* Bank Info Security. Retrieved from https://www.bankinfosecurity.com/instagram-shows-kids-contact-details-in-plain-sight-a-12631

67 *About professional accounts.* (n.d.). Instagram. Retrieved from https://help.instagram.com/138925576505882

68 Cowan, J. (2010). Why we'll never escape Facebook. *Canadian Business, 83*(10), 28–32.

69 Leclerc, B., Beauregard, E., & Proulx, J. (2008). Modus operandi and situational aspects in adolescent sexual offenses against children: A further examination. *International Journal of Offender Therapy and Comparative Criminology, 52*(1), 46–61.

70 Ramsey, G., & Venkatesan, S. (2010). Cybercrime strategy for social networking and other online platforms. *Licensing Journal, 30*(7), 23–7.

71 Javers. (2019). *Here's why LinkedIn is a 'gold mine' for foreign spies digging for corporate and government secrets.*

72 Balaouras, S. (2020, February 24). *Cybersecurity in 2020: The rise of the CISO.* MIT Technology Review. Retrieved from https://www.technologyreview.com/s/615092/cybersecurity-in-2020-the-rise-of-the-ciso/

73 Bajpai, P. (2019, December 11). *How Microsoft fights cyber threats.* Nasdaq. Retrieved from https://www.nasdaq.com/articles/how-microsoft-fights-cyber-threats-2019-12-11

74 Cadwalldr, C. (2020, January 4). Fresh Cambridge Analytica leak "shows global manipulation is out of control." *The Guardian.* Retrieved from https://www.theguardian.com/uk-news/2020/jan/04/cambridge-analytica-data-leak-global-election-manipulation

75 Agencies. (2018, March 28). *How Cambridge Analytica exploited Facebook users' data, and why it's a big deal.* South China Morning Post. Retrieved from https://www.scmp.com/news/world/united-states-canada/article/2139327/explainer-how-cambridge-analytica-exploited-facebook

76 Meredith, S. (2018, March 21). *Here's everything you need to know about the Cambridge Analytica scandal.* CNBC. Retrieved from https://www.cnbc.com/2018/03/21/facebook-cambridge-analytica-scandal-everything-you-need-to-know.html

77 Cadwalladr, C., & Graham-Harrison, E. (2018, March 17). Revealed: 50 million Facebook profiles harvested for Cambridge Analytica in major data breach. *The Guardian.* Retrieved from https://www.theguardian.com/news/2018/mar/17/cambridge-analytica-facebook-influence-us-election

78 Cadwalladr & Graham-Harrison. (2018). Revealed.

79 Chan, R. (2019, October 5). *The Cambridge Analytica whistleblower explains how the firm used Facebook data to sway elections.* Insider. Retrieved from https://www.businessinsider.com/cambridge-analytica-whistleblower-christopher-wylie-facebook-data-2019-10

80 Chan, R. (2019). *The Cambridge Analytica whistleblower explains how the firm used Facebook data to sway elections.*

81 Meredith, S. (2018, April 10). *Facebook-Cambridge Analytica: A timeline of the data hijacking scandal.* CNBC. Retrieved from https://www.cnbc.com/2018/04/10/facebook-cambridge-analytica-a-timeline-of-the-data-hijacking-scandal.html

82 Reley, M., & Frier, S. (2018, March 22). *Facebook data scandal: The key questions answered.* Irish Examiner. Retrieved from https://www.irishexaminer.com/ireland/facebook-data-scandal-the-key-questions-answered-468582.html
83 Meredith. (2018). *Facebook-Cambridge Analytica.*
84 Reley & Frier. (2018). *Facebook data scandal.*
85 Meredith. (2018). *Facebook-Cambridge Analytica.*
86 Meredith. (2018). *Facebook-Cambridge Analytica.*
87 *ICO issues maximum £500,000 fine to Facebook for failing to protect users' personal information.* (2018, October 25). Information Commissioner's Office. Retrieved from https://ico.org.uk/about-the-ico/news-and-events/news-and-blogs/2018/10/facebook-issued-with-maximum-500-000-fine/
88 Bruce, P. (2018, April 18). The Facebook controversy: Privacy is not the issue. *Scientific American.* Retrieved from https://blogs.scientificamerican.com/observations/the-facebook-controversy-privacy-is-not-the-issue/
89 Nieva, R. (2018, March 19). *Can Facebook be blamed for Cambridge Analytica?* CNet. Retrieved from https://www.cnet.com/news/can-facebook-be-blamed-for-cambridge-analytica/
90 Sherr, I. (2018, April 18). *Facebook, Cambridge Analytica and data mining: What you need to know.* CNet. Retrieved from https://www.cnet.com/news/facebook-cambridge-analytica-data-mining-and-trump-what-you-need-to-know/
91 *Scoop: Facebook's internal Cambridge Analytica document.* (2019, August 23). Byers Market. Retrieved from https://link.nbcnews.com/view/57c09634487ccd31218b6128amnse.5g5/46de4778/
92 Reley & Frier. (2018). *Facebook data scandal.*
93 Davies, H. (2015, December 11). Ted Cruz using firm that harvested data on millions of unwitting Facebook users. *The Guardian.* Retrieved from https://www.theguardian.com/us-news/2015/dec/11/senator-ted-cruz-president-campaign-facebook-user-data
94 Cadwalladr, C. (2018, March 18). "I made Steve Bannon's psychological warfare tool": Meet the data war whistleblower. *The Guardian.* Retrieved from https://www.theguardian.com/news/2018/mar/17/data-war-whistleblower-christopher-wylie-faceook-nix-bannon-trump
95 Chang, A. (2018, May 2). *The Facebook and Cambridge Analytica scandal, explained with a simple diagram.* Vox. Retrieved from https://www.vox.com/policy-and-politics/2018/3/23/17151916/facebook-cambridge-analytica-trump-diagram
96 Wong, J. (2019, August 23). Document reveals how Facebook downplayed early Cambridge Analytica concerns. *The Guardian.* Retrieved from https://www.theguardian.com/technology/2019/aug/23/cambridge-analytica-facebook-response-internal-document
97 *ICO issues maximum £500,000 fine to Facebook for failing to protect users' personal information.* (2018, October 25). Information Commissioner's Office. Retrieved from https://ico.org.uk/about-the-ico/news-and-events/news-and-blogs/2018/10/facebook-issued-with-maximum-500-000-fine/
98 Sherr. (2018). *Facebook, Cambridge Analytica and data mining.*
99 Zialcita, P. (2019, October 30). *Facebook pays $643,000 fine for role in Cambridge Analytica scandal.* NPR. Retrieved from https://www.npr.org/2019/10/30/774749376/facebook-pays-643-000-fine-for-role-in-cambridge-analytica-scandal
100 *ICO issues maximum £500,000 fine to Facebook for failing to protect users' personal information.* (2018). Information Commissioner's Office.

101 Swinhoe, D. (2019, August 5). *GDPR vs UK Data Protection Act 2018: What's the difference?* CSO. Retrieved from https://www.csoonline.com/article/3410039/gdpr-vs-uk-data-protection-act-2018-whats-the-difference.html

102 Zialcita. (2019). *Facebook pays $643,000 fine for role in Cambridge Analytica scandal.*

103 *FTC imposes $5 billion penalty and sweeping new privacy restrictions on Facebook.* (2019). Federal Trade Commission.

104 Hamilton, I. (2019, July 24). *Facebook just got clobbered with a record $5 billion penalty over the Cambridge Analytica data breach.* Business Insider. Retrieved from https://www.businessinsider.com/facebook-ftc-record-penalty-mark-zuckerberg-2019-5

105 Meredith. (2018). *Here's everything you need to know about the Cambridge Analytica scandal.*

106 Miller, M. (2019, December 30). Brazil fines Facebook $1.6 million over improper data sharing. *The Hill.* Retrieved from https://thehill.com/policy/cybersecurity/476274-brazil-fines-facebook-16-million-over-improper-data-sharing

107 Agencies. (2018). *How Cambridge Analytica exploited Facebook users' data, and why it's a big deal.*

108 Butcher, M. (2020, January 17). *Cambridge Analytica email chain with Facebook sheds new light on data misuse scandal.* TechCrunch. Retrieved from https://techcrunch.com/2020/01/17/cambridge-analytica-whistleblower-reveals-new-documents-showing-how-facebook-handled-the-data-misuse-scandal/

109 Reley & Frier. (2018). *Facebook data scandal.*

110 Sherr. (2018). *Facebook, Cambridge Analytica and data mining.*

111 Zialcita. (2019). *Facebook pays $643,000 fine for role in Cambridge Analytica scandal.*

Chapter 12

1 Ingham, L. (2020, March 11). *Public cloud will be home to 76% of IT infrastructure by 2025, but security concerns remain.* Verdict. Retrieved from https://www.verdict.co.uk/public-cloud-security/

2 *Dropbox's three-pronged defense against digital fraud.* (2020, February 28). PYMNTS. Retrieved from https://www.pymnts.com/fraud-prevention/2020/dropboxs-three-pronged-defense-against-digital-fraud/

3 Talbot, D. (2010). Security in the ether. *Technology Review, 113*(1), 36–42.

4 Ely, A. (2011, March 3). 5 steps to secure SaaS. *InformationWeek*, 17. Retrieved from https://www.networkcomputing.com/5-steps-secure-saas

5 Ricadela, A. (2011, September 1). *Cloud security is looking overcast.* Businessweek. Retrieved from http://www.businessweek.com/magazine/cloud-security-is-looking-overcast-09012011.html

6 Barnott, G. (2019, December 16). *McAfee reveals survey results and breaks down complexities behind cloud security migration.* DevOps Online. Retrieved from https://www.devopsonline.co.uk/mcafee-reveals-survey-results-and-breaks-down-complexities-behind-cloud-security-migration/

7 Moyer, S. (2019, December 17). *Majority of financial professionals don't trust cloud computing.* Financial Director. Retrieved from https://www.financialdirector.co.uk/2019/12/17/majority-of-financial-professionals-dont-trust-cloud-computing/

8 Braue, D. (2019, December 2). *Cloud security is taking a back seat to cloud enthusiasm.* CSO. Retrieved from https://www2.cso.com.au/article/669298/cloud-security-taking-back-seat-cloud-enthusiasm/

9 Bourne, J. (2020, January 28). *More sensitive data moves to the enterprise cloud – but the security risk widens with it*. Cloud Tech News. Retrieved from https://www.cloudcomputing-news.net/news/2020/jan/28/more-sensitive-data-moves-enterprise-cloud-security-risk-widens-it/

10 Ingham. (2020). *Public cloud will be home to 76% of IT infrastructure by 2025, but security concerns remain*.

11 Tillery, S. (2010, October 15). *How safe is the cloud?* Baseline. Retrieved from http://www.baselinemag.com/c/a/Security/How-Safe-Is-the-Cloud-273226

12 *Cloud data theft affects 1 in 4*. (2018). McAfee. Retrieved from https://www.mcafee.com/enterprise/en-us/solutions/lp/cloud-security-report.html

13 *Cloud data theft affects 1 in 4*. (2018). McAfee.

14 Sheridan, K. (2020, February 19). *44% of security threats start in the cloud*. Dark Reading. Retrieved from https://www.darkreading.com/cloud/44--of-security-threats-start-in-the-cloud/d/d-id/1337088

15 *Global cyber risk perception survey*. (2019). Marsh.

16 Spadafora, A. (2019, May 15). *China Telecom outage takes down Apple, Microsoft and AWS for 7 hours*. Techradar.pro. Retrieved from https://www.techradar.com/news/china-telecom-outage-takes-down-apple-microsoft-and-aws-for-7-hours

17 Larkin, E. (2010). Will cloud computing kill privacy? *PCWorld, 28*(3), 44.

18 Wittow, M.H., & Buller, D.J. (2010). Cloud computing: Emerging legal issues for access to data, anywhere, anytime. *Journal of Internet Law, 14*(1), 1–10.

19 Ingthorsson, O. (2011, August 29). *Regulations a barrier to cloud growth in Europe*. Data Center Knowledge. Retrieved from http://www.datacenterknowledge.com/archives/2011/08/29/enhancing-cloud-development-in-europe/

20 Crosman, P. (2009, December 1). Securing the clouds. *Wall Street & Technology*, 23.

21 Olavsrud, T. (2012). *Security in the cloud is all about visibility and control*. CIO. Retrieved from https://www.cio.com/article/2399225/security-in-the-cloud-is-all-about-visibility-and-control.html

22 McCann, D. (2010, March 24). *Accountants head to the cloud*. CFO. Retrieved from http://cfo.com/article.cfm/14484960/c_14485112?f=home_todayinfinance

23 Wittow & Buller. (2010). Cloud computing.

24 Allen, J.M. (2011). Cloud computing: Heavenly solution or pie in the sky? *Pennsylvania CPA Journal, 82*(1), 1–4.

25 Wilshusen, G.C. (2010). Information security federal guidance needed to address control issues with implementing cloud computing. *GAO Reports*. Retrieved from https://www.gao.gov/products/gao-10-513

26 Messmer, E. (2010, July 6). *Secrecy of cloud computing providers raises IT security risks*. NetworkWorld. Retrieved from https://www.networkworld.com/article/2214080/secrecy-of-cloud-computing-providers-raises-it-security-risks.html

27 Wilshusen. (2010). Information security federal guidance needed to address control issues with implementing cloud computing.

28 Crosman. (2009). Securing the clouds.

29 Wittow & Buller. (2010). Cloud computing.

30 Wittow & Buller. (2010). Cloud computing.

31 Coble, S. (2019, December 2). *Recommendations published for EU cloud service providers certification scheme*. InfoSecurity Group. Retrieved from https://www.infosecurity-magazine.com/news/eu-cloud-certification-scheme/

32 Zucker, L. (1986). Production of trust: Institutional sources of economic structure 1840–1920. *Research in Organizational Behaviour, 8*, 3–11.
33 Turner, K. (2016, September 7). Hacked Dropbox login data of 68 million users is now for sale on the dark Web. *The Washington Post*. Retrieved from https://www.washingtonpost.com/news/the-switch/wp/2016/09/07/hacked-dropbox-data-of-68-million-users-is-now-or-sale-on-the-dark-web/
34 Cox, J. (2016, August 30). *Hackers stole account details for over 60 million Dropbox users*. Vice. Retrieved from https://www.vice.com/en/article/nz74qb/hackers-stole-over-60-million-dropbox-accounts
35 Agarwal, A. (2012, July 31). *Security update and new features*. Dropbox. Retrieved from https://blog.dropbox.com/topics/company/security-update-new-features
36 Schwartz, M.J. (2016, May 18). *LinkedIn breach: Worse than advertised*. Bank Info Security. Retrieved from https://www.bankinfosecurity.com/linkedin-breach-worse-than-advertised-a-9113
37 Gibbs, S. (2016, August 31). Dropbox hack leads to leaking of 68m user passwords on the internet. *The Guardian*. Retrieved from https://www.theguardian.com/technology/2016/aug/31/dropbox-hack-passwords-68m-data-breach#:~:text=Popular%20cloud%20storage%20firm%20Dropbox,had%20been%20stolen%20as%20well
38 Cox. (2016). *Hackers stole account details for over 60 million Dropbox users*.
39 Newman, L.H. (2016, August 31). Hack brief: 4-year-old Dropbox hack exposed 68 million people's data. *Wired*. Retrieved from https://www.wired.com/2016/08/hack-brief-four-year-old-dropbox-hack-exposed-68-million-peoples-data/
40 Brewster, T. (2016, August 31). Why you shouldn't panic about Dropbox leaking 68 million passwords. *Forbes*. Retrieved from https://www.forbes.com/sites/thomasbrewster/2016/08/31/dropbox-hacked-but-its-not-that-bad/?sh=6fec27565576
41 Newman. (2016). Hack brief.
42 Ferguson, S. (2020, September 30). *Russian gets 7-year sentence for hacking LinkedIn, Dropbox*. Bank Info Security. Retrieved from https://www.bankinfosecurity.com/russian-gets-7-year-sentence-for-hacking-linkedin-dropbox-a-15092

Chapter 13

1 *Generating value from big data analytics* [White paper]. (2014). ISACA. Retrieved from http://www.isaca.org/Knowledge-Center/Research/ResearchDeliverables/Pages/Generating-Value-From-Big-Data-Analytics.aspx
2 Cohen, A. (2013, February 4). Will "stalking apps" be stopped? *Time*. Retrieved from http://ideas.time.com/2013/02/04/will-stalking-apps-be-stopped/
3 Truxillo, C. (2013, July 8). *Five myths about unstructured data and five good reasons you should be analyzing it*. SAS. Retrieved from http://blogs.sas.com/content/sastraining/2013/07/08/five-myths-about-unstructured-data-and-five-good-reasons-you-should-be-analyzing-it/
4 Hooper, J. (2010, February 24). Google executives convicted in Italy over abuse video, *The Guardian*. Retrieved from http://www.theguardian.com/technology/2010/feb/24/google-video-italy-privacy-convictions
5 Brill, J. (2012, March 2). *Big data, big issues, Fordham University School of Law*. Federal Trade Commission. Retrieved from http://www.ftc.gov/public-statements/2012/03/big-data-big-issues

6 Nevskaya, Y. (2012). *Consumer information asymmetry in online product reviews* [Working paper]. William E. Simon Graduate School of Business, University of Rochester.
7 Talbot, D. (2013, October 9). *Data discrimination means the poor may experience a different Internet.* MIT Technology Review. Retrieved from https://www.technologyreview.com/2013/10/09/112852/data-discrimination-means-the-poor-may-experience-a-different-internet/#:~:text=Expand%20menu-,Data%20Discrimination%20Means%20the%20Poor%20May%20Experience%20a%20Different%20Internet,analytics%20were%20used%20against%20them
8 Hudgins, V. (2019, November 11). *Big data creates big problems: Organizations mandating data minimization efforts.* Law.com. Retrieved from https://www.law.com/legaltechnews/2019/11/11/big-data-creates-big-problems-organizations-mandating-data-minimization-efforts/
9 Hudgins. (2019). *Big data creates big problems.*
10 Kshetri, N. (2016). Big data's role in expanding access to financial services in China. *International Journal of Information Management, 36*(3), 297–308.
11 *Online credit services in China accused of abusing personal data.* (2015). Wantchinatimes.com. Retrieved May 5, 2016 from http://www.wantchinatimes.com/news/content?id=20150927000087&cid=1203
12 Mayer-Schönberger, V., & Cukier, K. (2013). *Big data: A revolution that will transform how we live, work and think.* Houghton Mifflin Harcourt.
13 Mayer-Schönberger & Cukier. (2013). *Big data.*
14 Einav, L., & Levin, J. (2013). *The data revolution and economic analysis.* National Bureau of Economic Research. Retrieved from http://www.nber.org/papers/w19035
15 *Big data: What it is and why it matters.* (n.d.). SAS. Retrieved from https://www.sas.com/en_us/insights/big-data/what-is-big-data.html
16 IMF. (2013). Report finds data breaches mainly involve outsourced IT. *Information Management Journal, 47*(3), 16.
17 Wood, P. (2013, March 4). *How to tackle big data from a security point of view.* Computer Weekly. Retrieved from http://www.computerweekly.com/feature/How-to-tackle-big-data-from-a-security-point-of-view
18 Kshetri, N. (2013). Privacy and security issues in cloud computing: The role of institutions and institutional evolution. *Telecommunications Policy, 37*(4–5), 372–86.
19 Wood. (2013). *How to tackle big data from a security point of view.*
20 Mantelero, A. (2012). Cloud computing, trans-border data flows and the European Directive 95/46/EC: Applicable law and task distribution. *European Journal for Law and Technology, 3*(2).
21 Tung, L. (2013, June 14). *Sweden tells council to stop using Google Apps.* ZDNet. Retrieved from http://www.zdnet.com/sweden-tells-council-to-stop-using-google-apps-7000016850/
22 Kshetri, N. (2014). Big data's impact on privacy, security and consumer welfare. *Telecommunications Policy, 38,* 1134–45.
23 Varonis Systems. (2008). *Ponemon study.*
24 *Generating value from big data analytics* [White paper]. (2014). ISACA.
25 Ahmed, M. (2019, October 22). *Data lakes: Where big businesses dump their excess data, and hackers have a field day.* GCN. Retrieved from https://gcn.com/articles/2019/10/22/data-lake-security.aspx
26 *Do you actually need a data lake?* (2019, November 6). insideBIGDATA. Retrieved from https://insidebigdata.com/2019/11/06/do-you-actually-need-a-data-l
27 *Cyber criminals: Who they are and why they do it.* (2018, February 7). Vircom. Retrieved from https://www.vircom.com/blog/cybercriminals-who-they-are-and-why-they-do-it/

28 Crawford, K., & Schultz, J.M. (2013). *Big data and due process: Toward a framework to redress predictive privacy harms*. New York University Public Law and Legal Theory Working Papers. Paper 429. Retrieved May 5, 2014 from http://lsr.nellco.org/nyu_plltwp/429/
29 Duhigg, C. (2012, February 16). How companies learn your secrets. *The New York Times*. Retrieved from http://www.nytimes.com/2012/02/19/magazine/shopping-habits.html?_r=1&pagewanted=all
30 Brill. (2013). Demanding transparency from data brokers.
31 Teufel, H., III. (2008). *Privacy policy guidance memorandum*. Memorandum Number: 2008-01. The Privacy Office US Department of Homeland Security.
32 Skok, G. (2000). Establishing a legitimate expectation of privacy in clickstream data. *Michigan Telecommunications & Technology Law Review*. Retrieved from http://cyber.law.harvard.edu/privacy/PrivacyInClickstream%28Skok%29.htm
33 *CDT's guide to online privacy*. (2000). Center for Democracy & Technology. Retrieved May 5, 2010 from http://www.cdt.org/privacy/guide/start
34 Teufel. (2008). *Privacy policy guidance memorandum*.
35 King, N.J., & Jessen, P.W. (2010). Profiling the mobile customer – privacy concerns when behavioural advertisers target mobile phones – part I. *Computer Law and Security Review, 26*(6), 595–612.
36 Austin, S., & Dowell, A. (2012, March 31). "Girls around me" developer defends app after Foursquare dismissal. *The Wall Street Journal*. Retrieved from https://www.wsj.com/articles/BL-DGB-24194
37 Yan, Z. (2012, July 4). *Personal data crimes set to be defined*. China Daily. Retrieved from www.chinadaily.com.cn/china/2012-07/04/content_15546503.htm
38 Secure auditing of information systems. United States Patent Application 20030220940. (2003). FPO. Retrieved from http://www.freepatentsonline.com/y2003/0220940.html
39 Gens, F. (2011). *Predictions 2012: Competing for 2020*. IDC. Retrieved May 5, 2013 from http://cdn.idc.com/research/Predictions12/Main/downloads/IDCTOP10Predictions2012.p
40 Varonis Systems. (2008). *Ponemon study*.
41 Varonis Systems. (2008). *Ponemon study*.
42 Fonseca, B. (2008, July 1). *Unstructured data at risk in most firms, survey finds*. Computerworld. Retrieved from http://www.computerworld.com/article/2534496/data-center/unstructured-data-at-risk-in-most-firms--survey-finds.html
43 Rosenbush, S. (2014, April 2). Few businesses are focused on unstructured data. *The Wall Street Journal*. Retrieved from https://www.wsj.com/articles/BL-CIOB-4206
44 Pirlot, A. (2014, January 21). Big data: A tool for development or threat to privacy? Privacy International. Retrieved from https://privacyinternational.org/news-analysis/1434/big-data-tool-development-or-threat-privacy
45 *Generating value from big data analytics* [White paper]. (2014). ISACA.
46 Talbot. (2013). Data discrimination means the poor may experience a different Internet.
47 Brill. (2012). *Big data, big issues, Fordham University School of Law*.
48 Crawford, K., & Schultz, J.M. (2013). Big data and due process: Toward a framework to redress predictive privacy harms.
49 Teufel. (2008). *Privacy policy guidance memorandum*.
50 Manos, D. (2011, March 18). *Patients sue Walgreens for making money on their data*. HealthcareITnews. Retrieved from http://www.healthcareitnews.com/news/patients-sue-walgreens-making-money-their-data

51 Sullivan, B. (2003, August 26). Who profits from spam? Surprise. *NBC News*. Retrieved from https://www.nbcnews.com/id/wbna3078642
52 Sullivan. (2003). Who profits from spam?
53 Malim, G. (2018, June 18). *The arrival of 5G – do you have a fast data strategy?* Vanilla Plus. Retrieved from https://www.vanillaplus.com/2018/06/18/39505-arrival-5g-fast-data-strategy/
54 Dixon, M. (2018, December 21). *How 5G and data analytics will change the way we live*. Selerity. Retrieved from https://seleritysas.com/blog/2018/12/21/5g-data-analytics-change-the-way-we-live/
55 Kuo, M. (2019, January 15). *The quest for 5G technology dominance: Impact on US national security*. The Diplomat. Retrieved from https://thediplomat.com/2019/01/the-quest-for-5g-technology-dominance-impact-on-us-national-security/
56 Williams, R. (2019, July 15). *Securing 5G networks*. Council on Foreign Relations. Retrieved from https://www.cfr.org/report/securing-5g-networks
57 Williams. (2019). *Securing 5G networks*.
58 Reddig, G. (2018, December 27). *In 5G we trust. Why flexible security is a 5G business essential*. Nokia. Retrieved from https://www.nokia.com/blog/5g-we-trust-why-flexible-security-5g-business-essential
59 Smith, M., & Mulrain, G. (2017). Equi-failure: The national security implications of the Equifax hack and a critical proposal for reform. *Journal of National Security Law & Policy, 9*, 549–88.
60 Belmonte, A. (2019, July 25). Equifax settled its big data breach lawsuit. Here's how much consumers could get. *Huffington Post*. Retrieved from https://www.huffpost.com/entry/how-to-get-equifax-money-data-breach-lawsuit_l_5d38bd4be4b004b6adba579a
61 FBI. (2020). *Chinese PLA members, 54th research institute*. Retrieved from https://www.fbi.gov/wanted/cyber/chinese-pla-members-54th-research-institute
62 Hymes, C., & Becket, S. (2020, February 11). Justice Department charges 4 members of Chinese military for massive Equifax hack. *CBS News*. Retrieved from https://www.cbsnews.com/news/equifax-hack-chinese-military-members-charged-department-of-justice/
63 Corera, G. (2020, February 11). Equifax: US charges four Chinese military officers over huge hack. *BBC News*. Retrieved from https://www.bbc.com/news/world-us-canada-51449778
64 Corera. (2020). Equifax.
65 Belmonte. (2019). Equifax settled its big data breach lawsuit.
66 Belmonte. (2019). Equifax settled its big data breach lawsuit.
67 Hymes & Becket. (2020). Justice Department charges 4 members of Chinese military for massive Equifax hack.
68 Fruhlinger, J. (2019, February 12). *Equifax data breach FAQ: What happened, who was affected, what was the impact?* CSO. Retrieved from https://www.csoonline.com/article/3444488/equifax-data-breach-faq-what-happened-who-was-affected-what-was-the-impact.html
69 Hymes & Becket. (2020). Justice Department charges 4 members of Chinese military for massive Equifax hack.
70 Fruhlinger. (2019). *Equifax data breach FAQ.*
71 Andriotis, A. (2018, February 9). Equifax hack might be worse than you think. *The Wall Street Journal*. Retrieved from https://www.wsj.com/articles/equifax-hack-might-be-worse-than-you-think-1518191370?mod=searchresults&page=1&pos=3
72 Borak, D., & Vasel, K. (2018, February 10). *The Equifax hack could be worse than we thought*. CNN. Retrieved from https://money.cnn.com/2018/02/09/pf/equifax-hack-senate-disclosure/index.html
73 Fruhlinger. (2019). *Equifax data breach FAQ.*

Chapter 14

1 Dowden, M. (2014, March 13). *UK: Smart cities: The end of privacy or the key to active citizenship?* Mondaq. Retrieved from http://www.mondaq.com/x/299362/Data+Protection+Privacy/Smart+Cities+The+End+Of+Privacy+Or+The+Key+To+Active+Citizenship
2 Perlroth, N. (2015, April 21). Smart city technology may be vulnerable to hackers. *The New York Times*. Retrieved from http://bits.blogs.nytimes.com/2015/04/21/smart-city-technology-may-be-vulnerable-to-hackers/?_r=0
3 Kuranda, S. (2016, November 4). *Security experts: IoT will be biggest threat of the next decade*. CRN. Retrieved from http://www.crn.com/news/security/300082723/security-experts-iot-will-be-biggest-threat-of-the-next-decade.htm
4 Mann, T. (2019, October 23). *Nokia vp: 5G security risks are huge*. SdxCentral. Retrieved from https://www.sdxcentral.com/articles/news/nokia-vp-5g-security-risks-are-huge/2019/10/
5 *Global cyber risk perception survey*. (2019).
6 *How much do organizations understand the risk exposure of IoT devices?* (2019, August 1). Deloitte. Retrieved from https://www2.deloitte.com/us/en/pages/about-deloitte/articles/press-releases/deloitte-cyber-shares-top-cyber-risks-for-iot-environments.html
7 Chin, J. (2016). Security experts question Chinese webcam maker's response to cyberattack. *The Wall Street Journal*. Retrieved from http://www.wsj.com/articles/security-experts-question-chinese-webcam-makers-response-to-cyberattack-1477404426
8 *41.6 billion IoT devices will be generating 79.4 zettabytes of data in 2025*. (2019, June 21). Help Net Security. Retrieved from https://www.helpnetsecurity.com/2019/06/21/connected-iot-devices-forecast/
9 Wilson, C. (2016, February 12). *Level 3's Drew sees liability issues in IoT botnets*. LightReading. Retrieved from http://www.lightreading.com/security/security-strategies/level-3s-drew-sees-liability-issues-in-iot-botnets/d/d-id/728748
10 *Smart cities and big data: Challenges and opportunities.* (2014). European Utility Week. Retrieved February 5, 2015 from http://www.european-utility-week.com/SmartCitiesandBigData
11 Symantec. (2013). *Transformational "smart cities": Cyber security and resilience*. Retrieved from https://securingsmartcities.org/wp-content/uploads/2019/03/Transformational-Smart-Cities-Symantec-Executive-Report.pdf
12 Brewster, T. (2014, May 21). Smart or stupid: Will our cities of the future be easier to hack? *The Guardian*. Retrieved from http://www.theguardian.com/cities/2014/may/21/smart-cities-future-stupid-hack-terrorism-watchdogs
13 Kitchin, R. (2014). The real-time city? Big data and smart urbanism. *GeoJournal, 79*, 1–14.
14 Cecco, L. (2020, May 7). Google affiliate Sidewalk Labs abruptly abandons Toronto smart city project. *The Guardian*. Retrieved from https://www.theguardian.com/technology/2020/may/07/google-sidewalk-labs-toronto-smart-city-abandoned
15 China plans to track Beijing citizens through their mobiles. (2011, March 4). *The Guardian*. Retrieved from http://www.theguardian.com/world/2011/mar/04/china-tracking-beijing-citizens-mobiles
16 *A closer look at smart cities*. (2014, November 18). MIT Technology Review. Retrieved from https://www.technologyreview.com/2014/11/18/170383/a-closer-look-at-smart-cities/
17 Singh, I.B., & Pelton, J.N. (2013). Securing the cyber city of the future. *Futurist, 47*(6), 22–7.

18 Wendzel, S., et al. (2014). *Envisioning smart building botnets.* In Lecture Notes in Informatics (LNI), Proceedings – Series of the Gesellschaft fur Informatik (GI), pp. 319–29.
19 Cadzow, S. (2013). *Privacy and security in smart cities: The i-tour contribution to protecting the citizen.* Cadzow Communications Consulting Limited (C3L). Retrieved from https://docplayer.net/amp/3236243-Privacy-and-security-in-smart-cities-the-i-tour-contribution-to-protecting-the-citizen-scott-cadzow-c3l-for-i-tour.html
20 Wendzel et al. (2014). *Envisioning smart building botnets.*
21 Chipley, M. (2014, February 21). *Cybersecurity: Whole Building Design Guide.* Retrieved from http://www.wbdg.org/resources/cybersecurity.php?r=secure_safe
22 Ernst, A. (2015, February 5). *Is this the future of cyberwarfare? Experts warn that new malware called BlackEnergy could be used to sabotage America's most critical infrastructure.* Al Jazeera. Retrieved from http://america.aljazeera.com/watch/shows/america-tonight/articles/2015/2/5/blackenergy-malware-cyberwarfare.html
23 Carr, J. (2012, September 10). *Why wasn't Saudi Aramco's oil production targeted?* Digital Dao. Retrieved from http://jeffreycarr.blogspot.de/2012/09/why-wasnt-saudi-aramcos-oil-production.html
24 Cornwell, A. (2014, October 16). *Cyber attacks an increasing threat for Mideast oil and gas.* Business. Retrieved from http://gulfnews.com/business/oil-gas/cyber-attacks-an-increasing-threat-for-mideast-oil-and-gas-1.1399982
25 Chipley. (2014). *Cybersecurity.*
26 Kovacs, E. (2014, December 18). *Cyberattack on German steel plant caused significant damage: Report.* Security Week. Retrieved from http://www.securityweek.com/cyberattack-german-steel-plant-causes-significant-damage-report
27 Zetter, K. (2015, January 8). A cyberattack has caused confirmed physical damage for the second time ever. *Wired.* Retrieved from http://www.wired.com/2015/01/german-steel-mill-hack-destruction/
28 Ranger, S. (2015, January 27). *Internet of things: A security threat to business by the backdoor?* ZDNet. Retrieved from http://www.zdnet.com/article/internet-of-things-a-security-threat-to-business-by-the-backdoor/
29 Witter, B. (2013, Fall). *The move beyond building automation systems to a more secure energy infrastructure.* Area Development. Retrieved from http://www.areadevelopment.com/AssetManagement/Q4-2013/secure-facility-energy-management-systems-22627256.shtml
30 Storm, D. (2014, June 4). *Botnets coming soon to a smart home or automated building near you.* ComputerWorld. Retrieved from http://www.computerworld.com/article/2476386/cybercrime-hacking/botnets-coming-soon-to-a-smart-home-or-automated-building-near-you.html
31 Petock, M. (2013, March). *Cyber threats.* Retrieved from http://automatedbuildings.com/news/mar13/articles/lynxspring/130218033505lynxspring.html
32 King, R. (2013, April 5). Cyber attackers target building management systems. *The Wall Street Journal.* Retrieved from https://www.wsj.com/articles/BL-CIOB-1827
33 King. (2013). Cyber attackers target building management systems.
34 Woods, E. (2013) *Why smart cities need smart grids.* Retrieved May 5, 2014 from http://www.navigantresearch.com/blog/why-smart-cities-need-smart-grids
35 Perera, D. (2015, January 1). *Smart grid powers up privacy worries.* Politico. Retrieved from http://www.politico.com/story/2015/01/energy-electricity-data-use-113901.html
36 Perera. (2015). *Smart grid powers up privacy worries.*

37 Kitchin. (2014). The real-time city?

38 Martinez-Balleste, A., Perez-Martinez, P., & Solanas, A. (2013). The pursuit of citizens' privacy: A privacy-aware smart city is possible. *IEEE Communications Magazine, 51*(6), 136–41. Retrieved from http://ieeexplore.ieee.org/xpl/articleDetails.jsp?arnumber=6525606

39 Perera. (2015). *Smart grid powers up privacy worries.*

40 *Smart cities and big data.* (2014).

41 Murrill, B.J., Liu, E.C., & Thompson, R.M. (2012). *Smart Meter Data: Privacy and Cybersecurity.* Congressional Research Service.

42 Warning over smart meters privacy risk. (2012, June 12). *BBC News.* Retrieved from http://www.bbc.com/news/technology-18407340

43 Rebollo-Monedero, D., Bartoli, A., Hernández-Serrano, J., Forné, J., & Soriano, M. (2014). Reconciling privacy and efficient utility management in smart cities. *European Transactions on Telecommunications, 25*(1), 94–108. doi: 10.1002/ett.2708

44 *PwC's Global Information Security Survey – Japanese companies are behind the global standard.* PwC. (2015). Retrieved May 5, 2016 from http://www.pwc.com/jp/ja/advisory/press-room/news-release/2014/information-security-survey141105.jhtm

45 Bort, J. (2014, January 16). *For the first time, hackers have used a refrigerator to attack businesses.* Insider. Retrieved from http://www.businessinsider.com/hackers-use-a-refridgerator-to-attack-businesses-2014-1

46 Stanganelli, J. (2013, October 31). *They want your enterprise brains: Night of the botnet of things.* Enterprise Networking Planet. Retrieved from http://www.enterprisenetworkingplanet.com/netsecur/when-smart-devices-attack-the-botnet-of-things.html

47 *Internet of things will change cyber security forever.* (2015, September 2). Gartner. Retrieved from https://www.gartner.com/en/newsroom/press-releases/2015-09-02-gartner-says-that-the-internet-of-things-will-change-cybersecurity-forever

48 Simonite, T. (2013, August 1). *Hacking industrial systems turns out to be easy.* MIT Technology Review. Retrieved from https://www.technologyreview.com/2013/08/01/177124/hacking-industrial-systems-turns-out-to-be-easy/

49 Industrial Internet Consortium. (2016). *Industrial internet of things volume G4: Security framework.* Retrieved from https://www.iiconsortium.org/pdf/IIC_PUB_G4_V1.00_PB.pdf

50 Dickson, B. (2016, October 29). *Blockchain could help fix IoT security after DDoS attack.* VentureBeat. Retrieved from http://venturebeat.com/2016/10/29/blockchain-could-help-fix-iot-security-after-ddos-attack/

51 Leonard, J. (2016). *IoT-enabled botnet launches record 1.5Tbps DDoS attack.* The Inquirer. Retrieved May 5, 2017 from http://www.theinquirer.net/inquirer/news/2472432/iot-enabled-botnet-launches-record-15tbps-ddos-attack

52 Hill, K. (2015, September 2). *Watch out, new parents – internet-connected baby monitors are easy to hack.* Splinter. Retrieved from https://splinternews.com/watch-out-new-parents-internet-connected-baby-monitors-1793850489

53 Constantin, L. (2016, September 13). *Hackers found 47 new vulnerabilities in 23 IoT devices at DEF CON.* Computerworld. Retrieved from http://www.computerworld.com/article/3119766/security/hackers-found-47-new-vulnerabilities-in-23-iot-devices-at-def-con.html

54 *How much of a security problem is ransomware for IoT?* (2016). IDGAnswers. Retrieved May 5, 2017 from http://www.idganswers.com/question/29660/how-much-of-a-security-problem-is-ransomware-for-iot#src=ifw

55 Greenberg, A. (2016, August 1). The Jeep hackers are back to prove car hacking can get much worse. *Wired*. Retrieved from https://www.wired.com/2016/08/jeep-hackers-return-high-speed-steering-acceleration-hacks/
56 Schneier, B. (2016, October 16). *Security economics of the internet of things*. Schneier on Security. Retrieved from http://www.schneier.com/blog/archives/2016/10/security_econom_1.html
57 Baron, E. (2016, October 24). *Cybersecurity experts call for "internet of things" standards in wake of massive attack*. The Mercury News. Retrieved from http://www.mercurynews.com/2016/10/24/cybersecurity-experts-call-for-internet-of-things-standards-in-wake-of-massive-attack/
58 Herrod, S. (2016, November 2). IoT poses a unique challenge to the security industry. *Forbes*. Retrieved from http://www.forbes.com/sites/quora/2016/11/02/iot-poses-a-unique-challenge-to-the-security-industry/#7713a17a62df
59 Peckham, G. (2020, December 14). *More security threats for printers/MFP*. Office Systems of Texas. Retrieved from https://www.osot.com/more-security-threats-for-printers-mfp/
60 Wikholm, Z. (2016). *When vulnerabilities travel downstream*. Flashpoint. Retrieved from https://www.flashpoint-intel.com/cybercrime/when-vulnerabilities-travel-downstream/
61 Chin. (2016). Security experts question Chinese webcam maker's response to cyberattack.
62 Warren, J. (2016, October 21). IoT security is a mess that will take an age to fix. *Forbes*. Retrieved from http://www.forbes.com/sites/justinwarren/2016/10/21/iot-security-is-a-mess-that-will-take-an-age-to-fix/#6722ba6c2b8d
63 Taylor, A. (2016, September 29). *The internet of things, cyber-security and the role of the CIO*. ITProPortal. Retrieved from https://www.itproportal.com/features/the-internet-of-things-cyber-security-and-the-role-of-the-cio/
64 Corfield, G. (2016, September 27). *No wonder we're being hit by internet of things botnets: Ever tried patching a thing?* The Register. Retrieved from https://www.theregister.com/2016/09/27/akamai_chief_security_officer_andy_ellis_interview_internet_of_things/
65 Schneier, B. (2007). *Information security and externalities, social & economic factors shaping the future of the internet* [Speakers' position papers]. Retrieved from https://www.oecd.org/sti/ieconomy/37985707.pdf
66 Dickson, B. (2016, August 16). *How to prevent your IoT devices from being forced into botnet bondage*. TechCrunch. Retrieved from https://techcrunch.com/2016/08/16/how-to-prevent-your-iot-devices-from-being-forced-into-botnet-slavery/
67 Reyneri, R. (2018, August 31). *IoT Update: US wireless industry establishes IoT security certification program*. Covington. Retrieved from https://www.insidetechmedia.com/2018/08/31/iot-update-u-s-wireless-industry-establishes-iot-security-certification-program/
68 Eggerton, J. (2019, March 7). *First IoT device gets CTIA certification*. Broadcasting + Cable. Retrieved from https://www.broadcastingcable.com/news/first-iot-device-gets-ctia-certification
69 Combs, V. (2020, February 20). *Cisco adds 5G and machine learning to IoT monitoring platform*. TechRepublic. Retrieved from https://www.techrepublic.com/article/cisco-adds-5g-and-machine-learning-to-iot-monitoring-platform/
70 Lindsey, N. (2019, December 5). New IoT security ratings a positive development for internet of things. *CPO Magazine*. Retrieved from https://www.cpomagazine.com/cyber-security/new-iot-security-ratings-a-positive-development-for-internet-of-things/
71 *UL now offers CTIA cybersecurity certification for IoT devices*. (2018, October 16). PR Newswire. Retrieved from https://www.prnewswire.com/news-releases/ul-now-offers-ctia-cybersecurity-certification-for-iot-devices-300731583.html

72 Yaga, D., Mell, P., Roby, N., & Scarfone, K. (2018). *Blockchain technology overview*. National Institute of Standards and Technology Internal Report (NISTIR) 8202. U.S. Department of Commerce. Retrieved from https://nvlpubs.nist.gov/nistpubs/ir/2018/NIST.IR.8202.pdf

73 Banafa, A. (2017, January 10). *IoT and blockchain convergence: Benefits and challenges*. IEEE: Internet of Things. Retrieved from http://iot.ieee.org/newsletter/january-2017/iot-and-blockchain-convergence-benefits-and-challenges.html

74 Banafa, A. (2016, December 21). *A secure model of IoT with blockchain*. OpenMind. Retrieved from https://www.bbvaopenmind.com/en/a-secure-model-of-iot-with-blockchain/?utm_source=views&utm_medium=article06&utm_campaign=MITcompany&utm_content=banafa-jan07

75 Hackett, R. (2017, February 25). *How blockchains could save us from another Flint-like contamination crisis*. VentureBeat. Retrieved from http://venturebeat.com/2017/02/25/how-blockchains-could-save-us-from-another-flint-like-contamination-crisis/

76 Hackett. (2017). *How blockchains could save us from another Flint-like contamination crisis*.

77 Debucquoy-Dodley, D. (2016, January 14). *Did Michigan officials hide the truth about lead in Flint?* CNN. Retrieved from http://www.cnn.com/2016/01/14/us/flint-water-investigation/

78 *Drinking water requirements for states and public water systems*. (n.d.). United States Environmental Protection Agency. Retrieved from https://www.epa.gov/dwreginfo/lead-and-copper-rule

79 Coward, J. (2017, May 31). *Meet the visionary who brought blockchain to the industrial IoT*. Internet of Things World. Retrieved from https://www.iotworldtoday.com/2017/05/31/meet-visionary-who-brought-blockchain-industrial-iot/

80 Hackett. (2017). *How blockchains could save us from another Flint-like contamination crisis*.

81 Hackett. (2017). *How blockchains could save us from another Flint-like contamination crisis*.

82 Combs, V. (2020, February 20). *Cisco adds 5G and machine learning to IoT monitoring platform*. TechRepublic. Retrieved from https://www.techrepublic.com/article/cisco-adds-5g-and-machine-learning-to-iot-monitoring-platform/

83 O'Brien, C. (2020, February 11). *Is AI cybersecurity's salvation or its greatest threat?* VentureBeat. Retrieved from https://venturebeat.com/2020/02/11/is-ai-cybersecuritys-salvation-or-its-greatest-threat/

84 *Ericsson launches new AI-powered network services*. (2020, February 13). Ericsson. Retrieved from https://www.ericsson.com/en/press-releases/2020/2/ericsson-launches-new-ai-powered-network-services

85 Mann, T. (2020, February 20). *The role AI must play in 5G*. SDXcentral. Retrieved from https://www.sdxcentral.com/articles/news/the-role-ai-must-play-in-5g/2020/02/

86 Dickson, B. (2016, October 2). *What makes IoT ransomware a different and more dangerous threat?* TechCrunch. Retrieved from https://techcrunch.com/2016/10/02/what-makes-iot-ransomware-a-different-and-more-dangerous-threat/

87 Stavridis, J., & Weinstein, D. (2016, November 3). The internet of things is a cyberwar nightmare. *Foreign Policy*. Retrieved from http://foreignpolicy.com/2016/11/03/the-internet-of-things-is-a-cyber-war-nightmare/?utm_source=Sailthru&utm_medium=email&utm_campaign=New%20Campaign&utm_term=Flashpoints

88 Wong, J. (2016, October 26). *Good news: The hackers who broke the internet last week are less powerful than originally believed*. Quartz. Retrieved from https://qz.com/820003/dyn-dns-ddos-the-mirai-botnet-is-smaller-than-originally-thought/

89 Feingold, J. (2016, October 27). *Dyn issues analysis of "complex and sophisticated" cyberattacks*. New Hampshire Business Review. Retrieved from https://www.nhbr.com/dyn-issues-analysis-of-complex-and-sophisticated-cyberattacks/

90 Thielman, S., & Hunt, E. (2016, October 22). Cyber attack: Hackers "weaponised" everyday devices with malware. *The Guardian*. Retrieved from https://www.theguardian.com/technology/2016/oct/22/cyber-attack-hackers-weaponised-everyday-devices-with-malware-to-mount-assault

91 Geller, E., & Room, T. (2016, October 21). *WikiLeaks supporters claim credit for massive US cyberattack, but researchers skeptical*. Politico. Retrieved from https://www.politico.com/story/2016/10/websites-down-possible-cyber-attack-230145

92 Geller & Room. (2016). *WikiLeaks supporters claim credit for massive US cyberattack, but researchers skeptical*.

93 *Internet denied: What's behind the massive DDoS attacks*. (2016, October 22). RT. Retrieved from https://www.rt.com/usa/363714-ddos-dns-attacks-hackers/

94 Satter, R., & Bajak, F. (2016, October 21). Cyberattack disrupts access to major websites, worries experts. *Portland Press Herald*. Retrieved from https://www.pressherald.com/2016/10/21/cyber-attack-disrupts-access-to-popular-websites/

95 Liu, C. (2018, December 19). The forgotten object lesson of the Dyn DDoS attack. *Forbes*. Retrieved from https://www.forbes.com/sites/forbestechcouncil/2018/12/19/the-forgotten-object-lesson-of-the-dyn-ddos-attack/#10b58e752c06

96 Lanxon, N., Kahn, J., and Brustein, J. (2016, October 22). *Connected gadgets blamed as internet recovers from Friday attack*. Bloomberg Technology. Retrieved from https://www.bloomberg.com/news/articles/2016-10-22/connected-gadgets-blamed-as-internet-recovers-from-friday-attack

97 Roberts, P. (2017, February 3). *Exclusive: Mirai attack was costly for Dyn, data suggests*. The Security Ledger. Retrieved from https://tinyurl.com/yb7kg5gs

98 Reuters. (2016, October 22). Hackers used your vulnerable tech to throttle the internet. *Huffington Post*. Retrieved from http://www.huffingtonpost.com/entry/hackers-used-your-vulnerable-tech-to-throttle-the-internet_us_580b7ab8e4b0cdea3d87c4f0?section=

99 Woolf, N. (2016, October 26). DDoS attack that disrupted internet was largest of its kind in history, experts say. *The Guardian*. Retrieved from https://www.theguardian.com/technology/2016/oct/26/ddos-attack-dyn-mirai-botnet

100 Blumenthal, E., & Weise, E. (2016). Hacked home devices caused massive Internet outage. *USA Today*. Retrieved from http://www.usatoday.com/story/tech/2016/10/21/cyber-attack-takes-down-east-coast-netflix-spotify-twitter/92507806/

101 *From attacking IoT devices to Linux servers: Mirai is back!* (2018, December 7). Panda Security. Retrieved from https://www.pandasecurity.com/mediacenter/malware/new-mirai-attacks-linux-servers/

102 Feldman, B. (2017, December 13). *Three Minecraft players were behind the botnet that took down a chunk of the internet last year*. Intelligencer. Retrieved from https://nymag.com/intelligencer/2017/12/three-minecraft-players-created-the-webs-scariest-botnet.html

103 Mimoso, M. (2016, October 19). *Mirai bots more than double since source code release*. ThreatPost. Retrieved from https://threatpost.com/mirai-bots-more-than-double-since-source-code-release/121368/

104 Levi, R. (2016, February 11). *Beware of the internet of attacking things*. Finance Magnates. Retrieved from http://www.financemagnates.com/forex/bloggers/beware-internet-attacking-things/
105 Mimoso. (2016). *Mirai bots more than double since source code release*.
106 Waqaz. (2015, November 4). *Another country is under massive DDoS attacks – thanks to Mirai malware*. Hackread. Retrieved from https://www.hackread.com/ddos-attack-with-mirai-malware/
107 *3rd cyberattack "has been resolved" after hours of major outages: Company*. (2016, October 21). NBC New York. Retrieved from http://www.nbcnewyork.com/news/local/Major-Websites-Taken-Down-by-Internet-Attack-397905801.html
108 Woolf, N. (2016, October 26). DDoS attack that disrupted internet was largest of its kind in history, experts say. *The Guardian*. Retrieved from https://www.theguardian.com/technology/2016/oct/26/ddos-attack-dyn-mirai-botnet
109 Thielman & Hunt. (2016). Cyber attack.
110 Woolf. (2016). DDoS attack that disrupted internet was largest of its kind in history, experts say.
111 Perlroth, N. (2016, October 22). Hackers used new weapons to disrupt major websites across US. *The New York Times*. Retrieved from http://www.nytimes.com/2016/10/22/business/internet-problems-attack.html?_r=0
112 Soo, Z. (2016, October 24). *China's Xiongmai Tech admits product flaws contributed to cyberattack on US sites*. South China Morning Post. Retrieved from https://www.scmp.com/tech/article/2039584/chinas-xiongmai-tech-admits-product-flaws-contributed-cyberattack-us-sites
113 Horwitz, J., & Yinyin, E.H. (2016, October 31). *A collision of Chinese manufacturing, globalization, and consumer ignorance could ruin the internet for everyone*. Quartz. Retrieved from http://qz.com/819391/a-collision-of-chinese-manufacturing-globalization-and-consumer-ignorance-could-ruin-the-internet-for-everyone/
114 Wikholm. (2016). *When vulnerabilities travel downstream*.
115 Horwitz & Yinyin. (2016). *A collision of Chinese manufacturing, globalization, and consumer ignorance could ruin the internet for everyone*.
116 Chin. (2016). Security experts question Chinese webcam maker's response to cyberattack.
117 Lovelace, B. (2016, October 21). Dyn says cyberattack "resolved" after services shut down. *NBC News*. Retrieved from https://www.nbcnews.com/tech/tech-news/dyn-says-cyberattack-resolved-after-services-shut-down-n670926
118 Roach, C. (2017, February 9). *Lessons learned from the Dyn attack*. CFO. Retrieved from https://www.cfo.com/cyber-security-technology/2017/02/lessons-learned-dyn-attack/

Chapter 15

1 *The AI in cybersecurity market to generate $101.8 billion in 2030*. (2020, November 27). Help Net Security. Retrieved from https://www.helpnetsecurity.com/2020/11/27/ai-in-cybersecurity-market/
2 Mackay, M. (2020, February 6). *Adversarial artificial intelligence: Winning the cyber security battle*. Information Age. Retrieved from https://www.information-age.com/adversarial-artificial-intelligence-winning-cyber-security-battle-123487325/
3 Yang, C. (2020, May 13). *AI and ML in cybersecurity: Adoption is rising, but confusion remains*. Webroot. Retrieved from https://www.webroot.com/blog/2020/05/13/ai-and-ml-in-cybersecurity-adoption-is-rising-but-confusion-remains/

4 Gil, L. (n.d.). *Chronicle puts AI into action on security: The time has come for machine learning*. Techbeacon. Retrieved from https://techbeacon.com/security/chronicle-puts-ai-action-security-time-has-come-machine-learning
5 *Facebook targets misinformation, issues rules on "deepfake" videos*. (2020, January 7). Security Magazine. Retrieved from https://www.securitymagazine.com/articles/91506-facebook-targets-misinformation-issues-rules-on-deepfake-videos
6 *Martech interview with Director of Strategic Threat, Darktrace – Marcus Fowler*. (2020, October 6). MarTechCube. Retrieved from https://www.martechcube.com/martech-interview-with-director-of-ai-company-darktrace/
7 Jones, M. (2020, September 9). *How hackers are weaponizing artificial intelligence*. TechHQ. Retrieved from https://techhq.com/2020/09/how-hackers-are-weaponizing-artificial-intelligence/
8 Mauriello, J. (2019, September 5). *What's the real role of AI and ML in cybersecurity?* Security Magazine. Retrieved from https://www.securitymagazine.com/articles/90871-whats-the-real-role-of-ai-and-ml-in-cybersecurity
9 Winder, D. (2020, February 28). Google confirms new AI tool scans 300 billion Gmail attachments every week. *Forbes*. Retrieved from https://www.forbes.com/sites/daveywinder/2020/02/28/google-confirms-new-ai-tool-scans-300-billion-gmail-attachments-every-week/#68de9a8a3edd
10 Johnson, A. (2020, February 20). *Delivering on the promise of security AI to help defenders protect today's hybrid environments*. Microsoft. Retrieved from https://blogs.microsoft.com/blog/2020/02/20/delivering-on-the-promise-of-security-ai-to-help-defenders-protect-todays-hybrid-environments/
11 Columbus, L. (2019, November 4). 10 charts that will change your perspective of AI in security. *Forbes*. Retrieved from https://www.forbes.com/sites/louiscolumbus/2019/11/04/10-charts-that-will-change-your-perspective-of-ai-in-security/#7761818168b7
12 Capgemini Institute. (2019). *Reinventing cybersecurity with artificial intelligence: The new frontier in digital security*. Retrieved from https://www.capgemini.com/wp-content/uploads/2019/ Research 07/AI-in-Cybersecurity_Report_20190711_V06.pdf
13 Columbus. (2019). 10 charts that will change your perspective of AI in security.
14 Columbus, L. (2019, July 14). Why AI is the future of cybersecurity. *Forbes*. Retrieved from https://www.forbes.com/sites/louiscolumbus/2019/07/14/why-ai-is-the-future-of-cybersecurity/#2db7fee4117e
15 Metz, C. (2015, November 5). Baidu, the "Chinese Google," is teaching AI to spot malware. *Wired*. Retrieved from https://www.wired.com/2015/11/baidu-the-chinese-google-is-teaching-ai-to-spot-malware/
16 *The role of AI in cybersecurity*. (2019, December 13). EC-Council. Retrieved from https://blog.eccouncil.org/the-role-of-ai-in-cybersecurity/#:~:text=With%20AI%20stepping%20into%20cybersecurity,learn%20about%20new%20attack%20vectors
17 TD Ameritrade Institutional. (2019, January 8). *TD Ameritrade Institutional 2019 RIA sentiment survey*. Retrieved from https://s1.q4cdn.com/959385532/files/doc_downloads/research/2019/2019-RIA-Sentiment-Survey.pdf
18 Fadilpašić, S. (2020, November 20). *AI could be the next big defence against cybercrime*. ITProPortal. Retrieved from https://www.itproportal.com/news/ai-could-be-the-next-big-defence-against-cybercrime/
19 Perez, B. & Soo, Z. (2017, October 28). *China a fast learner when it comes to artificial intelligence-powered fintech, experts say*. South China Morning Post. Retrieved from https://www.scmp.com/tech/innovation/article/2117298/china-fast-learner-when-it-comes-artificial-intelligence-powered

20 Perez & Soo. (2017). *China a fast learner when it comes to artificial intelligence-powered fintech, experts say.*
21 *What is next-gen endpoint security?* (n.d.). McAfee. Retrieved from https://www.mcafee.com /enterprise/en-us/security-awareness/endpoint/what-is-next-gen-endpoint-protection.html
22 Martin, J.A., & Waters, J.K. (2018, October 9). *What is IAM? Identity and access management explained.* CSO. Retrieved from https://www.csoonline.com/article/2120384/what-is-iam-identity -and-access-management-explained.html
23 *Facebook targets misinformation, issues rules on "deepfake" videos.* (2020). Security Magazine.
24 Statt, N. (2020, March 4). *Facebook's new AI-powered moderation tool helps it catch billions of fake accounts.* The Verge. Retrieved from https://www.theverge.com/2020/3/4/21164695/facebook -ai-moderation-dec-deep-entity-classification-fake-accounts-spam-scams
25 *Community standards enforcement report: Facebook transparency.* (2020). Facebook. Retrieved from https://transparency.facebook.com/community-standards-enforcement#fake-accounts
26 *Deep learning.* (2019, January). JohnCReid.com. Retrieved from http://johncreid.com/2019/03 /deep-learning/
27 Jiang, F., Jiang, Y., Zhi, H., Dong, Y., Li, H., Ma, S., Wang, Y., Dong, Q., Shen, H., & Wang, Y. (2017). Artificial intelligence in healthcare: Past, present and future. *Stroke and Vascular Neurology, 2*(4), 230–43. Retrieved from https://svn.bmj.com/ content/2/4/230.full.pdf
28 Gills, A., & Steele, C. (n.d.). GPU (graphics processing unit). TechTarget. Retrieved from https:// searchvirtualdesktop.techtarget.com/definition/GPU-graphics-processing-unit
29 Au, D. (2019, May 14). *What deep learning means for cybersecurity.* Security Week. Retrieved from https://www.securityweek.com/what-deep-learning-means-cybersecurity
30 Lunden, I. (2020, February 12). *Deep Instinct nabs $43M for a deep-learning cybersecurity solution that can suss an attack before it happens.* TechCrunch. Retrieved from https://techcrunch.com/2020/02 /12/deep-instinct-nabs-43m-for-a-deep-learning-cybersecurity-solution-that-can-suss-an-attack -before-it-happens/
31 Ippolito, P. (2019, October 10). *Feature extraction techniques.* Towards Data Science. Retrieved from https://towardsdatascience.com/feature-extraction-techniques-d619b56e31be
32 Lunden. (2020). *Deep Instinct nabs $43M for a deep-learning cybersecurity solution that can suss an attack before it happens.*
33 Apruzzese, G., Ferretti, L., Marchetti, M., Colajanni, M., & Guido, A. (2018). *On the effectiveness of machine and deep learning for cyber security.* 10th International Conference on Cyber Conflict. Retrieved from https://ccdcoe.org/uploads/2018/10/Art-19-On-the-Effectiveness-of-Machine -and-Deep-Learning-for-Cyber-Security.pdf
34 *Top trends on the Gartner hype cycle for artificial intelligence, 2019.* (2019). Retrieved May 5, 2020 from www.gartner.com
35 *The role of AI in cybersecurity.* (2019).
36 Ismail, N. (2019, September 4). *Darktrace unveils the cyber AI analyst: A faster response to threats.* Information Age. Retrieved from https://www.information-age.com/darktrace-cyber-ai-analyst -faster-response-threats-123485089/
37 Williams, O. (2019, September 4). *Darktrace launches "Cyber AI Analyst" as skills shortage intensifies.* NS Tech. Retrieved from https://tech.newstatesman.com/security/darktrace-cyber-ai -analyst
38 *Darktrace Asia Pacific hits new milestone with over 500 customers.* (2019, January 11). Cambridge Network. Retrieved from https://www.cambridgenetwork.co.uk/news/darktrace-asia-pacific-hits -new-milestone-over-500-customers

39 Ismail. (2019). *Darktrace unveils the cyber AI analyst: A faster response to threats.*
40 Ismail. (2019). *Darktrace unveils the cyber AI analyst: A faster response to threats.*
41 The enterprise immune system. (n.d.). Darktrace. Retrieved from https://www.darktrace.com/en/products/enterprise/
42 *Finding the needle in the haystack: Accelerating threat remediation with AI.* (2020, November 9). Tech HQ. Retrieved from https://techhq.com/2020/11/ai-ml-cyber-security-protection-enterprise-jobs-cybersecurity-best-darktrace/
43 Darktrace. (2020). *Darktrace cyber AI analyst* [White paper]. Retrieved from https://www.darktrace.com/fr/resources/wp-cyber-ai-analyst.pdf
44 O'Donnell, L. (2020, March 6). *Critical Zoho zero-day flaw disclosed.* Threat Post. Retrieved from https://threatpost.com/critical-zoho-zero-day-flaw-disclosed/153484/
45 *Finding the needle in the haystack: Accelerating threat remediation with AI.* (2020). Tech HQ.
46 Apruzzese et al. (2018). *On the effectiveness of machine and deep learning for cyber security.*
47 Yang, C. (2020, May 13). *AI and ML in cybersecurity: Adoption is rising, but confusion remains.* Webroot. Retrieved from https://www.webroot.com/blog/2020/05/13/ai-and-ml-in-cybersecurity-adoption-is-rising-but-confusion-remains/
48 Mauriello. (2019). *What's the real role of AI and ML in cybersecurity?*
49 Press, G. (2019, December 3). 141 cybersecurity predictions for 2020. *Forbes.* Retrieved from https://www.forbes.com/sites/gilpress/2019/12/03/141-cybersecurity-predictions-for-2020/#482c468e1bc5
50 Ismail, N. (2019, September 17). *Data science cowboys are exacerbating the AI and analytics challenge.* Information Age. Retrieved from https://www.information-age.com/data-science-cowboys-ai-analytics-challenge-123484974/
51 Kahn, J. (2020, April 30). Cybercriminals adapt to coronavirus faster than the A.I. cops hunting them. *Fortune.* Retrieved from https://fortune.com/2020/04/30/cybercriminals-adapt-to-coronavirus-faster-than-the-a-i-cops-hunting-them/
52 Broomfield. (2017, December 14). *More than 90 percent of cybersecurity professionals concerned about cybercriminals using AI in attacks.* Webroot. Retrieved from https://www.webroot.com/us/en/about/press-room/releases/cybercriminals-using-ai-in-attacks
53 Ghosemajumder, S. (2020, September 6). *How AI will automate cybersecurity in the post-COVID world.* VentureBeat. Retrieved from https://venturebeat.com/2020/09/06/how-ai-will-automate-cybersecurity-in-the-post-covid-world/
54 Woollacott, E. (2020, May 27). *Machine learning techniques applied to crack CAPTCHAs.* Port Swigger. Retrieved from https://portswigger.net/daily-swig/machine-learning-techniques-applied-to-crack-captchas
55 Ghosemajumder. (2020). *How AI will automate cybersecurity in the post-COVID world.*
56 Kahn. (2020). Cybercriminals adapt to coronavirus faster than the A.I. cops hunting them.
57 Palmer, D. (2017, January 9). *AI will supercharge spear-phishing.* Darktrace. Retrieved from https://www.darktrace.com/en/blog/ai-will-supercharge-spear-phishing/
58 Jones. (2020). *How hackers are weaponizing artificial intelligence.*
59 Jones. (2020). *How hackers are weaponizing artificial intelligence.*
60 Choudhury, S. (2019, December 18). *Automated hacking, deepfakes are going to be major cybersecurity threats in 2020.* CNBC. Retrieved from https://www.cnbc.com/2019/12/18/automated-hacking-deepfakes-top-cybersecurity-threats-in-2020.html
61 Pollard, J. (2019, October 30). *Predictions 2020: This time, cyberattacks get personal.* Forrester. Retrieved from https://go.forrester.com/blogs/predictions-2020-cybersecurity/

62 Stupp, C. (2019, August 30). Fraudsters used AI to mimic CEO's voice in unusual cybercrime case. *The Wall Street Journal*. Retrieved from https://www.wsj.com/articles/fraudsters-use-ai-to-mimic-ceos-voice-in-unusual-cybercrime-case-11567157402
63 Laurens, C. (2020, January 10). *How to spot a deepfake*. World Economic Forum. Retrieved from https://www.weforum.org/agenda/2020/01/how-to-spot-a-deepfake/
64 Scroxton, A. (2020, January 29). *Fintechs fear deepfake fraud*. Computer Weekly. Retrieved from https://www.computerweekly.com/news/252477448/Fintechs-fear-deepfake-fraud
65 Press. (2019). 141 cybersecurity predictions for 2020.
66 Scroxton. (2020). *Fintechs fear deepfake fraud*.
67 Combs, V. (2020, October 7). *3 ways criminals use artificial intelligence in cybersecurity attacks*. TechRepublic. Retrieved from https://www.techrepublic.com/article/3-ways-criminals-use-artificial-intelligence-in-cybersecurity-attacks/
68 Moisejevs, I. (2019, July 15). *Poisoning attacks on machine learning*. Towards Data Science. Retrieved from https://towardsdatascience.com/poisoning-attacks-on-machine-learning-1ff247c254db
69 Best, K.L., Schmid, J., Tierney, S., Awan, J., Beyene, N., Holliday, M., Khan, R., & Lee, K. (2020). *How to analyze the cyber threat from drones*. RAND Corporation. Retrieved from https://www.rand.org/content/dam/rand/pubs/research_reports/RR2900/RR2972/RAND_RR2972.pdf
70 Artificial intelligence topic hub. (n.d.). SDXCentral. Retrieved from https://www.sdxcentral.com/big-data/artificial-intelligence/
71 Wheeler, T., & Simpson, D. (2019, September 3). *Why 5G requires new approaches to cybersecurity*. Brookings. Retrieved from https://www.brookings.edu/research/why-5g-requires-new-approaches-to-cybersecurity/
72 Kirk, J. (2018, January 31). *What's riding on 5G security? The internet of everything*. Bank Info Security. Retrieved from https://www.bankinfosecurity.com/whats-riding-on-5g-security-internet-everything-a-10618
73 Thornhill, J. (2019, July 22). Formulating values for AI is hard when humans do not agree. *Financial Times*. Retrieved from https://www.ft.com/content/6c8854de-ac59-11e9-8030-530adfa879c2
74 Feng, E. (2020, January 5). *In China, a new call to protect data privacy*. NPR. Retrieved from https://www.npr.org/2020/01/05/793014617/in-china-a-new-call-to-protect-data-privacy
75 Feng. (2020). *In China, a new call to protect data privacy*.
76 Mozur, P. (2019, April 14). One month, 500,000 face scans: How China is using A.I. to profile a minority. *The New York Times*. Retrieved from https://www.nytimes.com/2019/04/14/technology/china-surveillance-artificial-intelligence-racial-profiling.html
77 Mozur. (2019). One month, 500,000 face scans.
78 Zenz, A. (2019, December 11). Xinjiang's new slavery. *Foreign Policy*. Retrieved from https://foreignpolicy.com/2019/12/11/cotton-china-uighur-labor-xinjiang-new-slavery/
79 Mozur. (2019). One month, 500,000 face scans.
80 Palmer, J. (2015, February 16). *All-seeing, all-knowing*. Aeon. Retrieved from https://aeon.co/essays/will-information-open-up-china-or-close-the-ranks-for-good
81 Mandhana, N. (2019, February 20). Huawei's video surveillance business hits snag in Philippines. *The Wall Street Journal*. Retrieved from https://www.wsj.com/articles/huaweis-video-surveillance-business-hits-snag-in-philippines-11550683135
82 O'Rourke, B., & Choy, G. (2019, January 28). *Big brother Huawei kitted out this Philippine city. Is China watching?* South China Morning Post. Retrieved from https://www.scmp.com/week-asia/economics/article/2183540/big-brother-huawei-watches-philippine-city-does-china-too

83 Chinese facial recognition tech installed in nations vulnerable to abuse. (2019, October 16). *CBS News*. Retrieved from https://www.cbsnews.com/news/china-huawei-face-recognition-cameras-serbia-other-countries-questionable-human-rights-2019-10-16/
84 Rollet, C. (2018, August 9). Ecuador's all-seeing eye is made in China. *Foreign Policy*. Retrieved from https://foreignpolicy.com/2018/08/09/ecuadors-all-seeing-eye-is-made-in-china/
85 Metz, C. (2016, June 2). Google's training its AI to be Android's security guard. *Wired*. Retrieved from https://www.wired.com/2016/06/googles-android-security-team-turns-machine-learning/
86 Hurst, A. (2020, February 18). *Google revealed to have acquired the most AI startups since 2009*. Information Age. Retrieved from https://www.information-age.com/google-revealed-acquired-most-ai-startups-since-2009-123487752/
87 *Making the world's information safely accessible*. (n.d.). Google. Retrieved from https://safebrowsing.google.com/#:~:text=Safe%20Browsing%20launched%20in%202007,across%20desktop%20and%20mobile%20platforms
88 Protalinski, E. (2017, September 11). *Google Safe Browsing now protects over 3 billion devices*. VentureBeat. Retrieved from https://venturebeat.com/2017/09/11/google-safe-browsing-now-protects-over-3-billion-devices/
89 Protalinski, E. (2020, February 24). *Google Cloud beefs up Chronicle, reCaptcha Enterprise and Web Risk API hit general availability*. VentureBeat. Retrieved from https://venturebeat.com/2020/02/24/google-cloud-beefs-up-chronicle-recaptcha-enterprise-and-web-risk-api-hit-general-availability/
90 Newman, L.H. (2018, April 29). AI can help cybersecurity – if it can fight through the hype. *Wired*. Retrieved from https://www.wired.com/story/ai-machine-learning-cybersecurity/
91 Winder. (2020). Google confirms new AI tool scans 300 billion Gmail attachments every week.
92 Vincent, J. (2019, February 6). *Gmail is now blocking 100 million extra spam messages every day with AI*. The Verge. Retrieved from https://www.theverge.com/2019/2/6/18213453/gmail-tensorflow-machine-learning-spam-100-million
93 Marr, B. (2017, August 8). The amazing ways Google uses deep learning AI. *Forbes*. Retrieved from https://www.forbes.com/sites/bernardmarr/2017/08/08/the-amazing-ways-how-google-uses-deep-learning-ai/#212270f53204
94 Teller, A. (2018, January 24). *Graduation day: Introducing Chronicle*. X – The Moonshot Factory. Retrieved from https://blog.x.company/graduation-day-introducing-chronicle-318d34b80cce
95 Protalinski. (2020). Google Cloud beefs up Chronicle.
96 Protalinski. (2020). Google Cloud beefs up Chronicle.
97 Teller. (2018). *Graduation day*.
98 Walker, J. (2018, January 25). *Alphabet launches Chronicle cybersecurity business*. PortSwigger. Retrieved from https://portswigger.net/daily-swig/alphabet-launches-chronicle-cybersecurity-business
99 Brandom, R. (2018, January 1). *Google X is launching a cybersecurity company called Chronicle*. The Verge. Retrieved from https://www.theverge.com/2018/1/24/16929320/google-x-cybersecurity-chronicle-spin-off-alphabet
100 Brewster, T. (2019). Alphabet's Chronicle startup finally launches – it's like Google photos for cybersecurity. *Forbes*. Retrieved from https://www.forbes.com/sites/thomasbrewster/2019/03/04/alphabets-chronicle-startup-finally-launchesits-like-google-photos-for-cybersecurity/#6d81c0d179c5

101 Franklin Jr., C. (2019, June 27). *Chronicle folds into Google*. Dark Reading. Retrieved from https://www.darkreading.com/analytics/chronicle-folds-into-google/d/d-id/1335082

102 Condon, S. (2020, September 23). Google unveils new real-time threat detection tool from Chronicle. ZDNet. Retrieved from https://www.zdnet.com/article/google-unveils-new-real-time-threat-detection-tool-from-chronicle/

103 Winder. (2020). Google confirms new AI tool scans 300 billion Gmail attachments every week.

104 Winder. (2020). Google confirms new AI tool scans 300 billion Gmail attachments every week.

Chapter 16

1 *Managing the impact of increasing interconnectivity: Trends in cyber risk.* (2020, November 19). Allianz. Retrieved from https://www.agcs.allianz.com/news-and-insights/reports/cyber-risk-trends-2020.html

2 Lemos, R. (2020, November 30). *Driven by ransomware, cyber claims rise in number & value*. Dark Reading. Retrieved from https://www.darkreading.com/risk/driven-by-ransomware-cyber-claims-rise-in-number-and-value/d/d-id/1339557

3 Lemos. (2020). *Driven by ransomware, cyber claims rise in number & value.*

4 Ilascu, I. (2020, May 1). *NetWalker adjusts ransomware operation to only target enterprise.* BleepingComputer. Retrieved from https://www.bleepingcomputer.com/news/security/netwalker-adjusts-ransomware-operation-to-only-target-enterprise/

5 Seals, T. (2020, May 20). *NetWalker ransomware gang hunts for top-notch affiliates*. ThreatPost. Retrieved from https://threatpost.com/netwalker-ransomware-gang-top-notch-affiliates/155946/

6 Cluley, G. (2020, May 28). *NetWalker ransomware – what you need to know*. Tripwire. Retrieved from https://www.tripwire.com/state-of-security/featured/netwalker-ransomware-what-need-know/

7 Lieberman, M. (2020, December 3). *Schools aren't doing enough to protect their networks, top cybersecurity official warns*. EducationWeek. Retrieved from https://blogs.edweek.org/edweek/digitaleducation/2020/12/federal_cybersecurity_chief_most_schools_don.html

8 *Domain blocking and reporting (MDBR) newest service for US SLTTs*. (n.d.). Center for Internet Security. Retrieved from https://www.cisecurity.org/blog/malicious-domain-blocking-and-reporting-mdbr-newest-service-for-u-s-sltts/

9 Lieberman. (2020). *Schools aren't doing enough to protect their networks.*

10 *Global digital population as of October 2020.* (2020). Statista. Retrieved from https://www.statista.com/statistics/617136/digital-population-worldwide/#:~:text=Almost%204.66%20billion%20people%20were,percent%20of%20the%20global%20population

11 African News Agency. (2020, December 9). *More Africans concerned about cybercrime*. IOL. Retrieved from https://www.iol.co.za/news/africa/more-africans-concerned-about-cybercrime-report-848f4cb7-1e15-5e3b-9669-2d991237cb52

12 Mandia, K. (2020, December 8). *FireEye shares details of recent cyber attack, actions to protect community*. FireEye. Retrieved from https://www.fireeye.com/blog/products-and-services/2020/12/fireeye-shares-details-of-recent-cyber-attack-actions-to-protect-community.html

13 Gara, A. (2020, December 18). SolarWinds hack throws wrench in private equity's most profitable market. *Forbes*. Retrieved from https://www.forbes.com/sites/antoinegara/2020/12/18/solarwinds-hack-throws-wrench-in-private-equitys-most-profitable-market/?sh=764c1fa5ef73

14 Hern, A. (2020, December 14). Orion hack exposed vast number of targets – impact may not be known for a while. *The Guardian*. Retrieved from https://www.theguardian.com/world/2020/dec/14/solarwinds-breach-orion-hacked-cyber-espionage

15 Schneier, B. (2020, December 23). The US has suffered a massive cyberbreach. It's hard to overstate how bad it is. *The Guardian*. Retrieved from https://www.theguardian.com/commentisfree/2020/dec/23/cyber-attack-us-security-protocols

16 *SolarWinds hack could affect 18K customers*. (2020, December 15). KrebsonSecurity. Retrieved from https://krebsonsecurity.com/2020/12/solarwinds-hack-could-affect-18k-customers/

17 Patterson, D. (2020, December 23). Here is what we know – and don't know – about the suspected Russian hack. *CBS News*. Retrieved from https://www.cbsnews.com/news/solarwinds-fireeye-cyberattack-russia-hack-explained/

18 Schneier. (2020). The US has suffered a massive cyberbreach.

19 Christophe. (2020, December 14). *FireEye & SolarWinds incidents: What happened?* CYBR. Retrieved from https://cybr.com/cybersecurity/fireeye-solarwinds-incidents-what-happened/

20 Reuters. (2020, December 22). *Former SolarWinds adviser says he warned of lax security years before suspected Russian hack*. South China Morning Post. Retrieved from https://www.scmp.com/tech/enterprises/article/3114876/former-solarwinds-adviser-says-he-warned-lax-security-years

21 Zetter, K. (2020, December 24). *Solarwinds hack infected critical infrastructure, including power industry*. The Intercept. Retrieved from https://theintercept.com/2020/12/24/solarwinds-hack-power-infrastructure/

22 Allen, M. (2020, December 24). *Biden says Russian-linked cyberattack started "last year."* Axios. Retrieved from https://www.axios.com/biden-cyber-attack-russia-2019-289cfd66-f897-474c-a7bc-3ca35da8a6be.html

23 Nakashima, E. (2020, December 31). Microsoft says Russians hacked its network, viewing source code. *The Washington Post*. Retrieved from https://www.washingtonpost.com/national-security/microsoft-russian-hackers-source-coce/2020/12/31/a9b4f7cc-4b95-11eb-839a-cf4ba7b7c48c_story.html

24 Chappell, B., & Myre, G. (2020, December 15). *What we know about Russia's latest alleged hack of the US government*. MPRNews. Retrieved from https://www.mprnews.org/story/2020/12/15/npr-what-we-know-about-russias-latest-alleged-hack-of-the-u-s-government

25 Schneier. (2020). The US has suffered a massive cyberbreach.

26 Reuters. (2020). *Former SolarWinds adviser says he warned of lax security years before suspected Russian hack*.

27 Schneier. (2020). The US has suffered a massive cyberbreach.

28 Chappell & Myre. (2020). What we know about Russia's latest alleged hack of the US government.

29 Johnson, D. (2020, December 23). *Can SolarWinds survive? For breached companies it's a long, painful road to restoring trust*. SC. Retrieved from https://www.scmagazine.com/home/solarwinds-hack/can-solarwinds-survive-for-breached-companies-its-a-long-painful-road-to-restoring-trust/

30 Morgan, S. (2019, October 24). *Cybersecurity talent crunch to create 3.5 million unfilled jobs globally by 2021*. Cybercrime Magazine. Retrieved from https://cybersecurityventures.com/jobs/#:~:text=The%20US%20has%20a%20total,Standards%20and%20Technology%20(NIST)%20in

31 Knake, R. (2020, December 22). *No, the United States does not spend too much on cyber offense*. Council on Foreign Relations. Retrieved from https://www.cfr.org/blog/no-united-states-does-not-spend-too-much-cyber-offense

32 Kahn. (2020). Cybercriminals adapt to coronavirus faster than the A.I. cops hunting them.

33 Laurens. (2020). *How to spot a deepfake*.

34 Williams, C. (2017, January 10). *4 risk response strategies you will have to consider after assessing risks*. ERM Insights. Retrieved from https://www.erminsightsbycarol.com/risk-response-strategies/

35 Chen, S. (2010, May 12). *Workplace rants on social media are headache for companies*. CNN. Retrieved from http://www.cnn.com/2010/LIVING/05/12/social.media.work.rants/index.html

36 *China's cyber-theft jet fighter*. (2014, November 12). WSJ Opinion. Retrieved from http://www.wsj.com/articles/chinas-cyber-theft-jet-fighter-1415838777

37 Groll, E. (2016, June 10). Preventing a blackout by taking the power grid offline. *Foreign Policy*. Retrieved from http://foreignpolicy.com/2016/06/10/preventing-a-blackout-by-taking-the-power-grid-offline/

38 Parisi, R. (2019, October 20). *Staying atop the changing cyber insurance market is a necessity for boards*. Brink. Retrieved from https://www.brinknews.com/staying-atop-the-changing-cyber-insurance-market-is-a-necessity-for-boards/

39 *Cyber insurance claims on the rise*. (2020, November 27). Help Net Security. Retrieved from https://www.helpnetsecurity.com/2020/11/27/cyber-insurance-claims-rise/

40 Coker, J. (2020, December 23). *Cyber insurance market expected to surge in 2021*. InfoSecurity. Retrieved from https://www.infosecurity-magazine.com/news/cyber-insurance-market-surge-2021/

41 *21% of SMBs do not have a data backup or disaster recovery solution in place*. (2020, March 31). Help Net Security. Retrieved from https://www.helpnetsecurity.com/2020/03/31/data-backup

42 Ashford, W. (2014, March 17). *SMEs believe they are immune to cyber attack*. Computer Weekly. https://www.computerweekly.com/news/2240216202/SMEs-believes-it-is-immune-to-cyber-attack-study-shows

43 Wolf, J. (2020, March 6). *9 strategies for retaining women in cybersecurity and STEM in 2020*. Security Intelligence. Retrieved from https://securityintelligence.com/articles/9-strategies-for-retaining-women-in-cybersecurity-and-stem-in-2020/

44 Tarun, R. (2019, December 16). *Gender diversity in cybersecurity matters to the business*. CSO. Retrieved from https://www.csoonline.com/article/3490417/gender-diversity-in-cybersecurity-matters-to-the-business.html

45 Morgan, S. (2019, March 28). *Women represent 20 percent of the global cybersecurity workforce in 2019*. Cybercrime Magazine. Retrieved from https://cybersecurityventures.com/women-in-cybersecurity/

46 Ashford, W. (2017, June 15). *PaloAlto Networks partners with US Girl Scouts on security skills*. Computer Weekly. Retrieved from https://www.computerweekly.com/news/450420822/PaloAlto-Networks-partners-with-US-Girl-Scouts-on-security-skills

47 Sottile, C., & Kent, J. (2018, March 4). Girl Scouts fight cybercrime with new cybersecurity badge. *NBC News*. Retrieved from https://www.nbcnews.com/tech/tech-news/girl-scouts-fight-cybercrime-new-cybersecurity-badge-n852971

48 *Female cybersecurity experts take the floor at BBVA*. (2019, November 26). BBVA. Retrieved from https://www.bbva.com/en/female-cybersecurity-experts-take-the-floor-at-bbva/

49 Williams. (2017). *4 risk response strategies you will have to consider after assessing risks*.

50 Perrin, A. (2018, September 5). *Americans are changing their relationship with Facebook*. Pew Research Center. Retrieved from https://www.pewresearch.org/fact-tank/2018/09/05/americans-are-changing-their-relationship-with-facebook/

51 Perrin. (2018). *Americans are changing their relationship with Facebook*.

52 Saxena, R. (2018, September 27). *Will personal cyber insurance cover phishing, hacking and stalking?* Bloomberg. Retrieved from https://www.bloombergquint.com/law-and-policy/will-personal-cyber-insurance-cover-phishing-hacking-and-stalking

53 Ralph, O. (2017, November 8). Should individuals buy insurance against cyber attacks? *Financial Times*. Retrieved from https://www.ft.com/content/72e11ca6-98ad-11e7-8c5c-c8d8fa6961bb

54 Metz, J. (2020, October 21). Personal cyber insurance for the growing risk of cyberattacks. *Forbes*. Retrieved from https://www.forbes.com/advisor/homeowners-insurance/personal-cyber-insurance/
55 Ralph. (2017). Should individuals buy insurance against cyber attacks?
56 Pinsker, B. (2018, June 7). *Personal cyber insurance-security packages being tested with wealthy*. Insurance Journal. Retrieved from https://www.insurancejournal.com/news/national/2018/06/07/491496.htm
57 Home Cyber Protection. Retrieved from https://www.munichre.com/hsb/en/products/personal-lines-agents-and-brokers/home-cyber-protection.html
58 Ralph. (2017). Should individuals buy insurance against cyber attacks?
59 van Hooijdonk, R. (2017, March 10). *6 of the smartest smart cities in the world*. Retrieved from https://www.richardvanhooijdonk.com/blog/en/6-of-the-smartest-smart-cities-in-the-world/
60 Schuman, E. (2017, April 18). *How one personal cyber insurance policy stacks up*. ComputerWorld. Retrieved from https://www.computerworld.com/article/3190209/how-one-personal-cyber-insurance-policy-stacks-up.html
61 Kshetri, N. (2018, October 26). *As digital threats grow, will cyber insurance take off?* The Conversation. Retrieved from https://theconversation.com/as-digital-threats-grow-will-cyber-insurance-take-off-104371
62 Morgan. (2019). *Cybersecurity talent crunch to create 3.5 million unfilled jobs globally by 2021*.
63 Luxner, L. (2018, September 26). *This new program is recruiting Israeli girls for cyber warfare and high-tech futures*. Jewish Telegraphic Agency. Retrieved from https://www.jta.org/2018/09/26/israel/new-program-recruiting-israeli-girls-cyber-warfare-high-tech-futures
64 IBM Security. (2016, March 8). *IBM's women in security* [Video]. Youtube. Retrieved from https://www.youtube.com/watch?v=-LQmorNEnZ8
65 Ricciuto, H. (2017, March 15). *Representation of women in cybersecurity remains stagnant, despite recent efforts to balance the scales*. Security Intelligence. Retrieved from https://securityintelligence.com/representation-of-women-in-cybersecurity-remains-stagnant-despite-recent-efforts-to-balance-the-scales/

Index